Handbook of Behavioral and Cognitive Therapies with Older Adults

Handbook of Behavioral and Cognitive Therapies with Older Adults

Edited by

Dolores Gallagher-Thompson, Ph.D. ABPP
Ann M. Steffen, Ph.D.
Larry W. Thompson, Ph.D.

Springer

Dolores Gallagher-Thompson
Department of Psychiatry
and Behavioral Science
Stanford University
Stanford, CA 94305-5717
USA
dolorest@stanford.edu

Ann M. Steffen
Department of Psychology
University of Missouri-St. Louis
St. Louis, MO 63121
USA
ann_steffen@umsl.edu

Larry W. Thompson
Department of Medicine
Stanford University School of Medicine and
Pacific Graduate School of Psychology
Los Atlos, CA 94024
USA
larrywt@stanford.edu

Library of Congress Control Number: 2007926119

ISBN-13: 978-0-387-72006-7 eISBN-13: 978-0-387-72007-4

Printed on acid-free paper.

Printed in the United States of America.

9 8 7 6 5 4 3 2 1

springer.com

To my parents, and most especially my grandparents, who inspired me to achieve and excel in all of my life's endeavors

Dolores Gallagher-Thompson

To my beloved husband, George Gerules, for his abiding gifts to me of support and joy

Ann M. Steffen

To my patients and students who taught me more than I could ever imagine

Larry W. Thompson

Foreword

It gives me great pleasure to witness continued growth in the application of cognitive and behavioral theories and therapies to more diverse populations – including, in this volume, their application to the mental health problems of later life. Evidence continues to accumulate, demonstrating that these are effective in treating a broad range of elderly patient groups. This is the first book to examine a number of these evidence-based interventions currently in use with older adults. The editors have assembled chapters developed in many of the leading clinical and clinical research programs focusing on elderly patients, both in this country and in the UK. Since the emphasis of this volume is primarily on clinical application, each author group was asked to discuss the empirical data for the treatment strategies it is using with the specific patient group selected. Typically, this was followed by a detailed description of treatment procedures that were then illustrated by one or more clinical examples.

The book begins by examining the treatment of depressive and anxiety disorders and then moves on to more complicated and/or serious disorders, including schizophrenia and other psychoses, suicidal behavior, personality disorders, bipolar disorders, dementia, and complicated bereavement. A chapter on the problems and issues in training therapists to use evidence-based interventions effectively is also included, along with a chapter discussing the implications of Medicare policies and guidelines for service delivery.

The editors have recruited an impressive group of professionals to participate in the creation of this volume. It is noteworthy that several of the contributing authors, who over the years have published articles on the conceptual development, empirical testing, and refinement of various cognitive and behavioral treatment models, are now turning their attention to the growing mental health needs of this patient group. This book will provide the reader with a comprehensive picture of recent clinical work targeted specifically to the treatment of the most prevalent age-related mental health problems. Both clinicians and researchers in many professions concerned with the health and mental health of this population will find this book to be extremely valuable for obtaining up-to-date information about empirical evidence and its clinical application. As the need for services continues to increase with the onset of the "baby boomer" era, we can also expect this text to serve as a valuable resource for continued research in the development of more effective treatment programs.

Philadelphia, PA

Aaron T. Beck, MD

Preface

There was a time in the last century when professionals from any discipline involving mental health (e.g., nursing, psychiatry, psychology, social work) held the belief that the elderly would gain little from psychiatric treatment of any kind. Indeed, very little time and effort was invested in attempting to help older adults with mental health problems. As late as the mid 1960s, one could still occasionally hear arguments emphasizing that older persons simply could not benefit from "true" psychotherapy. The treatment of choice would thus have to be a trial with one of the latest "breakthrough" psychoactive medications, or ECT if they weren't too frail; otherwise, they would just have to settle for some type of supportive counseling. Even as these arguments were challenged with countering evidence, few clinicians moved beyond the languorous posture of "why bother."

Without making this a history lesson, a number of significant individuals and policies came to light during the 1960s that began to change this scene. Politicians began to feel pressure from their constituents that the elderly population was increasing, and something needed to be done to assure that older adults had adequate health care. Congress saw to it that federal funds were allocated for aging research; by the late 1960s, burgeoning gerontological and geriatric research activities stimulated numerous discussions at many different levels, leading to the creation of an independent Institute of Aging within the National Institutes of Health. As more funds were committed to research, so too, was the interest of the scientific and academic communities.

More importantly, theories focusing on the elements of change in psychiatric patients began to incorporate data and models from the psychological literature. Interesting comparisons between learning theories and psychoanalytic/psychodynamic models began to occur. The growing number of clinical psychologists, which started in earnest during World War II, quickly saw the value of applying these change models when working with mentally disturbed patients. Within short order, behavioral and cognitive intervention models were developed, refined, and empirically tested, leading to compelling arguments that there were more efficient ways of treating mental patients than psychoanalysis or vintage psychodynamic psychotherapy.

And so, we have had a wealth of prominent theorists to lead us, some of whom have become household names rivaling the reputation of Sigmund Freud. This list begins, perhaps, with Watson (a little before our time), but then moves on to Guthrie, Meehl, Dollard, Miller, Jacobs, Wolpe, Kelly, Lewinsohn, Beck, Jacobson, and many more, all of whom laid the groundwork with conceptual models and intervention strategies more favorably disposed to the treatment of the elderly.

Despite these developments, there were few nurses, psychiatrists, psychologists, and social workers in those early days who were interested in working with elderly psychiatric patients, let alone attempting to apply therapy techniques that were notably different than the traditional analytic/dynamic therapies. One such clinician in the trenches, who comes to mind, is Bob Kahn. When few were thinking of a clinical geropsychology profession, he was forging ahead in Chicago and training some of our leading geropsychologists; these individuals are active today in shaping the pathways that behavioral and cognitive interventionists, of whatever discipline, must traverse.

We provide the above narration to illustrate how much this group of papers symbolizes the rapidity with which the times have changed. Conceptualization, assessment, and interventions that rely heavily on behavioral and cognitive approaches (CBTs) have advanced significantly in the past four decades. The recent name change of the premier interdisciplinary international professional association devoted to the development, evaluation, and dissemination of these approaches, from *Association for the Advancement of Behavior Therapy (AABT)*, to the *Association for Behavioral and Cognitive Therapies (ABCT)*, signifies a number of interrelated changes in the field. Included is the transition of CBTs to mainstream professional practice, increased attention to the role of cognitive processes in behavior change, and acknowledgement of the wide range of theories and clinical practices that are covered by the terms "behavioral," "cognitive," and "cognitive-behavioral." For example, from earlier work on behavioral (Goldfried & Davidson, 1976) and cognitive (Beck, Rush, Shaw, & Emery, 1979) therapies, the range of populations and problems addressed by CBTs has expanded dramatically (Craighead, Craighead, Kazdin & Mahoney, 1994) and continues to grow. We are very pleased to be able to offer this handbook as an indicator of the ongoing progress being made in the application of CBTs and newer integrative approaches to understanding and ameliorating mental health problems in older adults.

We attempted to sample a broad range of CBT interventions that would reflect their use with a wide variety of patient populations. Authors were asked to include a discussion of the empirical support for their approach, a brief description of the intervention, followed by a case illustration. In each chapter that describes a specific intervention approach, we have also asked authors to address issues of cultural diversity (Hays & Iwamasa, 2006) when applying the conceptualizations and interventions with ethnically diverse older adults.

The chapters included can be viewed as falling within four general categories. The first section reviews a number of common mental health problems and the evidence base documenting the efficacy of each treatment. The topics covered in this section include depression, anxiety, insomnia, alcohol abuse, pain management, and chronic stress of caregiving. The second section focuses on treatment of patients with more severe mental illness, such as schizophrenia and other psychoses, suicidal behavior, personality disorders, and dementia. The third section includes patient groups where the evidence base is not yet strong, but the interest on the part of many clinicians in using CBTs is. These chapters describe issues in treating patients with severe bipolar disorder, stroke victims, patients suffering with complicated bereavement, the indigent, and patients with PTSD. The final section includes three chapters discussing several issues that have relevance for the development of future directions. While not exhaustive, issues in training and compensation warrant consideration. Finally, we have included a chapter that turns our attention to more positive features in aging that are deserving of attention as we consider the mental health needs of the elderly. It is noteworthy that in our short history of treating older patients, we have yet to build a treatment model that is uniquely relevant for the elderly. What we have done thus far is adapt models developed for use with other younger groups, and then tweak them in ways to make them applicable for work with the elderly. This makes *abundant good sense*, but many characteristics of importance may often get left by the wayside. The chapter on positive aging reminds us of important constructs we need to consider as we begin to develop intervention models specifically for use with this segment of our population.

Although varied in focus, behavioral and cognitive theories and interventions are generally characterized as utilizing basic research in learning, cognitive processes, and emotional regulation, as well as fostering the accompanying principle that learning is a life-long process. Thus, the CBTs are well suited to helping the field address the diagnostic (Jeste, Blazer & First, 2005) and treatment challenges of working with older adults (Gallagher-Thompson & Thompson, 1996). These interventions are grounded in coherent theories of psychopathology and change, and involve structured, often time-limited or time-efficient approaches that use guided mastery of behavioral, cognitive, and emotional self-regulation skills through instruction, in-session practice, and between-session assignments. Also, specific efforts are made so that the skills developed during treatment can generalize to future problems and challenges. Depending upon the severity of the condition, goals range from better symptom self-management and psychosocial functioning to the client being able to initiate and pursue self-interventions after treatment is over; essentially, individuals become their own "therapist" or interventionist.

Although chapters in this handbook describe a wide range of intervention approaches that are considered behavioral and/or cognitive in nature and designed for use with other specific groups, we recommend that professionals working with older adults also become familiar with recommendations for adapting interventions for work with older adults (Zeiss & Steffen, 1996). With significant interindividual differences in physical and cognitive functioning in late life, such recommendations should be viewed as general guidelines as opposed to rules. These suggestions reflect adaptations to better fit the learning style of older adults, including a slower pacing of material presented, multimodal training (i.e. "say it, show it, do it"), using memory aides (e.g., written homework reminders, providing tapes of sessions to listen to in between sessions, etc.), making use of strategies to stay on track during sessions (e.g., refocusing, keeping agenda visible, etc.), and planning for generalization of training. It is also important for clinicians to identify strengths of the older client that can be used to advance therapy, and consider the role of wisdom in responding to life's challenges. Scogin (2000) expands on these issues in a very nice discussion of skills needed for beginning clinical work with older adults. We also would like to emphasize the strongly multidisciplinary nature of work with older adults, and suggest that professionals become familiar with concepts and practices in interdisciplinary team functioning (Zeiss & Steffen, 1998).

Behavioral and cognitive approaches to conceptualization, assessment, and intervention are also characterized by a strong emphasis on empiricism; this is true for each clinician who uses an individual case formulation approach, and also for the field in demonstrating treatment efficacy and effectiveness. That is, a great deal of attention is paid by clinicians to ongoing assessment of targeted problems, identifying mechanisms of change for a specific client, and isolating the strategies leading to a successful treatment response. Because of the emphasis on documenting both intervention mechanisms and outcomes, behavioral and cognitive approaches have strong empirical support in the treatment literatures for many mental health issues, and are ideally suited to many mental health problems in later life. In this handbook, we have attempted to balance our coverage of topics that have led to the development of empirically supported therapies (Chambless & Hollon, 1998) with attention to newer areas of inquiry that are perhaps better viewed from an evidence-based approach that acknowledges the role of clinician judgment in the absence of strong empirical support for a specific therapy (APA, 2006; Goodheart, Kazdin, & Sternberg, 2006; Norcross, Beutler, & Levant, 2006).

We would also like to remind prospective investigators that, although considered the "gold standard" for demonstrating treatment efficacy, large and correspondingly expensive randomized clinical trials are not the only means of advancing the science of mental health interventions for older adults (Stiles et al., 2006). In their description of the criteria used to define "empirically supported treatments," Chambless and Hollon (1998) discuss the role of carefully controlled single case experiments and their group analogues. An intervention would be labeled "possibly efficacious" if shown to be beneficial to three or more participants in research conducted by a single group. Multiple replications of controlled single case experiments (with three or more participants) by two or more independent research groups are needed to demonstrate treatment efficacy. Thus, professionals unable or uninterested in doing large scale intervention trials still have much to contribute. In addition, whether an intervention is being tested in an RCT design or in a controlled single case experiment, Chambless and Hollon (1998) emphasize the essential need for independent replication in at least two studies (i.e., by investigators unaffiliated with the group where the intervention originated). Therefore, in addition to developing new interventions, we would all be well served by taking the time to replicate those interventions originally developed and tested by others.

Gallagher-Thompson, Steffen, and Thompson

References

APA Presidential Task Force on Evidence-Based Practice. (2006). Evidence-based practice in psychology. *American Psychologist, 61*, 271–285.

Beck, A. T., Rush, A. J., Shaw, B. F., & Emery, G. (1979). *Cognitive therapy of depression*. New York: Guilford.

Chambless, D. L., & Hollon, S. D. (1998). Defining empirically supported therapies. *Journal of Consulting and Clinical Psychology, 66*, 7–18.

Craighead, L. W., Craighead, W. E., Kazdin, A. E., & Mahoney, M. J. (1994). *Cognitive and behavioral interventions: An empirical approach to mental health problems*. Needham Heights, MA: Allyn & Bacon.

Gallagher-Thompson, D., & Thompson, L. W. (1996). Applying cognitive-behavioral therapy to the psychological problems of later life. In S. H. Zarit & B. G. Knight (Eds.), *A guide to psychotherapy and aging* (pp. 61–82). Washington, DC: American Psychological Association.

Goldfried, M. R., & Davidson, G. C. (1976). *Clinical behavior therapy*. New York: Holt, Rinehart, and Winston.

Goodheart, C. D., Kazdin, A. E., & Sternberg, R. J. (2006). *Evidence-based psychotherapy: Where practice and research meet*. Washington, DC: American Psychological Association.

Hays, P. A., & Iwamasa, G. Y. (Eds.). (2006). *Culturally responsive cognitive-behavioral therapy: Assessment, practice, and supervision*. Washington, DC: American Psychological Association.

Jeste, D. V., Blazer, D. G., & First, M. (2005). Aging-related diagnostic variations: Need for diagnostic criteria appropriate for elderly psychiatric patients. *Biological Psychiatry, 58*, 265–271.

Norcross, J. C., Beutler, L. E., & Levant, R. F. (Eds.). (2006). *Evidence-based practices in mental health: Debate and dialogue on the fundamental questions* (pp. 56–130). Washington, DC: American Psychological Association.

Scogin, F. (2000). *The first session with seniors*. San Francisco, CA: Jossey-Bass.

Stiles, W., Hurst, R., Nelson-Gray, R., Hill, C., Greenberg, L., Watson, J. C., Borkovec, T. D., Castonguay, L. G., & Hollon, S. D. (2006). What qualifies as research on which to judge effective practice? In J. C. Norcross, L. E. Beutler, & R. F. Levant (Eds.), *Evidence-based practices in mental health: Debate and dialogue on the fundamental questions* (pp. 56–130). Washington, DC: American Psychological Association.

Zeiss, A. M., & Steffen, A. (1996). Treatment issues with elderly clients. *Cognitive and Behavioral Practice, 3*, 371–389.

Zeiss, A. M., & Steffen, A. M. (1998). Interdisciplinary health care teams in geriatrics: An international model. In B. A. Edelstein (Ed.), *Vol. 7: Clinical geropsychology* (pp. 551–570) of A. S. Bellack & M. Hersen (Eds.), *Comprehensive clinical psychology*. London: Pergamon Press.

Acknowledgments

At the present time "Cognitive and Behavioral Therapies" (CBTs) arguably have the strongest evidence-base for effectiveness with the elderly when considering the wide variety of mental/behavioral disorders experienced by this population. It seemed to us that the "time had come" for this information to be shared with the larger community of mental health practitioners working with older adults. We would like to express our sincere appreciation to Antoinette Zeiss, Ph.D., whose vision for this book was what really initiated this project. Dr. Zeiss, past president of ABCT, is a psychologist who is well known for strongly promoting the dissemination of evidence-based and empirically supported mental health treatments. Unfortunately, due to her current pressing employment commitments, she was not able to continue with the project, but her inspiration and support have been there for us throughout this process. We would also like to give special acknowledgement to Peter Lewinsohn, Ph.D. University of Oregon and Aaron T. Beck, M.D. University of Pennsylvania, for their mentoring during the early years of the CBTs, and for their conceptual and empirical contributions that have provided a rational and generative foundation for the range of useful CBT modifications subsequently developed, many of which are represented in this book. Knowing these individuals personally, learning from them, and maintaining positive relationships with them over the years, have greatly enhanced our expertise in this field. We also wish to thank the authors for their enthusiasm about this project, and for sharing their wisdom and practical experience with all of us. Lastly, we wish to acknowledge our patients, from whom we have learned so much over the years, and our many other colleagues whose collaborations and lively discussions have informed our clinical thinking and practice. It is our sincere hope that this book will address many current gaps in the treatment of late-life mental health issues of older adults, and that it will "stand the test of time" in its usefulness to the field.

Contents

Contributors

George Alexopoulos
Cornell University, Ithaca, NY

Patricia A. Areán
University of California, San Francisco, CA

Aaron T. Beck
Department of Psychiatry, University of Pennsylvania, Philadelphia, PA

Judith S. Beck
Beck Institute for Cognitive Therapy and Research, Department of Psychiatry, University of Pennsylvania, Philadelphia, PA

J. Gayle Beck
University at Buffalo, State University of New York, Park Hall, Buffalo, NY

Sunil Bhar
Department of Psychiatry, University of Pennsylvania, Philadelphia, PA

Andrea K. Booth
Department of Psychology, University of Memphis, Memphis, TN

Gretchen A. Brenes
Wake Forest University School of Medicine, Winston-Salem, NC

Gregory K. Brown
Department of Psychiatry, University of Pennsylvania, Philadelphia, PA

Lisa M. Brown
Department of Aging and Mental Health, University of South Florida, Tampa, FL

Jennifer S. Cheavens
Duke University Medical Center, Durham, NC

Stacey Cherup
University of Nevada, Reno, NV

Joyce Chu
University of California, San Francisco, CA

Daisha J. Cipher
Department of Psychiatry, University of North Texas Health Science Center, Fort Worth, TX

P. Andrew Clifford
Mind Body Wellness, P.C. & Senior Connections of Dallas, Dallas, TX

Joseph M. Currier
University of Memphis, Memphis, TN

Kristina O. Dearborn-Harshman
Department of Aging and Mental Health, Florida Mental Health Institute, University of South Florida, Tampa, FL

Claudia Drossel
University of Nevada, Reno, NV

Larry W. Dupree
Department of Aging and Mental Health, Florida Mental Health Institute, University of South Florida, Tampa, FL

Andrea Feit
Pacific Graduate School of Psychology, Palo Alto, CA

Kyle Ferguson
University of Nevada, Reno, NV

Jane E. Fisher
University of Nevada, Reno, NV

David Folsom
University of California San Diego, Veterans Affairs San Diego Healthcare System, San Diego, CA

Dolores Gallagher-Thompson
Department of Psychiatry and Behavioral Sciences, Stanford University School of Medicine, Stanford, CA

Judith R. Gant
University of Missouri-St. Louis, St. Louis, MO

James M. Georgoulakis
Webster University, Fort Sam Houston Metro Campus (SANA)

E. Granholm
Department of Psychiatry, University of California, San Diego, VA San Diego Health Care System, La Jolla, CA

Paula E. Hartman-Stein
Center for Healthy Aging, Kent, OH

Robert D. Hill
University of Utah, Salt Lake City, UT

Jason M. Holland
University of Memphis, Memphis, TN

Lee Hyer
Georgia Neurosurgical Institute and Department of Psychiatry, Mercer Medical School, Macon, GA

Michele J. Karel
VA Boston Healthcare System, Harvard Medical School, Brockton, MA

David Kingdon
Department of Psychiatry, Royal South Hants Hospital, University of Southampton, Southampton, UK

Bob Knight
Andrus Gerontology Center, University of Southern California, Los Angeles, CA

Kenneth Laidlaw
University of Edinburgh, Edinburgh, UK

Kenneth L. Lichstein
Department of Psychology, The University of Alabama, Tuscaloosa, AL

Laurie Lindamer
University of California San Diego, Veterans Affairs San Diego Healthcare System, San Diego, CA

Catherine Loh
University of California San Diego, Veterans Affairs San Diego Healthcare System, San Diego, CA

Thomas R. Lynch
Duke University and Duke University Medical Center, Durham, NC

Nancy Lynn
Department of Aging and Mental Health, Florida Mental Health Institute, University of South Florida, Tampa, FL

E. Mansour
University of Utah, Salt Lake City, UT

Rene Marquett
Pacific Graduate School of Psychology, Palo Alto, CA

Meghan McCarthy
Department of Psychiatry at Dartmouth Medical School in Hanover, NH, Dartmouth Psychiatric Research Center, Concord, NH

Christine L. McKibbin
University of California San Diego, Veterans Affairs San Diego Healthcare System, San Diego, CA

Tara Mehta
University of Memphis, Memphis, TN

Johnathan Meyer
University of California San Diego, Veterans Affairs San Diego Healthcare System, San Diego, CA

Victor Molinari
Department of Aging and Mental Health, Louis de la Parte Florida Mental Health Institute, University of South Florida, Tampa, FL

Kathryn S. Moss
The University of Alabama, Tuscaloosa, AL

Kim T. Mueser
New Hampshire-Dartmouth Psychiatric Research Center, Concord, NH

Robert A. Neimeyer
University of Memphis, Memphis, TN

Tam Nguyen
Pacific Graduate School of Psychology, Palo Alto, CA

Nancy A. Pachana
School of Psychology, University of Queensland, Brisbane, Australia

Sarah I. Pratt
Department of Psychiatry at Dartmouth Medical School in Hanover, NH, Dartmouth Psychiatric Research Center, Concord, NH

Robert Reiser
Pacific Graduate School of Psychology, Palo Alto, CA

Kristi D. Roper
Mind Body Wellness, P.C. & Senior Connections of Dallas, Dallas, TX

Amanda Sacks
Moun Sinai School of Medicine, Department of Rehabilitation Medicine, New York, NY

Lawrence Schonfeld
Department of Aging and Mental Health, Florida Mental Health Institute, University of South Florida, Tampa, FL

Forrest R. Scogin
Department of Psychology, The University of Alabama, Tuscaloosa, AL

A'verria Sirkin
University of California San Diego, Veterans Affairs San Diego Healthcare System, San Diego, CA

A. Lynn Snow
Center for Mental Health and Aging, Department of Psychology, The University of Alabama, Tuscaloosa, AL

Melinda A. Stanley
Baylor College of Medicine, Winston-Salem, NC

Ann M. Steffen
Department of Psychology, University of Missouri-St. Louis, St. Louis, MO

Kristen C. Stone
Brown Medical School, Providence, RI

Maged Swelam
Department of Psychiatry, Royal South Hants Hospital, Southampton, UK

Merry Sylvester
University of Nevada, Reno, NV

Larry W. Thompson
Department of Medicine, Stanford University School of Medicine and Pacific Graduate School of Psychology, Los Atlos, CA 94024, USA

Diana Truong
Pacific Graduate School of Psychology, Palo Alto, CA

Wendi Wachsmuth
Pacific Graduate School of Psychology, Palo Alto, CA

Paula Wagener
Baylor College of Medicine, Winston-Salem, NC

1
Behavioral and Cognitive Treatments for Geriatric Depression: An Evidence-Based Perspective

Kathryn S. Moss and Forrest R. Scogin

Depressive disorders are common in late life and impact those who experience them on a number of levels. Although depressive disorders are among the most studied conditions in the older adult literature, their complex nature requires that continued research be conducted to determine the best methods to assess and treat them. The population of older persons is burgeoning, making it imperative that professionals become knowledgeable about the assessment and treatment of depressive disorders. This chapter provides an overview of the evidence-base for specific cognitive and behavioral treatments for late-life depression. Considerations for assessing depressive disorders in older adults as well as issues related to combination treatment (antidepressant medication + psychotherapy) are discussed. Suggestions for the provision of culturally competent treatment are covered as well. It is our hope that this chapter will both provide information for individuals seeking to provide evidence-based treatment to older adults as well as stimulate further research in this area.

Background/Prevalence

Major depressive disorder is one of the most commonly occurring mental health disorders in late life. Prevalence estimates of major depression in this population range from 3 to 5% in community-dwelling elders, to 6–8% in primary care settings, to approximately 13% in home health care recipients (Bruce, McAvay, & Raue, 2002) and those residing in assisted living facilities (e.g., Watson, Garrett, Sloane, Gruber-Baldini, & Zimmerman, 2003). The prevalence of depression (as identified via a Minimum Data Set (MDS) diagnosis, which is nonspecific in regard to formal diagnostic criteria) in nursing homes is even higher at approximately 20% (Jones, Marcantonio, & Rabinowitz, 2003). Rates of minor depression, dysthymia, or significant depressive symptoms are even higher among older adults. In fact, older adults have comparable or even higher rates in these categories than do younger persons (Blazer, 2002). Overall, these percentages translate to roughly 5 million US older adults (age 65 or older) who are impacted by depressive disorders. Some believe that these estimates may in fact be low given that our existing diagnostic techniques for psychiatric disorders are not fully applicable to older adults (e.g., Jeste, Blazer, & First, 2005).

It has been suggested that depressive disorders are differentially expressed in older versus younger adulthood. Older people are less likely than their younger counterparts to report specific affective symptoms (e.g., depressed mood) and more likely to endorse somatic symptoms (e.g., fatigue, insomnia, and anorexia), anxiety, memory loss, and cognitive impairment (Birrer & Vemuri, 2004; Gottfries, 1998). Risk factors for the development of a depressive illness include, but are not limited to, health conditions (e.g., cardiovascular diseases, arthritis, dementia); female sex; abuse of alcohol or other substances; functional disability; being single, divorced, or widowed; caregiving; lack of social support; a history of depression; and stressful life events. Also, although a controversial topic (e.g., Kotlyar, Dysken, & Adson, 2005), it is thought that certain medications may be associated with depressive symptoms in this population. Major classes of medication that have been implicated in relation to such symptoms include cardiovascular drugs, chemotherapeutics, antiparkinsonian drugs, antipsychotics, sedatives and antianxiety drugs, anticonvulsants, anti-inflammatory and anti-infective agents, stimulants, and hormones, among others (Birrer & Vemuri).

Depression is of particular concern to older adults for a number of reasons. First, clinically significant depression can accelerate the disease process of numerous health conditions and thus can increase disablement (van Gool et al., 2005). Depressive disorders increase the risk of several chronic health conditions, including Type 2 diabetes (Meyers, 2005), chronic pain (Currie & Wang, 2005), and ischemic heart disease (Glassman et al., 2003). In addition, depressive symptoms have been shown to predict nursing home admission (Harris & Cooper, 2006) and increased hospitalization (e.g., Rumsfeld et al., 2005). Linked to problems with health conditions and hospitalization is the cost of depression. Direct and indirect costs of the disorder are approximately US $43 billion annually in the United States alone (Hirschfeld et al., 1997) and are associated with high economic burden. This can be exceedingly difficult for older adults, who often live on a fixed income. It is estimated that annual direct excess costs average from US $1,000 to US $2,500 per individual with a depressive disorder (Luppa, Heinrich, Angermeyer, König, & Riedel-Heller, 2007). In addition, depression is an independent risk factor for early mortality. Romanelli, Fauerbach, Buch, and Ziegelstein (2002) found that, in older people who had experienced a myocardial infarction, the risk of death within 4 months of hospital discharge was four times greater for those who had a comorbid depressive disorder. Depressive disorders can affect health in a number of ways such that mortality is impacted (e.g., little social support, decreased motivation to manage comorbidities and medications, inadequate coping with life stressors). Further, suicide is a potential outcome of geriatric depression. Suicide rates among older adults are higher as a rule than they are for other age groups (e.g., Manthorpe & Iliffe, 2006). Indeed, depressive spectrum disorders have a great impact on older adults and thus present health care professionals with a number of challenges in terms of their assessment and treatment.

Evidence Base

Despite the challenges associated with the assessment and treatment of geriatric depression, the good news is that therapeutic treatments are available and are effective with this population. The examination of evidence-based psychotherapies is a relatively new area of study as it relates to older adult depression, with the majority of efforts being made within the last decade.

To our knowledge, the most recent efforts to catalog specific psychotherapies for late-life depression were undertaken by our research team in a critical review of evidence-based treatments, or EBTs (Scogin, Welsh, Hanson, Stump, & Coates, 2005). Using a coding system developed by Hawley and Weisz (2001), our team identified six treatments as being evidence-based, four of which are rooted in cognitive and/or behavioral theories: behavioral therapy, cognitive-behavioral therapy, cognitive bibliotherapy, and problem-solving therapy (PST) (brief psychodynamic and reminiscence therapies were also identified as evidence-based treatments; however they will not be covered in this review). Please refer to Scogin et al. for a thorough description of the criteria used to recognize these treatments as evidence-based.

Assessment Considerations

Assessment goes hand-in-hand with the cognitive and behavioral treatment of depressive disorders. It allows for proper screening, treatment planning, and monitoring change in symptoms over time in the depressed client. Depression spectrum disorders are typically assessed by multiple methods in older adults. The most common forms of assessment are self-report instruments, structured and unstructured clinical interviews, and interdisciplinary and other collateral source input.

Self-report measures are probably the most frequently used method of assessing late-life depression. Generally, this type of assessment works well for screening and monitoring change in symptoms. The most commonly used self-report measure of depression in older adults is the Geriatric Depression Scale (GDS; Yesavage et al., 1983). This instrument is preferred by many because it was developed specifically for use with older adults, has a simplified yes/no response format, and contains very few items related to somatic symptoms. Other measures that have strong psychometric properties but were not specifically tailored to measure late-life depression include the Beck Depression Inventory-II (BDI-II; Beck, Steer, & Brown, 1996) and the Center for Epidemiologic Studies Depression Scale (CES-D; Radloff, 1977).

Aside from depression instruments, self-report measures are also frequently used in cognitive and behavioral treatments for depression to monitor whether improvements in depressive symptoms can be linked to specific elements of a particular treatment. For example, CBT therapists may choose to use the Dysfunctional Attitudes Scale – A (DAS-A; Beck, Brown, Steer, & Weissman, 1991) to determine whether the treatment is successful in reducing the faulty attitudes that are maintaining their clients' depression. Research by Imber et al. (1990) revealed two DAS factors: perfectionism and need for social approval. Research in our lab with the DAS and depressed older adults (Floyd, Scogin, & Chaplin, 2004) suggests that the social approval factor seemed to hold reasonably well with older adults, whereas the perfectionism factor did not. Based on these findings, we suggest administering only the items comprising the social approval factor to older adult clients. Also, behavioral therapists may choose to use the pleasant events schedule – elderly (PES-E; Teri & Lewinsohn, 1982) to monitor changes in pleasant events throughout treatment. This measure examines the frequency of participation and level of enjoyment of events that are considered to be applicable to older adults.

Although brief self-report instruments serve as a solid method to screen and monitor late-life depressive disorders and fit well into the existing structure of cognitive and behavioral therapies for depression, we recommend more thorough assessment procedures when making formal diagnoses. We suggest that a thorough assessment be conducted during the initial clinical interview process that includes both structured and unstructured components. We recommend that the interviews be modified to facilitate a smoother assessment experience with the older adult client (e.g., simplified questions, shortened length, etc.; Snarski & Scogin, in press). The clinical assessment should also contain a suicide risk assessment, as depressive disorders are related to suicide and suicide occurs most frequently in older adults. In addition, clinicians should consult the DSM-IV-TR (American Psychiatric Association, 2000) to be sure that their clinical impression maps onto the symptoms required to meet a diagnosable depressive disorder. Finally, input from family members and other professionals is necessary in forming an opinion about client insight, denial, or potential memory impairment, among other things.

Indeed, assessment is key in the identification and treatment of late-life depression. Monitoring change over time in both process and outcome variables via self-report methods is common in cognitive and behavioral treatment. The use of more thorough assessment procedures prior to commencing treatment is suggested, as it allows clinicians to tailor treatment methods based on the needs of particular clients.

We remind our readers that this is simply a brief overview of the assessment of geriatric depression. Please refer to our recent chapter on the assessment of depression (Snarski & Scogin, in press) or other good sources (e.g., Edelstein, Kalish, Whipple Drozdick, & Rettig McKee, 1999) for further information on the measures/methodology we have presented as well as a more thorough description of issues encountered with older adults in the assessment process.

Treatment/Intervention Approach

Again, the specific cognitive, behavioral, and cognitive-behavioral treatments that have been identified as having sufficient evidence base to be supported as beneficial to older adults are behavioral therapy, cognitive-behavioral therapy, cognitive bibliotherapy, and PST. Below we briefly review the major tenets of each of these therapies independently, discussing the specific cognitive and/or behavioral strategies used in each as well as special considerations for mental health professionals in implementing these depression treatments with older adult clients. The special considerations named under each therapy are not exhaustive lists, nor are they necessarily exclusive only to their particular treatments. However, we feel our suggestions nonetheless offer clinicians interested in the provision of these therapies a good basis for informing their practice.

Behavioral Therapy

Behavioral therapy, or BT, is a structured means to increase access to positive reinforcement and decrease engagement in negative events and avoidance patterns. Many of the behavioral treatments of today are based on work originally conducted by Lewinsohn and colleagues (Lewinsohn, 1974; Lewinsohn, Biglan, & Zeiss, 1976). The origins of such treatments are rooted in learning theory models of depression (Ferster, 1973; Lewinsohn & Graf, 1973), which suggest that decreased engagement in positively reinforcing behaviors is key in the development of depression. More specifically, depressed mood is said to be a function of excesses in unpleasant experiences/negative reinforcement (e.g., death of a loved one; avoidant coping) and deficits in pleasant events/positive reinforcement (e.g., hobbies; socialization). Clients enrolled in BT are encouraged to engage in behavioral monitoring, which often involves keeping a daily schedule or list of any major activities conducted throughout the day and recording how those activities affect mood. Behavior therapists assist clients in reviewing behavioral monitoring forms to uncover any behaviors the client is engaging in that promote negative affect. Examining behavior using a functional analysis is also emphasized throughout the course of BT. In addition, therapists help the client either discover new pleasant events that may contribute to improved mood or rediscover pleasant events that the client has given up in response to the depression and build those reinforcing experiences into the client's daily life.

The five studies that supported BT as an EBT used a variety of manuals/protocols (e.g., Gallagher & Thompson, 1981; Lewinsohn et al., 1976; Rehm, 1977; Teri, 1991, 1994; Teri, Logsdon, Wagner, & Uomoto, 1994) all of which were shown to improve depressive symptoms above and beyond those in control groups. Please consult these materials for a more detailed description of how to conduct behavioral therapy.

Although it has been established that behavioral therapy can be applied successfully to older adults, a chief consideration therapists must take into account when using BT with the older adult client is that functional limitation/activity restriction arising from health conditions may preclude involvement in specific pleasant events. For example, an older client might enjoy bowling but is unable to bowl in the traditional sense due to functional limitation. In this case, adapting this behavior or finding a more suitable replacement is indicated. Following the aforementioned example, the behavioral therapist may suggest bowling with adaptive equipment (e.g., a bowling ramp) or finding an alternative that would be just as pleasurable (e.g., coaching a youth bowling league). It is important that the clinician using behavior therapy consider pleasant events to be more loosely defined (Coates & Scogin, 2005) and thus be creative in coming up with activities that severely activity-limited older adults can participate in. Another sensible consideration in administering BT involves checking for sensory or physical impairments that may prevent the older adult client from completing behavioral monitoring paperwork. Materials presented to older adult clients should have a larger font size. If the individual is unable to write, the therapist should consider completing the paperwork for the client in session, providing the client with a tape recorder to verbally describe the events of the day, suggesting the use of adaptive devices, or engaging the participation of a family member to assist with writing (with the client's consent and permission).

Although some special considerations exist for the delivery of BT to older adults, they are relatively minor and are generally easily addressed by the therapist. It is our thought that BT may be a very applicable treatment to older adult populations because of its reliance on pragmatic, personally relevant concepts and tasks that perhaps require less cognitive demand.

Cognitive-Behavioral Therapy

Cognitive-behavioral therapy, or CBT, has been widely studied in older adult populations and is the most researched therapy in the treatment of geriatric depression. Eight out of the nine studies our research team examined supported cognitive-behavioral therapy as a beneficial treatment for older adults with a depressive disorder or clinically significant depressive symptoms. A number of CBT manuals/protocols were used in the studies establishing CBT as an evidence-based treatment, many of which were based on the pioneering work conducted by Beck and colleagues (Beck, Rush, Shaw, & Emery, 1979). Since then, a stand-alone CBT manual specific to the treatment of late-life depression has been created by Laidlaw and colleagues that is based on many of those earlier studies (Laidlaw, Thompson, Dick-Siskin, & Gallagher-Thompson, 2003). This manual provides a very comprehensive set of instructions for therapists to follow as well as examples that can be used to facilitate older clients' understanding of various cognitive-behavioral concepts. We suggest that therapists consult this manual first as a "how-to," as it is more up-to-date and relevant to the needs of older clients than the older manuals used in the initial large-scale studies of CBT for geriatric depression.

Although cognitive adaptations are often necessary in applying CBT to older adults with a variety of mental health disorders, we would like to emphasize the importance of making such modifications in treating older adults with depression specifically. Memory impairment, executive dysfunction, and other cognitive limitations frequently occur in conjunction with late-life depression. An excellent collection of special considerations to take into account when conducting CBT with older adults has been recommended by Zeiss and Steffen (1996). Among other areas of discussion, these authors provide a set of suggestions describing how to adapt CBT to cognitive limitations experienced by some older adults. Due to the meta-cognitive skill required of clients engaged in CBT, we suggest that readers consult this article before applying the treatment to older adult clients with depression. Further, because CBT may continue to prove challenging to deliver despite making necessary adaptations, it has been recommended that the behavioral techniques of the treatment (which are purported to be less cognitively taxing) be emphasized in CBT in situations where cognitive techniques such as thought monitoring or disputation are too difficult to employ (Gallagher-Thompson & Thompson, 1996; Scogin, 2000).

Cognitive Bibliotherapy

Cognitive bibliotherapy is a self-administered or "self-help" treatment that involves reading and written exercises from a book based on cognitive-behavioral theory. The specific behavioral and cognitive strategies used in cognitive bibliotherapy closely mirror those used in CBT. This treatment is unique in comparison to the other evidence-based behavioral and cognitive therapies for geriatric depression in that it is self-paced and can be completed away from the therapist's office (i.e., at home or another convenient location) with minimal contact on the part of the clinician. In the studies we examined, therapist contact involved weekly telephone calls of 5 min or less to check in on client progress and answer any questions. There are several benefits of this treatment. It can be relatively brief and inexpensive. It can also serve as a means of treatment for older adults who would otherwise not be able to receive therapy due to geographical location, health conditions, transportation limitations, etc. Along those lines, it provides an alternative treatment to those who would rather not participate in face-to-face therapy for a variety of reasons (e.g., stigma, time constraints, etc.). Cognitive bibliotherapy could even be considered by primary care providers as an adjunctive treatment or perhaps a first step in primary care stepped-care models.

We examined four cognitive bibliotherapy intervention studies; all of which supported the treatment as evidence-based (Floyd, Scogin, McKendree-Smith, et al., 2004; Landreville & Bissonnette, 1997; Scogin, Hamblin, & Beutler, 1987; Scogin, Jamison, & Gochneaur, 1989). All of these studies used the *Feeling Good* book (Burns, 1980). We acknowledge that several other very good self-help publications based on cognitive theory exist. However, more research on these books is required to support them as evidence-based cognitive bibliotherapies for geriatric depression.

Although cognitive bibliotherapy is a very promising treatment for late-life depression, there are situations and/or participants for which it may not be well suited. Clinicians must be mindful of cognitive limitations or problems with literacy that could hinder a client's ability to adequately receive the information presented in bibliotherapy format. In addition, as visual limitations are common in older adulthood, reading materials should be presented in large font or clients should have adaptive devices readily available (e.g., glasses, a magnifying glass). Brief reading or vision tests to rule out significant deficits might be informally conducted prior to suggesting cognitive bibliotherapy to a client. Also, as cognitive bibliotherapy is more independent in nature than other therapies, the clinician has less control over client enactment. Because of this, clients expressing apathy and significant deficits in motivation may be inappropriate for this form of therapy due to the rather common recommendation that the material be read in a timely fashion (in the studies we examined, usually 4 weeks). However, for cognitively intact older clients who enjoy reading and may not otherwise have access to treatment, cognitive bibliotherapy is an avenue that can be explored for the management of depression spectrum disorders.

Problem-Solving Therapy

PST is a structured treatment that uses problem solving as an active coping strategy. In PST, therapists help clients through four key steps: (1) clearly defining and formulating a particular problem that is exacerbating their depressive disorder; (2) formulating a thorough list of possible solutions to the problem (usually alternatives to current coping methods); (3) examining potential consequences of each solution and determining one to make use of based on those consequences; and (4) evaluating the outcomes of the chosen solution after it has been implemented to determine whether or not to build the strategy into a routine. PST is praised for being a pragmatic, concrete, and easy to understand treatment. As such, it is generally accepted that it can be easily taught to and implemented by paraprofessionals (e.g., case managers, nurses) in a broad range of settings. Slightly modified versions of PST (as part of more comprehensive treatment services) have received a great deal of attention recently for proving successful in reducing depressive symptoms in primary care (Bruce et al., 2004; Lin, Katon, & Von Korff, 2003).

Three studies were examined in the EBT review, all of which supported PST as evidence-based. Two of these studies employed the commonly used protocol developed by Nezu, Nezu, and Perri (1989).

Because PST is very straightforward in its approach, we have very few recommendations concerning its delivery to depressed older adults. Because PST is cognitive-behavioral in its approach (Areán et al., 1993) we suspect that many of the suggestions offered by Zeiss and Steffen (1996) for CBT would be appropriate for PST as well.

Combination Treatment

Pharmacological treatment is often recommended in conjunction with psychotherapy in the treatment of geriatric depression. Both of these options have proven effective in treating late-life depression independently, and may result in an even more robust effect in certain individuals when administered concurrently (e.g., Baldwin et al., 2003; Hollon et al., 2005). The level of effectiveness of a particular monotherapy versus combination treatment may depend on the severity of depression experienced by an individual.

Baldwin et al. (2003), in forming guidelines for treating late-life depression in primary care, provided the following suggestions: (1) Mild to moderate depression should be treated with psychotherapy as a first line of treatment if the patient is amenable to psychological intervention. CBT and PST were highlighted in their article as cognitive-behavioral treatments that were promising in the treatment of late-life depression in primary care (along with other therapies – interpersonal (IPT) and brief focal dynamic treatments). Other research (e.g., Thompson et al., 2001) provides support for the suggestion that psychotherapy is an acceptable treatment for mild to moderate depression; (2) Combination treatment should be considered for moderate to severe depression given that it is associated with better outcome than either intervention alone in this severity level. In addition, combination therapy is effective in preventing recurrence of symptoms in high-risk patients; (3) In the case of sub-threshold levels of depression (e.g., adjustment disorder, minor depression), structured support may be preferable to either psychological or pharmacological treatment; and (4) Because of these findings, primary care teams should have access to individuals who can provide structured support and evidence-based therapies.

Hollon et al. (2005) have provided some other key points regarding the issue of combination treatment. They note that although medication is effective and prevents symptom return during the course of treatment, it has little if any effectiveness in reducing the risk of relapse once an individual discontinues taking the medication. They indicate that IPT and CBT are both as effective as antidepressant medication in treating depressed outpatients and that CBT can reduce subsequent risk following treatment. They further point out that combination therapy may improve overall treatment response.

Also, it is important to consider patient characteristics and treatment preference when deciding if combination treatment is appropriate for a particular older adult client. Several studies in the literature indicate that older adults have an equal or greater preference for psychological treatments as compared with pharmacological treatments (e.g., Gum et al., 2006; Landreville, Landry, Baillargeon, Guerette, & Matteau, 2001; Rokke & Scogin, 1995) although combination therapy may be preferred if depression is described as severe (Hanson & Scogin, 2007). This may be dependent upon patient characteristics (e.g., socioeconomic status), however. For example, a recent study (Cohen et al., 2006) found that older adults residing in a low-income census tract responded less favorably to combination treatment than did middle class older adults. Mental health professionals should also consider patient tolerance to the various antidepressant medications.

In sum, combination treatment does appear to be effective in general in treating depression in older adults. However, it is necessary to determine the level of depression severity, patient treatment preference, and other patient characteristics when determining if it is a suitable treatment for a particular older adult client.

Issues of Diversity

In general, the consensus is that cognitive and behavioral treatments are effective among diverse minority and cultural groups as long as the services are provided by culturally competent professionals (Chiriboga, Yee, & Jang, 2005). Chiriboga et al. cite several sources that all suggest the culturally competent professional should be able to exhibit mastery in the following areas: displaying respect for the client's cultural heritage, providing services in a client's language of heritage when English proficiency is limited, and understanding how a client's cultural background factors into symptom manifestation, significance, and treatment.

Reviewing the various strategies involved in making cognitive and behavioral therapists more culturally competent in their delivery of treatment is complex and could be written in volumes. Therefore, rather than to reiterate these recommendations, we instead have decided to point the reader in the direction of readings that can assist in developing cultural competence as a therapist in general as well as specifically in the administration of cognitive and behavioral therapies.

In developing general multicultural competency, we suggest Smith's (2004) *Practicing multiculturalism: Affirming diversity in counseling and psychology*. This book provides a good overview of understanding multicultural contexts, developing multicultural skill, and considerations for multicultural practice with specific populations and can assist in developing overall therapist cultural competency.

The treatment that has been most commonly studied cross-culturally in the literature is CBT. Thus, it has the most information available on its application to culturally diverse samples. Recently, a book discussing the integration of culture into CBT, Hays and Iwamasa's (2006) *Culturally responsive cognitive-behavioral therapy: Assessment, practice, and supervision*, was published. In this book, the authors discuss a variety of modifications therapists can make to increase the effectiveness of CBT with individuals representing a variety of cultures. Included in this volume is a chapter by Lau and Kinoshita that discusses the application of CBT specifically in relation to the needs of culturally diverse older adults. This chapter discusses issues of cultural identity in relation to the provision of treatment to older adults as well (e.g., how identifying with the majority, minority, or more than one culture map on to treatment procedures). Another good source that discusses the application of CBT to low-income elders is Areán and colleagues' (Areán et al. 2005) article, which discusses augmenting traditional cognitive-behavioral group therapy with clinical case management in an effort to better ensure the reduction of depressive symptoms.

Unfortunately, at this time no readings on culturally competent treatment provision specific to behavior therapy, PST, and cognitive bibliotherapy can be located using a computer-based literature search. However, because BT and PST are both components of CBT and cognitive bibliotherapy is essentially CBT in book format, we feel that many of the recommendations suggested for the provision of CBT to culturally diverse elders would generalize to these treatments. Despite this, it would be preferable to have empirical research backing up the applicability of these treatments to various cultures and ethnicities as well as providing any suggested modifications.

With numbers of ethnically and culturally diverse older adults steadily increasing in the United States, it is imperative that therapists become competent in approaching treatment from a multicultural standpoint. This is no different in the context of cognitive and behavioral therapies. It is our expectation that increasing work will be done to incorporate culturally based practice into the cognitive and behavioral therapies.

Case Example

Rose, a 68-year-old Caucasian female, was brought to the local psychology clinic by her two adult daughters. They recently visited their mother and took her to her primary care provider to seek help in understanding an apparent decline in their mother's memory and mood over the past year. The physician recommended that Rose begin taking antidepressants, which she was not receptive to; thus she was referred to the clinic.

Rose reported to her therapist that she had had a "bad couple of years": her husband died 2 years ago, from complications related to Alzheimer's disease; Rose had been his primary caregiver. She also had significant arthritic pain in her wrists that greatly reduced or eliminated her ability to engage in a variety of previously enjoyed activities, including a job as a waitress at the diner of a family friend. Rose also stated, "to top it all off I am becoming senile. Sometimes I wonder if I am getting Alzheimer's disease myself." She indicated that she often misplaced items in her house and sometimes forgot to attend her medical appointments.

When asked about her perceptions of her husband's death, Rose said she had done all she could for him while he was alive, and felt she had come to terms with his death. Rose said that keeping busy at the diner had helped her through the grieving process. However, now that she had lost that job, Rose reported that she was embarrassed about the decreased functioning in her wrists, and that having no structured activities to participate in made her feel "useless and alone." A friend from the diner had asked Rose to join her walking club, but Rose refused. She stated that she felt she should not try, given she was "getting old and decrepit" and because she had "no kind of energy." After this, she withdrew from her friends, spending most of her days sitting in a lounge chair in front of her television thinking about everything she had lost and how she had nothing to look forward to.

Rose's therapist used a semistructured clinical interview to get a detailed description of the presenting problems. Rose indicated no history of depressive illness and no family history of severe memory impairment. She reported a fairly average upbringing with no traumatic experiences and that she had completed high school with honors. She also indicated that her current relationship with her daughters was good and that, aside from the arthritis, she was remarkably healthy. The suicide risk assessment, which was included as part of the interview, yielded that Rose had occasionally wondered if she would be "better off dead." However, she expressed no suicidal intent or plan. The therapist also administered a Mini Mental State Exam (MMSE; Folstein, Folstein, & McHugh, 1975) and a GDS. Rose obtained a score of 25 out of 30 on the MMSE; 2 points higher than a score that would be indicative of cognitive impairment, but lower than would be expected for someone of her educational level. She also received a score of 20 out of 30 on the GDS, which is the cutoff score for severe depression. Based on these findings, Rose's therapist decided to do a more thorough cognitive assessment to help differentiate between a depressive disorder and dementia. Qualitative observations made during the testing suggested that Rose's symptoms were likely more closely related to a depressive disorder than significant memory impairment. For example, Rose demonstrated a lack of motivation to put effort into the tasks that were asked of her, but she was able to do the tasks with prompting. Also, depending on her level of effort at the time, Rose's scores on similar measures were quite variable, ranging from below average to above average. Taken together, this information suggested that Rose was a healthy, yet depressed and somewhat forgetful (but not demented), older adult.

Armed with these results, the therapist started a CBT regimen, meeting with Rose twice per week for the first 4 weeks. The behavioral strategies of CBT were emphasized first. The therapist emphasized involvement in meaningful activities that were manageable and within the client's capabilities, and educated Rose about the negative affects of avoidance. Rose started working on small goals, such as walking around the block each day and calling some of her friends from the diner. At each visit, Rose's therapist administered the short form of the GDS (GDS-SF; Sheikh & Yesavage, 1986) to monitor change in symptoms. After about a month, Rose's mood had observably improved. Another full GDS was administered, and her score had dropped to a total of 15 (out of 30), indicating improvement. Frequency of sessions was reduced to once per week. Depression assessment using the GDS-SF continued on a bi-monthly basis. The therapist then began integrating more cognitive techniques into treatment. Rose's negative perceptions about her self-worth, health, and misconceptions about aging were challenged and psychoeducation about depression, aging, memory, and pain was provided. Rose was also taught to engage in thought stopping and scheduled worry time to help with rumination. After about 3 months of treatment, Rose reported continued improvements in mood and also decreased forgetfulness. A full GDS and another MMSE were administered and she received scores of 9 and 29, respectively. These results supported the therapist's hypothesis that Rose's memory impairment was associated with her depression: as depression improved, so did cognitive function. Rose's walking and phone conversations had paid off. She felt more energetic and her friend again asked her to join her walking club. This time, Rose agreed. She began walking greater distances on a regular basis. She also began spontaneously contacting friends she had not talked with since her husband had become very ill. Soon, Rose's misconceptions about her health and aging greatly diminished. With the help of her primary care provider, she found an occupational therapist to help her adapt to her limitations caused by the arthritis. She was also prescribed medication to help with the pain. After 6 months, Rose reported feeling "happy as a clam." The GDS reflected this – she endorsed only two items. Together, Rose and her therapist decided, due to her marked improvement in symptoms, that she was no longer experiencing depression and that they could slowly taper the frequency of sessions for a few months until they were both comfortable that the disorder was in remission.

Commentary

Rose's case illustrates some of the issues and challenges typically seen by clinicians working with older adults with depression. Rose's initial symptom presentation and report were more consistent with somatic than psychologically-based complaints (e.g., fatigue, memory impairment, pain). In addition, because she experienced medical comorbidities and other life stressors (e.g., her husband's death), it was difficult to differentiate between a diagnosis of depression, grief, or a health condition. This was complicated by the fact that Rose first presented her symptoms to her family physician, which is more typical than seeking mental health services.

Treatment incorporated both cognitive and behavioral methods in an effort to treat as broad a range of symptoms as possible. However, behavioral methods were selected first because Rose was more willing to discuss behaviorally based issues than to discuss her negative perceptions. Regular assessment was key in determining if these treatment strategies were both on target and effective. In addition, like many other older adults with depression, Rose benefited from a multidisciplinary approach to her care. The intersection of psychology, medicine, and occupational therapy undoubtedly had a more profound impact on her outcomes than any one of those disciplines would have had alone. In sum, cognitive-behavioral methods facilitated the improvement in Rose's depressive symptoms with their inherent flexibility and customizability.

Conclusion

This chapter has highlighted the great impact depressive disorders have on older adults as well as the need for assessment and treatment strategies that are responsive to the needs of depressed older adults. Although the examination of cognitive and behavioral therapies as evidence-based treatments for geriatric depression is still in its preliminary stages, promising results have been found supporting CBT, BT, PST, and cognitive bibliotherapy as effective treatments. Combination therapy is an option for older adults that should be more frequently provided, especially if those with the disorder are experiencing moderate to severe symptoms and express a preference for combination treatment. Proper assessment is essential in diagnosing, treating, and monitoring the symptoms of depression in older adulthood. In addition, therapist competence in providing treatment from a multicultural standpoint is becoming increasingly vital in meeting older clients' needs. This information can serve as a basis from which researchers and clinicians can further explore treatment and assessment options that are applicable to the broadest range of older adults possible.

References

American Psychiatric Association. (2000). *The diagnostic and statistical manual of mental disorders: Text revision*. Washington, DC: American Psychiatric Association.

Areán, P. A., Gum, A., McCulloch, C. E., Bostrom, A., Gallagher-Thompson, D., & Thompson, L. (2005). Treatment of depression in low-income older adults. *Psychology and Aging, 20*(4), 601–609.

Areán, P. A., Perri, M. G., Nezu, A. M., Schein, R. L., Christopher, F., & Joseph, T. X. (1993). Comparative effectiveness of social problem-solving therapy and reminiscence therapy as treatments for depression in older adults. *Journal of Consulting and Clinical Psychology, 61*, 1003–1010.

Baldwin, R. C., Anderson, D., Black, S., Evans, S., Jones, R., Wilson, K., et al. (2003). Guideline for the management of late-life depression in primary care. *International Journal of Geriatric Psychiatry, 18*(9), 829–838.

Beck, A. T., Brown, G., Steer, R. A., & Weissman, A. (1991). Factor analysis of the Dysfunctional Attitudes Scale in a clinical population. *Psychological Assessment: A Journal of Consulting and Clinical Psychology, 3*, 478–483.

Beck, A. T., Rush, A. J., Shaw, B. F., & Emery, G. (1979). *Cognitive therapy of depression*. New York: Guilford.

Beck, A. T., Steer, R. A., & Brown, G. K. (1996). *Beck Depression Inventory – II*. San Antonio: The Psychological Corporation.

Birrer, R. B., & Vemuri, S. P. (2004). Depression in later life: A diagnostic and therapeutic challenge. *American Family Physician, 69*(10), 2375–2382.

Blazer, D. (2002). *Depression in late life* (3rd ed.). New York: Springer.

Bruce, M. L., McAvay, G. J., & Raue, P. J. (2002). Major depression in elderly home health care recipients. *American Journal of Psychiatry, 159*, 1367–1374.

Bruce, M. L., Ten Have, T. R., Reynolds, C. F., Katz, I. I., Schulberg, H. C., Mulsant, B. H., et al. (2004). Reducing suicidal ideation and depressive symptoms in depressed older primary care patients: A randomized controlled trial. *Journal of the American Medical Association, 291*(9), 1081–1091.

Burns, D. (1980). *Feeling good*. New York: Signet.

Chiriboga, D. A., Yee, B. W. K., & Jang, Y. (2005). Minority and cultural issues in late-life depression. *Clinical Psychology: Science and Practice, 12*(3), 358–363.

Coates, A., & Scogin, F. (2005). *Behavioral activation as a treatment for geriatric depression*. Manuscript submitted for publication.

Cohen, A., Houck, P. R., Szanto, K., Dew, M. A., Gilman, S. E., & Reynolds, C. F. (2006). Social inequalities in response to antidepressant treatment in older adults. *Archives of General Psychiatry, 63*(1), 50–56.

Currie, S. R., & Wang, J. (2005). More data on major depression as an antecedent risk factor for first onset of chronic back pain. *Psychological Medicine, 35*(9), 1275–1282.

Edelstein, B., Kalish, K. D., Whipple Drozdick, L., & Rettig McKee, D. (1999). Assessment of depression and bereavement in older adults. In P. A. Lichtenberg (Ed.), *Handbook of assessment in clinical gerontology* (pp. 11–58). New York: Wiley.

Ferster, C. B. (1973). A functional analysis of depression. *American Psychologist, 28*, 857–870.

Floyd, M., Scogin, F., & Chaplin, W. F. (2004). The Dysfunctional Attitudes Scale: Factor structure, reliability and validity with older adults. *Aging and Mental Health, 8*, 153–160.

Floyd, M., Scogin, F., McKendree-Smith, N. L., Floyd, D. L., & Rokke, P. D. (2004). Cognitive therapy for depression: A comparison of individual psychotherapy and bibliotherapy for depressed older adults. *Behavior Modification, 28*, 297–318.

Folstein, M. F., Folstein, S. E., & McHugh, P. R. (1975). Mini-mental state: A practical method for grading the cognitive state of patients for the clinician. *Journal of Psychiatric Research, 12*, 189–198.

Gallagher, D. E., & Thompson, L. W. (1981). *Depression in the elderly: A behavioral treatment manual*. Los Angeles: University of Southern California Press.

Gallagher-Thompson, D., & Thompson, L. W. (1996). Applying cognitive-behavioral therapy to the psychological problems of later life. In S. H. Zarit & B. G. Knight (Eds.), *A guide to psychotherapy and aging* (pp. 61–82). Washington, DC: American Psychological Association.

Glassman, A., Shapiro, P. A., Ford, D. E., Culpepper, L., Finkel, M. S., Swenson, J. R., et al. (2003). Cardiovascular health and depression. *Journal of Psychiatric Practice, 9*(6), 409–421.

van Gool, C. H., Kempen, G. I., Penninx, B. W., Deeg, D. J., Beekman, A. T., & van Eijk, J. T. (2005). Impact of depression on disablement in late middle aged and older persons: Results from the Longitudinal Aging Study Amsterdam. *Social Science & Medicine, 60*(1), 25–36.

Gottfries, C. G. (1998). Is there a difference between elderly and younger patients with regard to the symptomatology and aetiology of depression? *International Clinical Psychopharmacology, 13*(Suppl. 5), S13–S18.

Gum, A. M., Areán, P. A., Hunkeler, E., Tang, L., Katon, W., Hitchcock, P., et al. (2006). Depression treatment preferences in older primary care patients. *Gerontologist, 46*(1), 14–22.

Hanson, A. E., & Scogin, F. (2007). Older adults' acceptance of psychological, pharmacological, and combination treatments for geriatric depression. Manuscript under review.

Harris, Y., & Cooper, J. K. (2006). Depressive symptoms in older people predict nursing home admission. *Journal of the American Geriatrics Society, 54*(4), 593–597.

Hawley, K. M., & Weisz, J. R. (2001). *Procedural and coding manual for identification of evidence-based treatments*. Unpublished manual, University of California at Los Angeles.

Hays, P. A., & Iwamasa, G. Y. (Eds.). (2006). *Culturally responsive cognitive-behavioral therapy: Assessment, practice, and supervision*. Washington, DC: American Psychological Association.

Hirschfeld, R. M., Keller, M. B., Panico, S., Arons, B. S., Barlow, D., Davidoff, F., et al. (1997). The National Depressive and Manic-Depressive Association consensus statement on the undertreatment of depression. *Journal of the American Medical Association, 277*, 333–340.

Hollon, S. D., Jarrett, R. B., Nierenberg, A. A., Thase, M. E., Trivedi, M., & Rush, A. J. (2005). Psychotherapy and medication in the treatment of adult and geriatric depression: Which monotherapy or combined treatment? *Journal of Clinical Psychiatry, 66*(4), 455–468.

Imber, S. D., Pilkonis, P. A., Sotsky, S. M., Elkin, I., Watkins, J. T., Collins, J. F., et al. (1990). Mode-specific effects among three treatments for depression. *Journal of Consulting and Clinical Psychology, 58,* 352–359.

Jeste, D. V., Blazer, D. G., & First, M. (2005). Aging-related diagnostic variations: Need for diagnostic criteria appropriate for elderly psychiatric patients. *Biological Psychiatry, 58*(4), 265–271.

Jones, R. N., Marcantonio, E. R., & Rabinowitz, T. (2003). Prevalence and correlates of recognized depression in U. S. nursing homes. *Journal of the American Geriatrics Society, 51*(10), 1404–1409.

Kotlyar, M., Dysken, M., & Adson, D. E. (2005). Update on drug-induced depression in the elderly. *American Journal of Geriatric Pharmacotherapy, 3*(4), 288–300.

Laidlaw, K., Thompson, L. W., Dick-Siskin, L., & Gallagher-Thompson, D. (2003). *Cognitive behaviour therapy with older people*. New York: Wiley.

Landreville, P., & Bissonnette, L. (1997). Effects of cognitive bibliotherapy for depressed older adults with a disability. *Clinical Gerontologist, 17*, 35–55.

Landreville, P., Landry, J., Baillargeon, L., Guerette, A., & Msatteau, E. (2001). Older adults' acceptance of psychological and pharmacological treatments for depression. *Journal of Gerontology: Psychological Sciences, 56B*, P285–P291.

Lewinsohn, P. M. (1974). A behavioral approach to depression. In Friedman, R. J., & Katz, M. M. (Eds.). *The psychology of depression: Contemporary theory and research*, (pp. 157–176). Oxford, England: John Wiley & Sons.

Lewinsohn, P., Biglan, A., & Zeiss, A. (1976). Behavioral treatment of depression. In P. Davidson (Ed.), *Behavioral management of anxiety, depression, and pain*. New York: Brunner/Mazel.

Lewinsohn, P. M., & Graf, M. (1973). Pleasant activities and depression. *Journal of Consulting and Clinical Psychology, 41*, 261–268.

Lin, E. H., Katon, W., & Von Korff, M. (2003). Effect of improving depression care on pain and functional outcomes among older adults with arthritis: A randomized controlled trial. *Journal of the American Medical Association, 290*(18), 2428–2434.

Luppa, M., Heinrich, S., Angermeyer, M. C., König, H., & Riedel-Heller, S. G. (2007). Cost-of-illness studies of depression: A systematic review. *Journal of Affective Disorders, 98*, 29–43.

Manthorpe, J., & Iliffe, S. (2006). Suicide among older people. *Nursing Older People, 17*(10), 25–29.

Meyers, L. (2005). Depression increases type 2 risk. *Diabetes Forecast, 58*(9), 38.

Nezu, A. M., Nezu, C. M., & Perri, M. G. (1989). *Problem-solving therapy for depression: Theory, research, and clinical guidelines*. Oxford: Wiley.

Radloff, L. S. (1977). The CES-D scale: A self-report depression scale for research in the general population. *Applied Psychological Measurement, 1*, 385–401.

Rehm, L. P. (1977). A self-control model of depression. *Behavior Therapy, 8*, 787–804.

Rokke, P. D., & Scogin, F. R. (1995). Depression treatment preferences in younger and older adults. *Journal of Clinical Geropsychology, 1*, 243–257.

Romanelli, J., Fauerbach, J., Buch, D., & Ziegelstein, R. (2002). The significance of depression in older patients after myocardial infarction. *Journal of the American Geriatrics Society, 50*(5), 817–822.

Rumsfeld, J. S., Jones, P. G., Whooley, M. A., Sullivan, M. D., Pitt, B., Weintraub, W. S., et al. (2005). Depression predicts mortality and hospitalization in patients with myocardial infarction complicated by heart failure. *American Heart Journal, 150*(5), 961–967.

Scogin, F. (2000). *The first session with seniors*. San Francisco, CA: Jossey-Bass.

Scogin, F., Hamblin, D., & Beutler, L. (1987). Bibliotherapy for depressed older adults: A self-help alternative. *The Gerontologist, 27*, 383–387.

Scogin, F., Jamison, C., & Gochneaur, K. (1989). Comparative efficacy of cognitive and behavioral bibliotherapy for mildly and moderately depressed older adults. *Journal of Consulting and Clinical Psychology, 57*, 403–407.

Scogin, F., Welsh, D., Hanson, A., Stump, J., & Coates, A. (2005). Evidence-based psychotherapies for depression in older adults. *Clinical Psychology: Science and Practice, 12*(3), 222–237.

Sheikh, J. I., & Yesavage, J. A. (1986). Geriatric Depression Scale (GDS): Recent evidence and development of a shorter version. In T. L. Brink (Ed.), *Clinical gerontology: A guide to assessment and intervention* (pp. 165–173). New York: Haworth Press.

Smith, T. B. (Ed.). (2004). *Practicing multiculturalism: Affirming diversity in counseling and psychology*. Boston: Pearson Education, Inc.

Snarski, M., & Scogin, F. (2006). Assessing depression in older adults. In S. Qualls & B. Knight (Eds.), *Psychotherapy for depression in older adults* (pp. 45–78). Hoboken, NJ: John Wiley & Sons, Inc.

Teri, L. (1991). Behavioral assessment and treatment of depression in older adults. In P. A. Wisocki (Ed.), *Handbook of clinical behavior therapy with the elderly client* (pp. 225–243). New York: Plenum Press.

Teri, L. (1994). Behavioral treatment of depression in patients with dementia. *Alzheimer's Disease and Related Disorders, 8*, 66–73.

Teri, L., & Lewinsohn, P. M. (1982). Modification of the pleasant and unpleasant events schedules for use with the elderly. *Journal of Consulting Psychology, 50*(3), 444–445.

Teri, L., Logsdon, R., Wagner, A., & Uomoto, J. (1994). The caregiver role in behavioral treatment of depression in dementia patients. In E. Light & G. Niederehe (Eds.), *Stress effects on family caregivers of Alzheimer's patients: Research and interventions* (pp. 185–204). New York, : Springer Publishing Co.

Thompson, L. W., Coon, D. W., Gallagher-Thompson, D., Sommer, B. R., & Koin, D. (2001). Comparison of desipramine and cognitive-behavioral therapy in the treatment of elderly outpatients with mild-to-moderate depression. *American Journal of Geriatric Psychiatry, 9*(3), 225–240.

Watson, L. C., Garrett, J. M., Sloane, P. D., Gruber-Baldini, A. L., & Zimmerman, S. (2003). Depression in assisted living: Results from a four-state study. *American Journal of Geriatric Psychiatry, 11*(5), 534–542.

Yesavage, J. A., Brink, T. L., Rose, T. L., Lum, O., Huang, V., Adey, M. B., et al. (1983). Development and validation of a geriatric depression screening scale: A preliminary report. *Journal of Psychiatric Research, 17*, 37–49.

Zeiss, A. M., & Steffen, A. (1996). Treatment issues with elderly clients. *Cognitive and Behavioral Practice, 3*(2), 371–389.

2
Treating Generalized Anxiety in a Community Setting

J. Gayle Beck

Despite the fact that older adults are more prevalent than ever before in history, our knowledge about mental health disorders in this segment of the population is surprisingly thin. Anxiety disorders are a case in point. Although our knowledge about anxiety in the elderly is growing steadily, progress has been slow which has impacted knowledge about how to best treat these problems among older adults. Additionally, steadily accumulating evidence suggests that traditional definitions of anxiety disorders, as exemplified by the Diagnostic and Statistical Manual of Mental Disorders (*DSM*; American Psychiatric Association, 2000), may not best capture the symptom profiles reported by seniors, which could potentially result in inappropriate treatment (e.g., Jeste, Blazer, & First, 2005). This chapter begins by reviewing what current research indicates with respect to the symptom profile of generalized anxiety among community elders, with particular reference to the interrelationship between anxiety and depression. An overview of available assessment strategies will be included, with emphasis on the importance of conceptualizing the patient's problem using a functional analysis, in order to understand the interplay between specific symptoms in the context of the older adult's life circumstances. Current information about treating generalized anxiety in community-dwelling elders will be reviewed, to pinpoint potentially useful treatment avenues. Although we have made some significant strides towards understanding how to treat generalized anxiety in older adults, the empirical literature does not yet point to specific treatments that have established support for obtaining good outcomes. As such, the practitioner needs to be mindful in constructing treatment so as to maximize potential benefits, which may result in treatment models that involve considerable individualization. In many respects, it is an exciting era to work clinically with older adults, as the potential for discovery and innovation is high.

Generalized Anxiety in Older Adults – Prevalence, Definitions, and Conundrums

The available knowledge concerning the prevalence of anxiety disorders in general and generalized anxiety disorder (GAD) in particular presents a complex picture. This complexity derives from a number of factors. For example, the

Epidemiological Catchment Area (ECA) studies (e.g., Regier et al., 1988) reported an overall 1-month prevalence rate of 5.5% for the anxiety disorders, which is lower than the 7.3% rate reported for younger adults. With respect to GAD, the 6-month and lifetime prevalence rates for older adults were 1.9% and 4.6%, respectively (Blazer, George, & Hughes, 1991). Although these rates are lower than those obtained with younger adult samples, the ECA studies relied on hierarchical exclusionary criteria, as suggested by the *DSM-III* (American Psychiatric Association, 1980), wherein a diagnosis of GAD was not given if another Axis I disorder was diagnosed. Thus, it is difficult to extrapolate from these prevalence figures to estimate the "true" prevalence of GAD in the community. It is very likely that these rates may underestimate the prevalence of GAD in the community. In support of this speculation, Beekman et al. (1998) reported a 7.3% prevalence rate of GAD among community-dwelling older adults in the Netherlands, who were assessed without any diagnostic exclusionary criteria. As cogently discussed by Hyberls and Blazer (2004), the prevalence of anxiety disorders (and GAD) among older adults and the subsequent demand for mental health services is likely to increase, particularly as life expectancies increase.

In considering the epidemiology of GAD among older adults, an additional concern about most of the available data deserves mention. As discussed by Fisher, Zeiss, and Carstensen (2001), the ECA studies did not differentiate among older adults; individuals aged 65 or older were included in a single group. However, related studies suggest there are age-related differences in the prevalence of anxiety. As noted by Himmelfarb and Murrell (1984) and Feinson (1985), anxiety-based complaints appear to decrease from middle age to the beginning of older adulthood but then increase with advancing age, with individuals reporting more difficulties with symptoms beginning around age 75. This finding has been confirmed by Teachman (2006) in a cross-sectional community sample of 335 adults aged 18–93. Thus, it is difficult to convincingly state the prevalence of GAD among older adults, in light of the non-linear pattern that has been noted by these authors.

To compound this issue further, some writers have argued that the current diagnostic criteria do not fit well for older adults (e.g., Jeste et al., 2005). There are several aspects of this issue. First, differences have been observed between younger and older patients who suffer from the same anxiety disorder (e.g., Sheikh, Swales, Carlson, & Lindley, 2004). This would suggest that although the diagnostic criteria may accommodate both younger and older adults, there may be some phenotypic differences in symptomatology across age groups. Second, in considering the diagnostic criteria, there can be difficulty in distinguishing between "true" anxiety symptomatology and anxiety-related behaviors that are secondary to other factors (such as fear of falling in a frail older adult or reluctance to venture outside of the house owing to visual limitations, Jeste et al.). Although at a descriptive level, anxiety-based behaviors are indistinguishable from one another and from diagnosable disorders, the function of these behaviors is central for treatment formulation. As will be discussed in a later section, it is possible that the success of treatment can hinge on addressing the specific functional relationships among anxiety-based behaviors, particularly in the context of aging-related health and lifestyle changes.

The notable prevalence of subthreshold presentations of anxiety disorders also suggests that our diagnostic criteria might not apply well to older adults. For example, Wetherell, LeRoux, and Gatz (2003) and Diefenbach et al. (2003) have noted that considerable distress is reported by individuals reporting subthreshold levels of GAD, suggesting that despite failure to meet DSM criteria for the disorder, these individuals may benefit from clinical attention. Related work has suggested that subthreshold anxiety symptoms are common among older adults (e.g., deBeurs et al., 1999, Schuurmans et al., 2005) and potentially reflect a different nosology of anxiety disorders among the aged. In considering subthreshold presentations of anxiety, it is notable that longitudinal studies have documented that anxiety symptoms may progress over time into diagnosable depression, either alone or with comorbid GAD among older adults (Schoevers, Deeg, vanTilburg, & Beekman, 2005; Wetherell, Gatz, & Pedersen, 2001). The prevalence and apparently fluctuating course of subclinical anxiety symptoms suggests that refinement in our understanding of generalized anxiety in community-dwelling elders could inform future revisions to available diagnostic systems.

Additionally, an interesting study involving 167 younger–older family member pairs suggested that older adults are significantly less likely to report anxiety symptoms, both in themselves and in others (Levy, Conway, Brommelhoff, & Merikengas, 2003). Related conclusions have been drawn by Gallo, Anthony, and Muthén (1994). The tendency for older adults to minimize reporting of anxiety symptoms contributes to concern about the specific language used to define anxiety syndromes within the *DSM*. It is possible that these findings are unique to the specific cohort of older adults who are being studied and that the next generation of elders will be more likely to report anxiety symptoms. Notwithstanding this possibility, the fact that older adults are less likely to report anxiety symptoms suggests an additional complication in use of the current diagnostic system with this age group.

What emerges from these data and dialogs is a sense that although diagnosable GAD is lower among older adults, less specific anxiety complaints may actually be more prevalent. Given concerns about the fit of our current diagnostic system in describing anxiety disorders in senior adults, it may be reasonable not to adhere rigidly to the DSM criteria when assessing an older adult with anxiety complaints. Additionally, the interplay between generalized anxiety and depression becomes relevant, when considering clinical conundrums in diagnosing older adults. It is entirely possible that differential diagnosis may be considerably more complex with older adults, as articulated by Sheikh (1992). As well, we have very little information about racial and ethnic differences in the prevalence of GAD among older adults. Peng, Navaie-Waliser, and Feldman (2003), in a study of service utilization and health outcomes, reported that Caucasian older adults had the highest rates of anxiety and depressive symptoms, relative to all other racial and ethnic groups. Clearly, it would seem important for future epidemiological studies to more closely examine racial differences in the occurrence of GAD and the anxiety disorders in general.

Understanding Diversity Issues

As the reader might suspect, examination of cultural and ethnic influences on anxiety among older adults has barely begun. Mehta et al. (2003) examined the prevalence and correlates of anxiety symptoms in 3,075 African-American and Caucasian individuals, aged 70–79. Among people without concurrent depressive symptoms, Caucasian women reported the highest rate of anxiety symptoms (20%), followed by African-American women (17%). No racial differences were noted among men in reporting anxiety symptoms (12% for both African-American and Caucasian samples). Importantly, there were no ethnic differences in correlates of anxiety symptoms, suggesting that race does not seem to differentially impact the associated features of anxiety among older adults. More recently, Tolin, Robison, Gaztambide, and Blank (2005) examined anxiety disorders among 303 older Puerto Rican patients within a primary care setting. The rates of GAD were relatively equal among men (15%) and women (10%), with considerable comorbidity noted. Clearly, greater examination of the impact of cultural and ethnic factors on the prevalence and associated features of generalized anxiety is warranted. As will be noted in the next section, where issues pertaining to assessment of older adults for generalized anxiety complaints will be presented, there are preliminary data about diversity issues for several measures. As our awareness of cultural diversity expands, this is an area that is ripe for further work.

Assessment Strategies

Generally speaking, two different approaches have been used in considering strategies for assessing anxiety in older adults. One approach involves determination of the measurement properties of measures that have established utility for younger adults. Although this approach builds upon the existing research base for each of these measures, it may miss phenomenological features of anxiety that are especially salient among older adults. The second approach involves the development of anxiety measures specifically for older adults.

Clinician-Administered Measures

The most commonly used type of clinician-administered measure is the structured diagnostic interview. To date, two interviews have been used successfully to reliably diagnose anxiety disorders in older adults, the Anxiety Disorders Interview Schedule (ADIS-R; DiNardo, Brown, & Barlow, 1994) and the Structured Interview for DSM (SCID; First, Spitzer, Gibbon, & Williams, 2002). Both of these interviews contain structured questions that are designed to help the clinician to determine both primary and secondary diagnoses. The SCID assesses a broader range of diagnoses relative to the ADIS, although the ADIS contains more detailed questions with respect to the various anxiety disorders.

Both interviews have been shown to be useful with older adults (e.g., Beck et al., 2003; Segal, Hersen, Van Hasselt, Kabacoff, & Roth, 1993), although each can be time-consuming to administer. In a clinical setting, these structured interviews can facilitate the initial evaluation of a patient, particularly for a practitioner who is not experienced with older adults.

Other clinician-rated measures that are useful with older adults include general ratings of anxiety symptoms. The advantage of these measures is that they can be used to gauge general levels of anxiety, irrespective of diagnosis. Included in this category are the Hamilton Anxiety Rating Scale (HAM-A, Hamilton, 1959) and the Short Anxiety Screening Test (Sinoff, Ore, Zlotogorsky, & Tamir, 1999), both of which are relatively short and do not require administration by a trained professional. Properties of the HAM-A have been examined in older adults with and without GAD, with some support for their utility (Beck, Stanley, & Zebb, 1999; Diefenbach et al., 2001). These clinician-administered rating scales seem well-suited for tracking a patient's progress during treatment.

Self-Report Measures

Self-report measures are by far the easiest way to assess generalized anxiety in a clinical practice setting. A number of measures have been examined to determine their psychometric properties and utility with older adults, including the Penn State Worry Questionnaire (PSWQ, Meyer, Miller, Metzger, & Borkovec, 1990), the State Trait Anxiety Inventory (STAI, Spielberger, Gorsuch, Lushene, Vagg, & Jacobs, 1983), the Beck Anxiety Inventory (BAI, Beck, Epstein, Brown, & Steer, 1988), and the Worry Scale (WS; Wisocki, Handen, & Morse, 1986).

The PSWQ, BAI, and STAI are established measures within the anxiety literature. The PSWQ was designed to assess perceived controllability of worry and as such, reflects a salient dimension of GAD as defined within the current *DSM*. As reported by Watari and Broadbeck (1997), no differences have been noted in scores reported by Japanese American and European Americans on the PSWQ. Stanley, Beck, and Zebb (1996) and Stanley, Novy, Bourland, Beck, and Averill (2001) reported that the PSWQ shows good internal consistency and evidence of convergent validity, although test–retest reliability was not strong for the scale (unlike similar data with younger adults). The BAI is regarded as a well-validated measure of the severity of anxiety symptomatology. Several studies have suggested that this scale is useful in discriminating anxiety from depression in older adults (e.g., Kabacoff, Segal, Hersen, & VanHasselt, 1997; Morin et al., 1999). Importantly, Wetherell and Arean (1997) documented that scores on the BAI are unrelated to ethnicity, gender, and educational level. The STAI assesses both state and trait levels of anxiety and has been widely used for a number of years within the anxiety literature. Stanley et al. (1996, 2001) documented good internal consistency and evidence of convergent validity for both subscales of the STAI, although test–retest reliability was questionable. Related studies (e.g., Fuentes & Cox, 2000; Kabacoff et al., 1997) also suggest the utility of these measures for use with older adults.

The WS was developed expressly to assess worries of particular significance to older adults (specifically concerns about financial, health, and social issues). Initial normative data and support for concurrent validity have been provided (Wisocki et al., 1996) and replicated by Stanley et al. (1996, 2001). A subsequent revision to the WS expanded this measure to include more domains of worry (Wisocki, 2000). Unlike the PSWQ, the WQ focuses on the content of worry and is designed expressly for older adults. Given this focus, the scale can assist the practitioner in an initial assessment of specific worry domains

Recently, two additional measures have been introduced that are designed expressly for older adults. Although these scales are too new to have extensive psychometric support, the clinician may wish to consider their use, particularly given their topical focus. Kogan and Edelstein (2004) presented preliminary information on the Fear Survey Schedule – II for older adults, which is designed to assess self-reported fears. Initial data support the internal consistency, test–retest reliability, and convergence with related anxiety questionnaires. This measure may be particularly useful for assessing aging-related fears in the context of heightened worry among individuals with GAD. As well, an older adult version of the Adult Manifest Anxiety Scale (termed the Adult Manifest Anxiety Scale – Elderly version, Lowe & Reynolds, 2006) has been recently published. This scale has four factors (fear of aging, physiological anxiety, worry/oversensitivity, and a lie scale) and appears to have preliminary support for its temporal stability, construct validity, and factor invariance across gender. It will be interesting to determine how this scale performs with individuals with diagnosable GAD.

In assessing older adults with generalized anxiety, it is important to evaluate related aspects of functioning. Inclusion of a measure of depression is necessary, given the overlap between anxiety and depressive symptoms. Two measures are available, the Beck Depression Inventory (BDI, Beck, Ward, Mendelson, Mock, & Erbaugh, 1961) and the Geriatric Depression Scale (GDS, Yesavage & Brink, 1983). The GDS was developed to fit the depressive symptom profile of older adults and so, eliminates the assessment of somatic symptoms which may reflect normal aging. The BDI is arguably the most widely used self-report of depression. In considering which measure to select, the strengths and limitations of both the GDS and the BDI are described by Gallagher (1986) and Sheikh and Yesavage (1986). Snyder, Stanley, Novy, Averill, and Beck (2000) report that both measures possess adequate internal consistency and provide support for convergent, discriminant, construct, and discriminative validity for both scales, when administered to older adults with GAD. As well, clinical assessment of general lifestyle factors are essential for a thorough evaluation of older adults with generalized anxiety, including the presence and availability of social support, dietary and exercise habits, alcohol consumption, and financial circumstances. A thorough assessment of the individual's medical conditions and their associated medications is critical, particularly given the fact that aging-related medical problems may create anxiety-like symptoms.

As suggested previously, the clinician's job in evaluating an older adult's complaint of generalized anxiety is not simple. As cogently discussed by Fisher and Noll (1996), anxiety can be conceptualized as a syndrome, as a sign, or as a symptom. In particular, anxiety-like symptoms can be the result of medication or a physical disease process, can be related to aging-related changes in perception or physical ability, or can represent a form of generalized anxiety complaint or GAD. In drawing together assessment sources, it is essential for the clinician to establish functional associations between anxiety problems and related aspects of the individual's presentation. In particular, establishing links between specific anxiety complaints and issues within medical, social, financial, and family spheres requires careful questioning. Without understanding the functional associations between specific symptom reports and functioning within these domains, it is likely that treatment may not be formulated accurately. As will be discussed in the next section, standardized treatments do not appear to have the same efficacy for older adults with GAD as has been reported for younger adults with the disorder. As such, current clinical and research trends are moving towards more individualized treatments, which require a solid conceptualization of the individual's symptoms when placed in context.

Psychosocial Treatments for Generalized Anxiety

Efforts to develop and test psychosocial treatments for generalized anxiety in older adults have increased exponentially in the past 10 years. As reviewed by Mohlman (2004), these efforts have focused on cognitive behavioral therapy (CBT), conducted in individual and group formats. Within controlled trials of CBT with older adults, supportive group therapy and a general discussion group intervention have also been examined, which typically are conceptualized as comparison conditions.

In this literature, the emphasis on CBT stems from two sources. First, there is a general consensus within the larger field of anxiety disorders that CBTs are of documented efficacy (e.g., Barlow, 2002). With particular reference to GAD, CBT has produced superior results when contrasted with supportive psychotherapy and antianxiety medications in younger adults (e.g., Barlow, Rapee, & Brown, 1992; Gould, Otto, Pollak, & Yap, 1997). Given this research background, it seems natural to examine the effect of CBT for GAD in older adults, expecting similar levels of efficacy. A second reason behind this focus on CBT is that this approach to therapy fits well with the needs of older adults. As articulated by Zeiss and Steffen (1996), CBT does not require considerable "psychological mindedness" from a client. As well, this treatment approach does not pathologize the client, can be modified to accommodate age-related changes in memory, and is time-limited, fitting well with the economic constraints that some older adults live within. CBT can be used anywhere, including primary care offices or the patient's home (see the chapter by Brenes, Wagener, and Stanley in this volume).

Based on a small collection of well-designed studies, CBT has been shown to be somewhat effective in the treatment of older patients with GAD (see Mohlman, 2004 for a detailed review). CBT has been shown to lead to decreases in anxiety, worry, and depression. Response rates, however, are lower than expected and often are less marked than what is seen in younger adults. Importantly, group CBT has not been shown to be significantly more effective than supportive therapy or a discussion group intervention, suggesting that some aspect of the group setting might be salient in producing these positive outcomes. In response to this and related concerns, Wetherell, Sorrell, Thorp, and Patterson (2005) have presented a modified CBT protocol that is oriented towards addressing the specific needs of older adults. In this modified intervention, an individual treatment format is used, allowing the therapist to tailor the specific elements of treatment to a specific patient's needs. Although this approach still can be characterized as a form of CBT, the amount of emphasis on cognitive interventions is reduced, relative to previously examined CBTs. Fourteen skills-oriented modules are contained within this treatment approach, each included to address a specific feature of generalized anxiety that has been observed to occur within anxious older individuals. These modules include psychoeducation, progressive muscle relaxation (PMR) training, problem-solving skills training, mindfulness, stimulus control and thought-stopping, cognitive restructuring, exposure, and relapse prevention. As well, modules are presented which address sleep hygiene, behavioral activation (to address depression), life review (to help foster an alternative view of self and the world), assertiveness training, pain management, and time management, interventions which address associated features of generalized anxiety. As noted by Wetherell and colleagues, these interventions are not typically included in CBTs for anxious older adults but are intended to address features that may enhance treatment outcome. The therapist selects specific modules, based upon comorbid diagnoses, the patient's perceptions of their own problems, and responses to self-report questionnaires. In considering this newer form of CBT, it is flexible and can be modified to address individual concerns. As acknowledged by Wetherell and colleagues, greater work needs to be undertaken to determine how to select specific modules and how to integrate these various approaches in the treatment of generalized anxiety.

In considering treatments for generalized anxiety in older adults, it seems that the field has made a solid beginning with the initial research on CBT. At present, this literature has not evolved to the point where comparison of anxiolytic medication and CBT has occurred. Although promising, this approach to CBT is clearly not a panacea. The approach taken by Wetherell and colleagues exemplifies what easily could represent the next generation of CBTs for anxious older adults. In particular, individualized treatment which is constructed to address specific complaints and symptoms, appear potentially more efficacious than inflexible "package" treatments. Given the amount of heterogeneity among older adults, a functional analysis can aid in the selection of specific interventions for specific individuals and in conceptualizing treatment in its entirety. This approach will be illustrated in the following case example.

The Case of Evie and Her Angst

Evie[1] presented for individual treatment with the opening statement "I am full of angst – can you make it stop?" At age 77, Evie was seeking treatment for anxiety which had pervaded her life for the past 20 years. Careful evaluation using the ADIS-R indicated that Evie reported elevated levels of generalized anxiety and uncontrollable worry, an occasional panic attack, and intervals of extreme dysphoria and lack of interest in her usual activities. From a diagnostic perspective, Evie met criteria for GAD although she did not meet criteria for Dysthymic Disorder, owing to the intermittent nature of her depression. Evie was a self-described "Jewish princess" who had been widowed 10 years before presenting for treatment. A strong, active woman, Evie was immersed in the arts community and played a central role on the board of director's of the local gallery. She enjoyed an active social life and maintained good relationships with her two grown children, both of whom lived in other cities. She reported a number of medical problems, including high blood pressure, hypothyroid, an ulcer, and recent treatment for skin cancer. Most of these conditions were managed well by her physicians.

A functional assessment of Evie's "angst" indicated daily worry that lasted most of the day, with acute exacerbations if she was confronted with one of her worry triggers (see below). She reported difficulty in concentrating, inability in making decisions, irritability, muscle tension, and restlessness when worried. Her worry focused on health-related topics, minor matters such as being on time, traveling far from home, particularly during inclement weather, being surrounded by large crowds in areas that were unfamiliar to her, and money, as assessed using the WQ, as well as self-monitoring records. Several of these worry domains were grounded in actual concerns (e.g., her health, her financial situation); others appeared to stem from aging-related concerns, such as being lost in a strange place or surrounded by individuals whom she did not know if she were lost. This distinction had implications for treatment planning, as will be discussed further. She reported becoming depressed when faced with problems that "wouldn't go away," including uncertainty about whether she would outlive her savings and persistent symptoms from her ulcer. At these moments, she indicated that she felt overwhelmed, helpless, and demoralized by the recurrent nature of these problems. These feelings were reflected in an elevated score on the BDI at the beginning of treatment. Occasionally, she would worry to such an extent that she would experience a panic attack. She was not concerned about the panic attacks and understood them as a part of her larger anxiety problem.

Evie's GAD symptoms were conceptualized as a stemming from two sources, the first being areas of actual concern that she had not addressed effectively (e.g., her financial situation, her health problems) and the second being areas where her worry was clearly centered on aging-based concerns (e.g., traveling in unfamiliar

[1]This individual's name and identifying information have been modified to protect her confidentiality and privacy.

areas, being surrounded by unfamiliar people) and was excessive and debilitating. Treatment was formulated around these two foci, with different interventions being targeted at each. In particular, it was felt that worry which was centered on concrete issues would best be approached with more concrete, behavioral strategies, whereas worry that was more ethereal would be better addressed using exposure and cognitive strategies. Treatment began with psychoeducation about the nature of GAD and its treatment. Although Evie was fairly psychologically minded, she was not certain that the generalized anxiety symptoms were not part of her personality and so, was given several self-help books to read, in order to learn more about the disorder. Evie was taught how to use PMR, in order to provide a concrete skill to begin to control her anxiety. Evie reported that she had been using meditation for years and so, PMR was adapted in order to be integrated into her usual form of meditation. Throughout treatment, Evie was intermittently asked to complete the PSWQ, WQ, and BDI, in order to track her progress. These measures provided a concrete index of Evie's progress.

In considering which aspect of her worry to start treatment with, Evie reported greater anxiety about concerns that were based on actual problems (her health and finances). Because these topics were extremely anxiety-provoking, it was preferable to begin with interventions designed to address less anxiety-provoking worries (e.g., travel, being in unfamiliar places, being late, car repairs), much as one does with any form of gradual exposure. Evie was taught worry exposure (Craske, Barlow, & O'Leary, 1992) and was compliant with using this technique at home. She was able to notice reductions in the degree of anxiety generated by minor issues, travel, and meeting unfamiliar people within 3–4 weeks of steady exposure-based practice. In one of these sessions, Evie expressed some frustration at the need to target these concerns, as they had not characterized her thoughts when she was younger. In exploring this concern, Evie revealed that she had occasionally worried when she was younger, albeit about different topics. This was used as an entry to teach Evie some basic cognitive skills, including evidence gathering and how to recognize and correct logical errors. Evie acquired these cognitive skills easily and indicated that they helped her to put the worries into a better perspective.

Given Evie's success with basic exposure and cognitive therapy skills, it was felt appropriate to move on to those worries that had a concrete facet, focusing on health and financial concerns. Evie reported having a difficult time discussing either of these issues, given the level of anxiety that they would generate, and so, she was taught how to use mindfulness as a way of "staying in the moment" during treatment and at-home homework. Using a problem-solving framework, Evie explored these topics, with particular focus on defining the problem and developing a strategy for finding effective solutions. This led to her contacting a financial advisor to assist with financial planning and switching to work with a primary care physician who had particular expertise in treating ulcers in older adults. Her use of problem solving was so successful that she extended this approach to a range of other issues (not all of which had been the focus of worry). Included in this problem-solving spree was making plans for liquidating her art collection as her children were not interested in inheriting this, negotiating a new lease on a car, and structuring her time so that she could begin to write the story of her life, a goal that she wanted to achieve while she was still active and vibrant. In some respects, this goal shares some features with focused reminiscence (e.g., Zauszniewski et al., 2004) and other forms of life review, which are designed to help the older adult to attain a positive understanding of their life and accomplishments. It was interesting to see Evie pursuing this goal on her own, as it appeared to be important for her to evaluate her life with a different (nonanxious) perspective.

At the termination of treatment, Evie reported considerably less worry. This report was augmented by significantly reduced scores on the PSWQ and the WQ. She acknowledged that she would have moments of angst, which would serve as a cue to use one of the skills that she gained from therapy. It is salient that no interventions were included to target her depressed mood. Rather, considerable generalization was noted; as she grew more skilled in managing her worry, the frequency of dysphoric mood decreased, which was also reflected on the final BDI that she completed. As such, it was felt unnecessary to add additional sessions to target depression.

Commentary

This case highlights some of the issues discussed throughout this chapter. As noted, Evie reported a panoply of symptoms which included both anxiety and depressive symptoms. Technically, Evie met diagnostic criteria for GAD, without any secondary mental health diagnoses. Although the specific features of depression did not meet diagnostic criteria for Dysthymic Disorder, this aspect of her presentation was significant and had the potential to reduce her motivation during treatment. Treatment was structured around two foci, one being more "pure" worry (which was addressed with PMR, exposure-based techniques, and cognitive therapy approaches) and the second being a mixture of "real-life" problems to which Evie responded with worry and depression. Problem-solving techniques were used in an effort to help Evie seek alternative solutions to these problems. Importantly, these techniques provided this patient with an action-oriented approach to real-life problems, which seemed important in addressing her feelings of being overwhelmed and subsequent dysphoria. Although it is salient to recognize that the interventions used in Evie's treatment are all contained within package CBT approaches, their application was different and was tailored to meet the unique facets of this woman's situation. As noted, this approach to treatment was successful and importantly, showed radiating effects throughout several domains of her life.

Conclusion

As highlighted at the beginning of this chapter, it is an exciting time to be working clinically with older adults, particularly given increased awareness of the mental health needs of this segment of the population and greater emphasis on establishing mental health policy that is age-sensitive (e.g., Bartels, Dums, & Shea, 2004). In many respects, one of the more challenging features of clinical work with older adults is developing a framework for understanding when to use existing, empirically supported treatments, when to modify these treatments, and when to start anew in the development of age-specific interventions. When considering GAD, the field is just now beginning to examine the second option (modification of existing treatments), with the hope of improving efficacy for older patients. As we gain further knowledge about the psychopathology of GAD and begin to understand further the nuances of this diagnosis in older adults, it is likely that our treatments will be refined and improved. Certainly, this is an ideal point for clinicians and researchers to learn from one another towards the shared goal of enhancing outcomes of anxiety treatments.

References

American Psychiatric Association. (1980). *Diagnostic and statistical manual of mental disorders* (3rd ed.). Washington, DC: American Psychiatric Association.

American Psychiatric Association. (2000). *Diagnostic and statistical manual of mental disorders* (4th ed., text revision). Washington, DC: American Psychiatric Association.

Barlow, D. H. (2002). *Anxiety and its disorders*. New York: Guilford.

Barlow, D. H., Rapee, R., & Brown, T. A. (1992). Behavioral treatment of generalized anxiety disorder. *Behavior Therapy, 23*, 551–570.

Bartels, S. J., Dums, A. R., & Shea, D. G. (2004). Mental health policy and aging. In B. L. Levin, J. Petrila, & K. D. Hennessy (Eds.), *Mental health services: A public health perspective* (2nd ed., pp. 232–254). New York: Oxford University Press.

Beck, A. T., Epstein, N., Brown, G., & Steer, R. A. (1988). An inventory for measuring clinical anxiety. *Journal of Consulting and Clinical Psychology, 56*, 893–897.

Beck, J. G., Novy, D. M., Diefenbach, G. J., Stanley, M. A., Averill, P. M., & Swann, A. C. (2003). Differentiating anxiety and depression in older adults with Generalized Anxiety Disorder. *Psychological Assessment, 15*, 184–192.

Beck, J. G., Stanley, M. A., & Zebb, B. J. (1999). The effectiveness of the Hamilton Anxiety Rating Scale with older Generalized Anxiety Disorder patients. *Journal of Clinical Geropsychology, 5*, 281–290.

Beck, A. T., Ward, C., Mendelson, M., Mock, J., & Erbaugh, J. (1961). An inventory for measuring depression. *Archives of General Psychiatry, 4*, 53–63.

Beekman, A. T. F., Bremmer, M. A., Deeg, D. J. H., van Balkom, A. J. L. M., Snut, J. H., de Beurs, E., et al. (1998). Anxiety disorders in later life: A report from the Longitudinal Aging Study Amsterdam. *International Journal of Geriatric Psychiatry, 13*, 717–726.

deBeurs, E., Beekman, A. T. F., vanBalkom, A. J. L., Deeg, D. J. H., vanDyck, R. & vanTilburg, W. (1999). Consequences of anxiety in older persons: Its effects on disability, well-being, and use of health services. *Psychological Medicine, 29*, 583–593

Blazer, D., George, L. K., & Hughes, D. (1991). The epidemiology of anxiety disorders: An age comparison. In C. Salzman & B. D. Lebowitz (Eds.), *Anxiety in the elderly: Treatment and research* (pp. 17–30). New York: Springer.

Craske, M. G., Barlow, D. H., & O'Leary, T. A. (1992). *Mastery of your anxiety and worry*. Albany, NY: Graywind Publications.

Diefenbach, G. J., Hopko, D. R., Feigon, S., Stanley, M. A., Novy, D. M., Beck, J. G., et al. (2003). 'Minor GAD': Characteristics of subsyndromal GAD in older adults. *Behaviour Research and Therapy, 41*, 481–487.

Diefenbach, G. J., Stanley, M. A., Beck, J. G., Novy, D. M., Averill, P. M., & Swann, A. C. (2001). Examination of the Hamilton Scales in assessment with anxious older adults: A replication and extension. *Journal of Psychopathology and Behavioral Assessment, 23*, 117–124.

DiNardo P. A., Brown T. A., & Barlow D. H. (1994). *Anxiety disorders interview schedule for DSM-IV*. Boston. MA. Center for Stress and Anxiety Related Disorders, Boston University.

Feinson, M. (1985). Aging and mental health: Distinguishing myth from reality. *Research on Aging, 7*, 155–174.

First, M. B., Spitzer, R. L., Gibbon, M., & Williams, J. B. W. (2002). *Structured clinical interview for DSM-IV-TR Axis I disorders, research version, patient edition (SCID-I/P)*. New York: Biometrics Research, New York State Psychiatric Institute.

Fisher, J. E., & Noll, J. P. (1996). Anxiety disorders. In L. Carstensen, B. Edelstein, & L. Dornbrand (Eds.), *The practical handbook of clinical gerontology* (pp. 304–323). Thousand Oaks, CA: Sage Publications.

Fisher, J. E., Zeiss, A. M., & Carstensen, L. L. (2001). Psychopathology in the aged. In P. B. Sutker & H. E. Adams (Eds.), *Comprehensive handbook of psychopathology* (3rd ed., pp. 921–951). New York: Kluwer Academic/Plenum Publishers.

Fuentes, K. & Cox, B. (2000). Assessment of anxiety in older adults: A community-based survey and comparison with younger adults. *Behaviour Research and Therapy, 38*, 297–309.

Gallagher, D. (1986). The Beck Depression Inventory and older adults. In T. L. Brink (Ed.), *Clinical Gerontology: A guide to assessment and intervention* (pp. 149–163). New York: Haworth Press.

Gallo, J. J., Anthony, J. C., & Muthén, B. O. (1994). Age differences in the symptoms of depression: A latent trait analysis. *Journal of Gerontology: Psychological Sciences, 49*, 251–264.

Gould, R. A., Otto, M. N., Pollack, M. H., & Yap, L. (1997). Cognitive behavioral and pharmacological treatment of generalized anxiety disorder: A preliminary meta-analysis. *Behavior Therapy, 28*, 285–305.

Hamilton, M. (1959). The assessment of anxiety states by rating. *British Journal of Medical Psychology, 32*, 50–55.

Himmelfarb, S., & Murrell, S. (1984). The prevalence and correlates of anxiety symptoms in older adults. *Journal of Psychology, 116*, 159–167.

Hyberls, C. F., & Blazer, D. G. (2004). Epidemiology of mental disorders in older adults. In B. L. Levin, J. Petrila, and K. D. Hennessy (Eds.), *Mental health services: A public health perspective* (2nd ed., pp. 211–231). New York: Oxford University Press.

Jeste, D. V., Blazer, D. G., & First, M. (2005). Aging-related diagnostic variations: Need for diagnostic criteria appropriate for elderly psychiatry patients. *Biological Psychiatry, 58*, 265–271.

Kabacoff, R., Segal, D., Hersen, M., & VanHasselt, V. (1997). Psychometric properties and diagnostic utility of the Beck Anxiety Inventory and the State-Trait Anxiety Inventory with older adult psychiatric outpatients. *Journal of Anxiety Disorders, 11*, 33–47.

Kogan, J. N., & Edelstein, B. A. (2004). Modification and psychometric examination of a self-report measure of fear in older adults. *Journal of Anxiety Disorders, 18*, 397–409.

Levy, B. R., Conway, K., Brommelhoff, J., & Merikengas, K. (2003). Intergenerational differences in the reporting of elders' anxiety. *Journal of Mental Health and Aging, 9*, 233–241.

Lowe, P. A., & Reynolds, C. R. (2006). Examination of the psychometric properties of the Adult Manifest Anxiety Scale – Elderly version scores. *Educational and Psychological Measurement, 66*, 93–115.

Mehta, K. M., Simonsick, E. M., Penninx, B. W. J. H., Schulz, R., Rubin, S. M., Satterfield, S., et al. (2003). Prevalence and correlates of anxiety symptoms in well-functioning older adults: Findings from the Health Aging and Body Composition Study. *Journal of the American Geriatric Society, 51*, 499–504.

Meyer, T., Miller, M., Metzger, R., & Borkovec, T. (1990). Development and validation of the Penn State Worry Questionnaire. *Behaviour Research and Therapy, 28*, 487–495.

Mohlman, J. (2004). Psychosocial treatment of late-life generalized anxiety disorder: Current status and future directions. *Clinical Psychology Review, 24*, 149–169.

Morin, C., Landreville, P., Colecchi, C., McDonald, K., Stone, J., & Ling, W. (1999). The Beck Anxiety Inventory: Psychometric properties with older adults. *Journal of Clinical Geropsychology, 5*, 19–29.

Peng, T. R., Navaie-Waliser, M., & Feldman, P. H. (2003). Social support, home health service use, and outcomes among four racial-ethnic groups. *The Gerontologist, 43*, 503–513.

Regier, D., Boyd, J., Burke, J., Rae, D., Myers, J., Kramer, M., et al. (1988). One-month prevalence of mental disorders in the United States. *Archives of General Psychiatry, 45*, 977–986.

Schoevers, R. A., Deeg, D. J. H., vanTilburg, W., & Beekman, A. T. F. (2005). Depression and generalized anxiety disorder. *American Journal of Geriatric Psychiatry, 13*, 31–39.

Schuurmans, J., Comijs, H. C., Beekman, A. T. F., deBeurs, E., Deeg, D. J. H., Emmelkamp, P. M. G., et al. (2005). The outcome of anxiety disorders in older people at 6-year follow-up: Results from the longitudinal aging study Amsterdam. *Acta Psychiatrica Scandaniva, 111*, 420–428.

Segal, D., Hersen, M., VanHasselt, V., Kabacoff, R., & Roth, L. (1993). Reliability of diagnosis in older psychiatric patients using the Structured Clinical Interview for DSM-III-R. *Journal of Psychopathology and Behavioral Assessment, 15*, 347–356.

Sheikh, J. I. (1992). Anxiety and its disorders in old age. In J. E. Birren, R. B. Sloane, & G. D. Cohen (Eds.), *Handbook of mental health and aging* (2nd ed., pp. 409–432). San Diego, CA: Academic Press.

Sheikh, J. I., Swales, P. J., Carlson, E. B., & Lindley, S. E. (2004). Aging and panic disorder: Phenomenology, comorbidity, and risk factors. *American Journal of Geriatric Psychiatry, 12*, 102–109.

Sheikh, J. I. & Yesavage, J. A. (1986). Geriatric Depression Scale (GDS). In T. L. Brink (Ed.) *Clinical Gerontology: A guide to assessment and intervention* (pp. 165–173). New York: Haworth Press.

Sinoff, G., Ore, L., Zlotogorsky, D., & Tamir, A. (1999). Short anxiety screening test – A brief instrument for detecting anxiety in the elderly. *International Journal of Geriatric Psychiatry, 14*, 1062–1071.

Snyder, A. G., Stanley, M. A., Novy, D. M., Averill, P. M., & Beck, J. G. (2000). Measures of depression in older adults with Generalized Anxiety Disorder: A psychometric evaluation. *Depression and Anxiety, 11*, 114–120.

Spielberger, C. D., Gorsuch, R., Lushene, R., Vagg, P. R., & Jacobs, G. A. (1983). *STAI manual for the State-trait anxiety inventory*. Palo Alto, CA: Consulting Psychologists Press.

Stanley, M. A., Beck, J. G., & Zebb, B. J. (1996). Psychometric properties of four anxiety measures in older adults. *Behaviour Research and Therapy,* 34, 827–838.

Stanley, M. A., Novy, D. M., Bourland, S. L., Beck, J. G., & Averill, P. M. (2001). Assessing older adults with generalized anxiety: A replication and extension. *Behaviour Research and Therapy, 39*, 221–235.

Teachman, B. A. (2006). Aging and negative affect: The rise and fall and rise of anxiety and depression symptoms. *Psychology and Aging, 21*, 201–207.

Tolin, D. F., Robison, J. T., Gaztambide, S., & Blank, K. (2005). Anxiety disorders in older Puerto Rican primary care patients. *American Journal of Geriatric Psychiatry, 13*, 150–156.

Watari, K. & Brodbeck, C. (2000). Culture, health, and financial appraisals: Comparison of worry in older Japanese Americans and European Americans. *Journal of Clinical Geropsychology, 6*, 25–39.

Wetherell, J. & Arean, P. (1997). Psychometric evaluation of the Beck Anxiety Inventory with older medical patients. *Psychological Assessment, 9*, 136–144.

Wetherell, J. L., Gatz, M., & Pedersen, N. L. (2001). A longitudinal analysis of anxiety and depressive symptoms. *Psychology and Aging, 16*, 187–195.

Wetherell, J. L., LeRoux, H., & Gatz, M. (2003). *DSM-IV* criteria for generalized anxiety disorder in older adults: Distinguishing the worried from the well. *Psychology and Aging, 18*, 622–627.

Wetherell, J. L., Sorrell, J. T., Thorp, S. R., & Patterson, T. L. (2005). Psychological interventions for late-life anxiety: A review and early lessons from the CALM study. *Journal of Geriatric Psychiatry and Neurology, 18*, 72–82.

Wisocki, P. (2000). *Worry Scale – Revised.* Unpublished manuscript, University of Massachusetts.

Wisocki, P., Handen, B., & Morse, C. (1986). The Worry Scale as a measure of anxiety among homebound and community active elderly. *The Behavior Therapist, 5*, 91–95.

Yesavage, J., & Brink, T. L. (1983). Development and validation of a Geriatric Depression Screening Scale: A preliminary report. *Journal of Psychiatry Research, 17*, 37–49.

Zauszniewski, J. A, Eggenschwiler, K., Preechawong, S., Chung, C-W., Airey, T. F., Wilke, P. A., et al., (2004). Focused reflection reminiscence group for elders: Implementation and evaluation. *Journal of Applied Gerontology, 23*, 429–442.

Zeiss, A. M., & Steffen, A. (1996). Behavioral and cognitive-behavioral treatments: An overview of social learning. In S. H. Zarit & B. G. Knight (Eds.), *A guide to psychotherapy and aging: Effective clinical interventions in a life-stage context* (pp. 35–60). Washington, DC: American Psychological Association.

3
Treatment of Late-Life Generalized Anxiety Disorder in Primary Care Settings

Gretchen A. Brenes, Paula Wagener, and Melinda A. Stanley

Late-life anxiety disorders are twice as prevalent as mood disorders and 4–8 times more prevalent than major depressive episodes (Regier et al., 1988; Weissman et al., 1985). Generalized Anxiety Disorder (GAD) is second only to specific phobias in prevalence among older adults (Krasucki, Howard, & Mann, 1999). Furthermore, the most common anxiety disorder within the primary care setting is GAD (Wittchen et al., 2002), with prevalence estimates ranging from 2.8 to 11.2% (Ansseau, Fischler, Dierick, Mignon, & Leyman, 2005; Roy-Byrne & Wagner, 2004; Tolin, Robison, Gaztambide, & Blank, 2005). Among community samples of older adults, GAD is associated with poor quality of life (Wetherell, Gatz, & Craske, 2003), increased comorbidity (Krasucki, Howard, & Mann, 1998), and increased service use (Stanley, Roberts, Bourland, & Novy, 2001). In primary care settings, GAD in later life is associated with increased worry, depression, and sleep difficulties, as well as decreased quality of life and reduced perceptions of general health, mental health, and vitality (Stanley et al., 2003). GAD in older adults is also associated with greater physical disability (Astrom, 1996). Because older adults tend to somaticize psychiatric problems (Lenze et al., 2005) and because of the overlap between GAD, depression, and comorbid medical problems, GAD can be difficult to diagnose in older adults.

Brief Review of Empirically Supported Treatments of Anxiety in Older Adults

Most of the psychotherapeutic interventions used with anxious older adults are based on treatments that are efficacious with younger adults (Wetherell, Lenze, & Stanley, 2005). These interventions traditionally are based on a cognitive-behavioral approach and include: psychoeducation, awareness training, diaphragmatic breathing, relaxation, cognitive therapy, and exposure. Some modifications have been made to adapt treatment for use with older patients. These include using visual aids to assist persons with hearing difficulties, frequently summarizing information to aid in learning, presenting material at a slower pace, providing more psychoeducational information, and using more lay terms rather than psychological terms. Although cognitive-behavioral interventions have been successful at reducing anxiety (Wetherell et al.), studies conducted in an academic clinical setting may not generalize to the population of older adults with GAD. First, older adults are most likely to go to their primary care physician (PCP) for help rather than a mental health provider (Gallo, Marino, Ford, & Anthony, 1995; Klap, Unroe, & Unutzer, 2003; Pearson, 1998). Second, cognitive-behavioral therapy (CBT) trials thus far have been relatively rigid with a fixed number, duration, and content of sessions leaving little room for flexible attention to the needs of heterogeneous populations of older adults. Third, the majority of participants in academic clinical trials to date have been white, well-educated women who are relatively healthy. This only represents a portion of the older adult population.

Models of Integration of Mental Health Treatment: Primary Care Setting

Traditionally, primary care physicians have dealt with mental health problems by treating the problem themselves and referring the patient to a mental health services provider when necessary, such as with serious mental illness. This model is problematic because it assumes that anxiety disorders are appropriately assessed and diagnosed by PCPs. However, PCPs have a very limited amount of time and competing demands. Also, they may not have the training necessary to recognize the sometimes subtle differences among anxiety disorders as is reflected in the common use of "anxiety disorder NOS" among older adult primary care patients (Stanley et al., 2001). As a result, anxiety disorders are underrecognized among older adults in the primary care setting, even more so than depressive disorders (Ormel et al., 1997; Stanley et al.). When anxiety is recognized, PCPs often do not have the time and resources to determine if the patient followed up with a referral, and treatment is usually with medication of an inadequate duration (Roundy et al., 2005). Primary care treatment also is based on an acute disease model, but many anxiety disorders, particularly GAD, tend to be chronic problems. Acute disease models do not allow time for the education that older adults need regarding the importance of treatment for mental health problems. Many older adults may fail to perceive a need for mental health treatment (Klap et al., 2003; Mickus, Colenda, & Hogan, 2000), but those who recognize the need or are offered care typically prefer individual psychotherapy over pharmacotherapy (Arean, Alvidrez, Barrera, Robinson, & Hicks, 2002; Gum et al., 2006; Landreville, Landry, Baillargeon, Guerette, & Matteau, 2001; Wetherell et al., 2004).

Nonetheless, 75% of patients who get treatment for a mental health problem do so in the primary care setting (Klap et al., 2003). Recent efforts have focused on improving mental health treatment in medical settings by using a collaborative care model that integrates mental health providers into the primary care clinic.

In this model the PCP remains involved in the treatment, but a care manager coordinates care. Some of the research addressing the value of this treatment approach has focused on younger adults with anxiety disorders; another small collection of studies has used this model to treat depression in older medical patients.

Collaborative care models that incorporate psychotherapy for the treatment of panic disorder have been integrated into the primary care setting (Craske et al., 2002; Roy-Byrne et al., 2005). In this work, behavioral health specialists delivered a brief cognitive-behavioral treatment and coordinated pharmacological treatment with the PCP. Although 29% of patients in collaborative care were in remission (0 panic attacks, minimal anticipatory anxiety, minimal phobic avoidance) at 12 months compared with 16% of the usual care group, the magnitude of the treatment effects are smaller than what is seen in psychological/psychiatric clinic settings.

This model has been extended to include patients with GAD (Rollman, Belnap, Mazumdar, Houck, et al., 2005). Again, treatment included cognitive-behavioral and pharmacological components, but the CBT was delivered through the use of a workbook with minimal contact with the behavioral care manager. Findings indicate that collaborative care was more successful in patients with panic disorder with or without GAD than patients with GAD alone.

A third study has examined a collaborative care model targeted to older adults. The Primary Care Research in Substance Abuse and Mental Health for Elderly (PRISM-E; Levkoff et al., 2004) compared an enhanced system for mental health referrals with an integrated model of care for patients with depression, anxiety, and at risk drinking. Although there were differences in time to first appointments and number of follow-up appointments, there were no significant differences in diagnostic outcomes for anxious patients according to the model of care received; however, only 3% of the sample had a primary anxiety disorder. Significant differences were observed for participants who were depressed or at risk of drinking (Bartels et al., 2004).

Thus, while collaborative care for panic disorder is feasible and results in better outcomes for young patients, GAD may be more difficult to treat in a primary care setting with very limited sessions. Further adjustments need to be made to have a meaningful impact on anxiety in older patients, particularly those with GAD. GAD patients may require a more intensive treatment to demonstrate substantial reductions in anxiety and using less trained professionals may not be as effective for GAD.

Although not limited to just GAD or depression, another recent study evaluated the impact of treatment for anxiety and/or depressive symptoms in older patients with confirmed diagnoses of chronic obstructive pulmonary disease (Veazey, Stanley, Hopko, Densmore, & Kunik, 2003). Patients were randomized to eight sessions of either CBT or education. CBT included education, relaxation training, cognitive therapy, problem-solving techniques, increasing activity, and sleep management skills. Education included strategies for breathing and airway management, pathophysiology of lung disease, medications, use of oxygen, avoidance of environmental irritants, nutrition, exercise, smoking cessation, and end-of-life planning. Both interventions reduced anxiety and depressive symptoms and improved quality of life. However, no significant differences between the groups emerged. There are a few possible reasons for the lack of differences between groups. First, only 169 of the 238 participants randomized to the study attended the first session of the intervention. Thus, the least motivated and lowest functioning patients may have dropped out before the intervention even began. Second, the intervention was delivered in a group setting with a fixed and regular schedule, which may have hindered some patients from attending and/or benefiting maximally from the interventions. Third, both interventions included breathing exercises as a key component. Finally, it may be that CBT does not provide substantial benefits over and above COPD education for patients with coexistent anxiety and/or depression.

Extending Psychotherapy for Late-Life GAD into Primary Care

Three ongoing programs are examining how to successfully integrate psychotherapy as a treatment for late-life GAD into the primary care setting. A small pilot compared CBT with usual care in 12 patients diagnosed with GAD (Stanley et al., 2003). Patients were older adults who were recruited from a primary care clinic through physician and self-referrals. The treatment consisted of eight individual sessions with the option of two additional sessions for patients who needed more time to learn the techniques or who were in crisis. The topics addressed in the sessions included psychoeducation, relaxation, cognitive therapy, problem solving, exposure, and sleep management. Therapy was delivered by a master's level clinician, usually in the primary care office. However, sessions were flexible and could be delivered in the person's home. Improvements in GAD severity, worry, and depressive symptoms were noted in patients who received the CBT. A larger study is currently underway, with positive preliminary results (Stanley et al., 2005).

Another investigation of treatment for late-life GAD using a modular approach is currently underway. A preliminary paper describes the outcomes following a cognitive-behavioral intervention with two older primary care patients with GAD (Wetherell, Sorrell, Thorp, & Patterson, 2005). The modular treatment consisted of 14 possible components. Both patients received a module on psychoeducation and relapse prevention in conjunction with five additional modules selected from the following options: relaxation, sleep, problem solving, acceptance, worry control, life review, increasing pleasure, cognitive therapy, exposure, coping with pain, assertiveness training, and time management. Both patients reported improvements in anxiety, worry, and depression.

Finally, an ongoing trial for the treatment of late-life GAD is comparing CBT delivered through bibliotherapy and by telephone with usual care (Brenes, under review). Participants in the CBT condition receive ten workbook chapters by mail that address: psychoeducation, awareness training, breathing and relaxation techniques, cognitive therapy, behavioral therapy, increasing pleasurable activity, problem solving, sleep, pain, and relapse prevention. Each week, the participants receive one chapter in the mail. A study therapist then calls the participants to review the chapter, answer questions, and review the techniques. Preliminary results from four pilot participants indicate the acceptability of this treatment by older anxious patients. Moreover, three out of the four participants demonstrated meaningful declines in symptoms of anxiety and/or worry (Brenes, under review).

A Clinical Perspective on Integrating Psychotherapy into the Primary Care Setting

Collaborative care models hold promise for the treatment of late-life mental health problems. Large trials such as IMPACT and PRISM-E demonstrate that successful treatment of late-life depression can be adapted for use within the primary care setting (Levkoff et al., 2004; Unutzer et al., 2002). However, the treatment of late-life anxiety within such a model has received little attention. Although researchers have been successful in adapting treatment for panic disorder into primary care, albeit with reduced response rates (Roy-Byrne et al., 2005), GAD has received less attention. One study of young and middle aged adults demonstrated that medication in conjunction with a treatment involving minimal psychotherapeutic contact was not a sufficient treatment for GAD (Rollman, Belnap, Mazumdar, Houck, et al. 2005). Given older adults' preference for nonpharmacological treatments, psychotherapeutic interventions require further refinement.

Nonpharmacological approaches to treatment are appealing to older adults for a number of reasons. "I don't want to take another pill" is a commonly heard phrase. Anxious older adults are particularly concerned about side effects (Ayalon, Arean, & Alvidrez, 2005) and may have had previous bad experiences with medications. Others are resistant because of possible drug interaction concerns, while still others do not want to become "dependent" on a medication (Brown et al., 2005; Horne, Weinman, & Hankins, 1999). Given that worry is the primary component of GAD, it is not surprising that patients would be concerned about medications. Some patients do not experience enough symptom relief from medications and want an adjunctive treatment.

Fewer sessions are encouraged in a collaborative care model yet traditional psychotherapy for GAD involves 15–20 sessions. The need for brevity must be balanced with the complexity and chronicity of late-life GAD as well as the need for older adults to have more time to process information. One way of balancing these demands is to use written materials and telephone contacts in conjunction with in-person visits. Although bibliotherapy without accompanying psychotherapeutic contact is likely ineffective in treating late-life GAD (Rollman, Belnap, Mazumdar, Houck, et al., 2005), written materials can be a useful supplement. Providing information in a written format allows patients to read, reread, highlight, take notes, and process the information at their own speed. Used in conjunction with regular psychotherapeutic contact, bibliotherapy may be a cost-effective and satisfactory treatment option.

Regularly scheduled telephone contacts are another way of adapting psychotherapeutic interventions to the primary care setting. Although some therapists may be reluctant, patients seem to like this format. Two studies of telephone-based CBT for depressed patients found very low attrition rates, suggesting that patients like this format (Mohr et al., 2005; Simon, Ludman, Tutty, Operskalski, & von Korff, 2004). In the ongoing Brenes study, all psychotherapy sessions are held by telephone. Despite the offer to have occasional face-to-face sessions, no one has requested this. The use of telephone sessions also reduces barriers such as stigmatization and transportation, both for those who do not have access to transportation and those who do not like to drive in poor conditions (rain, snow, dark, highway travel). In order to have a successful therapeutic relationship by telephone, the therapist needs to explain that he/she will give the patient his/her undivided attention and takes this seriously. By engaging in active listening (repeating what the patient has said) and referring to previous comments made by the patient, the patient is able to determine that the therapist truly is interested in his/her problems.

Given that primary care appointments tend to be short and older adults do not perceive the need for mental health treatment, the initial contact with the anxious older adult is very important. Carefully explaining the treatment rationale and the treatment options available to the patient are critical for the patient to "buy in" to the need for treatment. Furthermore, because older adults tend to focus on somatic rather than psychological symptoms and some patients may be on medications causing side effects, a cognitive-behavioral model that explicitly addresses physical symptoms will be necessary. Training in relaxation techniques may be particularly useful for older adults in this setting (Ayers, Sorrell, Thorp, & Wetherell, in press). Relaxation can be taught fairly easily and it allows patients to experience success before proceeding on to other techniques that may be more difficult to learn (such as cognitive restructuring of worries). The choice of other treatment components should be based on the individual patient's symptoms, as the integration of GAD treatment into primary care settings requires flexibility in the content of the intervention. Modular treatment (Chorpita, Taylor, Francis, Moffitt, & Austin, 2004; Henin, Otto, & Reilly-Harrington, 2001) from which the therapist and patient choose items to focus on seems to be an innovative approach that allows for the tailoring of evidence-based treatment.

Communication with the patient's PCP is a key component of integration of mental health care into this setting. Electronic medical records and e-mail make it quick and easy to share patient information. This is a way to provide information regarding symptom management to the physician who may only see a patient every 3 months. Although psychologists cannot make recommendations regarding medications, they can certainly keep the PCP apprised of changes (or lack thereof) in symptoms.

Finally, psychologists working in the primary care setting will need to reexamine their own roles in the clinic. They may find themselves in a more supervisory position. Within the collaborative care framework, care managers or behavioral care specialists are used to deliver the psychotherapeutic interventions under the supervision of a psychologist. Furthermore, psychologists may be called upon for "curbside consultation" during clinics.

Assessment of Anxiety in Primary Care Settings

In a traditional psychology setting, the diagnosis of GAD is usually made based on a structured clinical interview, such as the structured clinical interview for DSM-IV (SCID; First, Spitzer, Gibbon, & Williams, 2002) or the anxiety disorders interview schedule for DSM-IV (ADIS; Brown, DiNardo, & Barlow, 1994). However, these interviews can take up to 90 min to complete and must be administered by a trained professional, limiting their usefulness in the primary care setting. In these settings, anxiety is typically diagnosed in a two stage process. First, the patient is administered a self-report measure of anxiety symptoms. If the patient scores above the cutoff, a shortened clinical interview is then administered.

One measure that was designed specifically for the diagnosis of psychiatric disorders in the primary care setting is the primary care evaluation of mental disorders (PRIME-MD; Spitzer et al., 1994), a two-part instrument that assesses for depression, anxiety, somatoform disorders, alcohol abuse, and eating disorders based on DSM-IV criteria. Patients complete 25 yes/no questions and if the responses are positive, the clinician interviews the patient. This clinician evaluation takes 8–9 min to complete. The self-report questions of the PRIME-MD have also been found to be a useful case-finding tool for GAD (Rollman, Belnap, Mazumdar, Zhu, et al., 2005). This measure also has good positive predictive value in identifying anxiety disorders in older COPD patients (Kunik et al., 2005). In an attempt to further increase its utility in the clinical setting, an entirely self-report version was developed. The patient health questionnaire (PHQ) assesses for both panic disorder and GAD. Questions related to GAD symptoms assess worry, restlessness, fatigue, muscle tension, sleep disturbance, impaired concentration, and irritability. In a large sample of adults ranging in age from 18 to 99 years, Spitzer reported that the sensitivity of the PHQ for diagnosing an anxiety disorder is 63%; the specificity is 97%; the overall accuracy is 91% (Spitzer, Kroenke, & Williams, 1999).

Another interview that is suitable for the primary care setting is the mini international neuropsychiatric interview (MINI; Sheehan et al., 1998). The MINI is a short structured interview that assesses for depressive and anxiety disorders based on DSM-IV criteria, and takes approximately 15 min to complete. Each disorder is represented by a module, so an individual module rather than the entire interview can be given. A recent study reported a 78% raw agreement between a

GAD diagnosis on the MINI and on the longer ADIS-IV clinical interview in a sample of older adults in a primary care setting (Wetherell, Birchler, Ramsdell, & Unutzer, in press).

The Penn State Worry Questionnaire (PSWQ) is a self-report measure of the frequency and controllability of worry and has been shown to distinguish patients with GAD from those without GAD (Hopko et al., 2003; Webb et al., 2005). It is also a reliable measure of worry in older GAD patients. A recent abbreviated version may be more user-friendly for older adults and more practical for use in primary care (Hopko et al.). Because the PSWQ only assesses worry, a measure like the GAD-Q-IV, a self-report questionnaire that assesses all of the symptoms of GAD based on DSM-IV criteria, may be preferable. Using a cutoff of 4.5 in an older primary care sample correctly classified 67% of patients with GAD (Senior et al., 2005). Furthermore, the PSWQ maintains its psychometric properties when administered by telephone (Senior et al.).

Diversity

A better understanding of anxiety in nonwhite patients who present in the primary care setting is needed. In general, minority patients tend to seek mental health treatment from primary care clinics (Gallo et al., 1995). In one primary care sample of Puerto Rican patients, 11.2% had GAD (Tolin et al., 2005). Latinos and Puerto Ricans with anxiety disorders present with more somatic symptoms than psychiatric complaints (Kirmayer, 2001; Zinbarg et al., 1994). In addition, older black primary care patients report more somatic symptoms than white patients of the same age even though the two groups reported similar levels of worry, depressive symptoms, quality of life, and functional status (Kraus et al., 2005). Thus, the very sparse literature suggests that nonwhite anxious older adults may have an even stronger tendency to experience anxiety through somatic symptoms. Much more research is needed to determine if there are other important differences, and if there are racial differences in treatment outcomes.

Case Study

An example of the implementation of psychotherapy in the primary care setting can be found in an ongoing large randomized controlled trial of the effectiveness of CBT for older adults diagnosed with GAD (Stanley et al., 2005). One patient who took part in this study is a self-referred 65-year-old married white female with a college education who had received an informational letter about the study from her primary care clinic. Administration of the SCID indicated that she met criteria for GAD, social phobia, and a specific phobia of injections, and her score of 43 on the PSWQ indicated moderate worry. At the time she entered the program, the patient led a very active lifestyle, volunteering for several organizations and traveling all over the world. Although she had always considered herself a worrier, she had become increasingly tense over the previous year and began to cancel some of her volunteer activities and other commitments.

One of the major topics of her worry was her health. She had some concerns about arthritis and polyps, but found most of the worries about her health to be manageable. After she had been diagnosed with chronic hepatitis C the previous year, however, her concerns about her health were difficult to control. Worry about her illness and its management began to occupy more of her time. She worried that the hepatitis would progress quickly and her health would deteriorate until she would be unable to care for herself. In addition, she worried that she would develop cancer of the liver as a result of the hepatitis. Her physician had recommended a course of treatment that involved a series of 24 injections. This treatment was quite expensive, costing thousands of dollars, and the patient worried about her ability to pay for the medication. She also feared the pain of repeated injections. She had researched the outcomes of this form of treatment and felt that the response rate of 50% was low, especially in light of the risk of unpleasant side effects and high cost. She thought that there was little chance the treatment would be effective for her. She was also particularly worried that the treatment would cause severe fatigue and limit her activities to a greater degree than her illness currently did. As a result of all of these concerns, she had procrastinated in making a decision about treatment for months. A letter was sent to the patient's PCP to inform him of her inclusion in the study before she entered treatment.

The patient's worries and fears about hepatitis management and the associated avoidance were then addressed with various components of CBT over the course of nine sessions. All treatment sessions were held in the primary care clinic. A chart note was placed in the research section of the patient's medical record after each session. The patient found relaxation training, which included deep breathing and progressive muscle relaxation, to be generally helpful when she began to worry about her health and became tense. Cognitive therapy components of the intervention were used to address her worries, beliefs, and expectations regarding her hepatitis and the recommended treatment. She was able to begin to examine these thoughts and to recognize that she was making logical errors, expecting the worst outcomes for both the course of her hepatitis and the injection therapy. This allowed her to then acknowledge that while these outcomes were possible, they were not the only potential outcomes and not necessarily the most probable. She, in turn, was able to think about the possible outcomes more realistically and estimated that the chance of responding well to the treatment was as likely as not responding.

Her worries about the cost of the injections were addressed through the use of problem solving. By using this process, she was able to look at the expense of the treatment as a situation that could potentially be resolved. She generated and evaluated different possible means to pay for the treatment and found that there were several methods she could use to pay for the injections. She was able to choose a method that worked best for her and developed a plan to pay for the portion of the hepatitis treatment that was not covered by her medical insurance.

The final barrier to beginning the recommended injection therapy was her fear of the pain of the frequent injections. She weighed the possible benefits of the treatment against the pain of the injections and decided that she wanted to go forward with the treatment. As the first step of exposure, she called her doctor and requested the prescription for the injections. She developed a plan to use relaxation and coping statements to manage her anxiety before and during receiving the injections.

At termination of CBT, she was scheduled to begin the course of injection therapy. Throughout the next year of telephone follow-up conducted within the study, she reported that the treatment was progressing very well. She was able to use the skills she gained during CBT to cope with the injections. Her response to the medication was good, and she was experiencing very few problems with side effects. She was able to continue her volunteer work and travel out of the country with almost no interference from her illness or the treatment. The results of the posttreatment SCID indicated that she no longer met criteria for GAD. The patient's scores on the PSWQ also reflected a significant decrease in worry throughout the year of follow-up assessments. Immediately following treatment, her score was 29, indicating mild worry. Her scores remained in the mild range over the follow-up period, with a score of 27 one year after CBT completion.

This case is a good representation of one of the primary challenges in treating older GAD patients: many of their worries relate to realistic problems (such as, declining health), rather than highly improbable events (such as, dying in a plane crash). In this case, the patient was worried about her health and finances, both of which were very realistic problems. The clinician must be careful not to imply that the patient should be unconcerned about these realistic worries, but at the same time needs to help the patient recognize the excessiveness of the worry. Relatedly, a true health problem may be present and require action, yet the clinician needs to help the patient recognize and manage worry that restricts the ability to seek appropriate treatment. In the case described above, this patient did have hepatitis that was potentially treatable, yet her worries initially prevented her from seeking treatment. Through the use of cognitive therapy, the patient was able to recognize that her worries, although based in reality, were excessively negative. Once the patient recognized this and agreed that she needed treatment, problem solving was used to help the patient manage her financial worries.

The other primary challenge to working with older GAD patient relates to cognitive capacity. Declines in cognitive capacity interfere with a person's ability to benefit from a complex and multifaceted intervention, such as CBT. Complex techniques, such as cognitive restructuring, can be difficult for these patients and lead to frustration with treatment. In these cases, the therapist should focus on more basic anxiety management techniques, such as deep breathing, relaxation, and the use of coping statements. Involvement of the patient's family members in the therapy can be helpful, as they can encourage and assist the patient in practicing the anxiety management techniques in their daily lives. Furthermore, recent evidence suggests that problem-solving therapy can be beneficial for older adults with comorbid depression and executive dysfunction (Alexopoulos, Raue, & Arean, 2003), although this has not be tested with anxious patients.

Summary and New Directions in the Treatment of GAD

GAD represents a problem for older adults that has significant implications for other areas of functioning. Recent studies have examined the integration of treatment for mental health problems, including GAD, into the primary care setting. One pilot study and three ongoing trials have suggested the feasibility and potential benefits of providing cognitive-behavioral psychotherapy for the treatment of late-life GAD into the primary care setting. The results of these trials will provide valuable data allowing the comparison of traditional psychotherapy with alternative methods of delivery. Furthermore, the integration of treatment into the primary care setting requires an adaptation in the way anxiety is traditionally diagnosed. The use of a two-stage screening tool is much more feasible than a longer, traditional structured psychiatric interview.

Two new approaches to the treatment of GAD focus on tolerance of uncertainty (Ladouceur, Leger, Dugas, & Freeston, 2004) and acceptance of anxiety (Roemer & Orsillo, 2002). Both of these approaches may have particular relevance for anxiety in older adults in the primary care setting. For example, an older adult recently diagnosed with cancer may experience an increase in her anxiety. Although behavioral management of her health is recommended (good nutrition, exercise, etc.) the patient may have very little control over the course of this disease. Therefore, strategies focused on tolerating and accepting this lack of control may be particularly useful in this situation.

References

Alexopoulos, G. S., Raue, P., & Arean, P. (2003). Problem-solving therapy versus supportive therapy in geriatric major depression with executive dysfunction. *American Journal of Geriatric Psychiatry, 11*, 46–52.

Ansseau, M., Fischler, B., Dierick, M., Mignon, A., & Leyman, S. (2005). Prevalence and impact of generalized anxiety disorder and major depression in primary care in Belgium and Luxemburg: the GADIS study. *European Psychiatry, 20*, 229–235.

Arean, P. A., Alvidrez, J., Barrera, A., Robinson, G. S., & Hicks, S. (2002). Would older medical patients use psychological services? *Gerontologist, 42*, 392–398.

Astrom, M. (1996). Generalized anxiety disorder in stroke patients. A 3-year longitudinal study. *Stroke, 27*, 270–275.

Ayalon, L., Arean, P. A., & Alvidrez, J. (2005). Adherence to antidepressant medications in black and Latino elderly patients. *American Journal of Geriatric Psychiatry, 13*, 572–580.

Ayers, C. R., Sorrell, J. T., Thorp, S. R., & Wetherell, J. L. (in press). Evidence-based treatments for late-life anxiety. *Psychology and Aging, 22*, 52–55.

Bartels, S. J., Coakley, E. H., Zubritsky, C., Ware, J. H., Miles, K. M., Arean, P. A., et al. (2004). Improving access to geriatric mental health services: A randomized trial comparing treatment engagement with integrated versus enhanced referral care for depression, anxiety, and at-risk alcohol use. *American Journal of Psychiatry, 161*, 1455–1462.

Brenes, G. A. (2007, March). *Acceptability of bibliotherapy with telephone contact as a treatment*. Paper presented at the annual meeting of the Anxiety Disorders Association of America, St. Louis, MO.

Brown, C., Battista, D. R., Bruehlman, R., Sereika, S. S., Thase, M. E., & Dunbar-Jacob, J. (2005). Beliefs about antidepressant medications in primary care patients: Relationship to self-reported adherence. *Medical Care, 43*, 1203–1207.

Brown, T. A., DiNardo, P. A., & Barlow, D. H. (1994). *Anxiety disorders interview schedule for DSM IV*. Albany, NY: Graywinds.

Chorpita, B. F., Taylor, A. A., Francis, S. E., Moffitt, C., & Austin, A. A. (2004). Efficacy of modular cognitive behavior therapy for childhood anxiety disorders. *Behavior Therapy, 35*, 263–287.

Craske, M. G., Roy-Byrne, P., Stein, M. B., Sherbourne, C. D., Bystritsky, A., Katon, W., et al. (2002). Treating panic disorder in primary care: A collaborative care intervention. *General Hospital Psychiatry, 24*, 148–155.

First, M. B., Spitzer, R. L., Gibbon, M., & Williams, J. B. W. (2002). *Structured clinical interview for DSM-IV-TR axis I disorders, research version, patient edition (SCID-I/P)*. New York: Biometrics Research, New York State Psychiatric Institute.

Gallo, J. J., Marino, S., Ford, D., & Anthony, J. C. (1995). Filters on the pathway to mental health care II: Sociodemographic factors. *Psychological Medicine, 25*, S79–S83.

Gum, A. M., Arean, P. A., Hunkeler, E., Tang, L., Katon, W., Hitchcock, P., et al. (2006). Depression treatment preferences in older primary care patients. *Gerontologist, 46*, 14–22.

Henin, A., Otto, M. W., & Reilly-Harrington, N. A. (2001). Introducing flexibility in manualized treatments: Application of recommended strategies to the cognitive-behavioral treatment of bipolar disorder. *Cognitive and Behavioral Practice, 8*, 317–327.

Hopko, D. R., Stanley, M. A., Reas, D. L., Wetherell, J. L., Beck, J. G., Novy, D. M., et al. (2003). Assessing worry in older adults: Confirmatory factor analysis of the Penn State Worry Questionnaire and psychometric properties of an abbreviated model. *Psychological Assessment, 15*, 173–183.

Horne, R., Weinman, J., & Hankins, M. (1999). The beliefs about medicines questionnaire: The development and evaluation of a new method of assessing the cognitive representation of medication. *Psychology and Health, 14*, 1–24.

Kirmayer, L. J. (2001). Cultural variations in the clinical presentation of depression and anxiety: Implications for diagnosis and treatment. *Journal of Clinical Psychiatry, 62*, 22–28.

Klap, R., Unroe, K. T., & Unutzer, J. (2003). Caring for mental illness in the United States: A focus on older adults. *American Journal of Geriatric Psychiatry, 11*, 517–524.

Krasucki, C., Howard, R., & Mann, A. (1998). The relationship between anxiety disorders and age. *International Journal of Geriatric Psychiatry, 13*, 79–99.

Krasucki, C., Howard, R., & Mann, A. (1999). Anxiety and its treatment in the elderly. *International Psychogeriatrics, 11*, 25–45.

Kraus, C. A., Kunik, M., Rhoades, H., Novy, D. M., Wilson, N., & Stanley, M. A. (2005). *Cross cultural differences in older adults with GAD*. To be presented at the 39th Annual Convention of the Association for Advancement of Behavior Therapy, Washington, DC, November.

Kunik, M. E., Roundy, K., Veazey, C., Souchek, J., Richardson, P., Wray, N. P., et al. (2005). Surprisingly high prevalence of anxiety and depression in chronic breathing disorders. *Chest, 127*, 1205–1211.

Ladouceur, R., Leger, E., Dugas, M., & Freeston, M. H. (2004). Cognitive-behavioral treatment of generalized anxiety disorder (GAD) for older adults. *International Psychogeriatrics, 16*, 195–207.

Landreville, P., Landry, J., Baillargeon, L., Guerette, A., & Matteau, E. (2001). Older adults' acceptance of psychological and pharmacological treatments for depression. *Journal of Gerontology B: Psychological Science and Social Science, 56*, 285–291.

Lenze, E. J., Karp, J. F., Mulsant, B. H., Blank, S., Shear, M. K., Houck, P. R., et al. (2005). Somatic symptoms in late-life anxiety: treatment issues. *Journal of Geriatric Psychiatry and Neurology, 18*, 89–96.

Levkoff, S. E., Chen, H., Coakley, E., Herr, E. C., Oslin, D. W., Katz, I., et al. (2004). Design and sample characteristics of the PRISM-E multisite randomized trial to improve behavioral health care for the elderly. *Journal of Aging and Health, 16*, 3–27.

Mickus, M., Colenda, C. C., & Hogan, A. J. (2000). Knowledge of mental health benefits and preferences for type of mental health providers among the general public. *Psychiatric Services, 51*, 199–202.

Mohr, D. C., Hart, S. L., Julian, L., Catledge, C., Honos-Webb, L., Vella, L., et al. (2005). Telephone-administered psychotherapy for depression. *Archives of General Psychiatry, 62*, 1007–1014.

Ormel, J., Kempen, G. I., Penninx, B. W., Brilman, E. I., Beekman, A. T., & van Sonderen, E. (1997). Chronic medical conditions and mental health in older people: disability and psychosocial resources mediate specific mental health effects. *Psychological Medicine, 27*, 1065–1077.

Pearson, J. L. (1998). Summary of a National Institute of Mental Health workshop on late-life anxiety. *Psychopharmacology Bulletin, 34*, 127–130.

Regier, D. A., Boyd, J. H., Burke, J. D., Jr., Rae, D. S., Myers, J. K., Kramer, M., et al. (1988). One-month prevalence of mental disorders in the United States. Based on five Epidemiologic Catchment Area sites. *Archives of General Psychiatry, 45*, 977–986.

Roemer, L., & Orsillo, S. M. (2002). Expanding our conceptualization of and treatment for generalized anxiety disorder: Integrating mindfulness/acceptance-based approaches with existing cognitive-behavioral models. *Clinical Psychology Science and Practice, 9*, 54–68.

Rollman, B. L., Belnap, B. H., Mazumdar, S., Houck, P. R., Zhu, F., Gardner, W., et al. (2005). A randomized trial to improve the quality of treatment for Panic and Generalized Anxiety Disorders in Primary care. *Archives of General Psychiatry, 62*, 1332–1341.

Rollman, B. L., Belnap, B. H., Mazumdar, S., Zhu, F., Kroenke, K., Schulberg, H. C., et al. (2005). Symptomatic severity of PRIME-MD diagnosed episodes of panic and generalized anxiety disorder in primary care. *Journal of General Internal Medicine, 20*, 623–628.

Roundy, K., Cully, J. A., Stanley, M. A., Veazey, C., Souchek, J., Wray, N. P., et al. (2005). Are anxiety and depression addressed in primary care patients with COPD? *Primary Care Companion Journal of Clinical Psychiatry, 715*, 213–218.

Roy-Byrne, P. P., Craske, M. G., Stein, M. B., Sullivan, G., Bystritsky, A., Katon, W., et al. (2005). A randomized effectiveness trial of cognitive-behavioral therapy and medication for primary care panic disorder. *Archives of General, 62*, 290–298.

Roy-Byrne, P. P., & Wagner, A. (2004). Primary care perspectives on generalized anxiety disorder. *Journal of Clinical Psychiatry, 65*(Suppl. 13), 20–26.

Senior, A. C., Kunik, M., Rhoades, H. M., Wilson, N., Novy, D. M., Williams, C. W., et al. (2005). *Psychometric properties of self-report measures administered over the telephone*. Presented at the 39th Annual Convention of the Association for Advancement of Behavior Therapy, Washington, DC, November.

Sheehan, D. V., Lecrubier, Y., Sheehan, K. H., Amorim, P., Janavs, J., Weiller, E., et al. (1998). The Mini-International Neuropsychiatric Interview (MINI): The development and validation of a structured diagnostic psychiatric interview for DSM-IV and ICD-10. *Journal of Clinical Psychiatry, 59*(Suppl. 20), 22–33.

Simon, G. E., Ludman, E. J., Tutty, S., Operskalski, B., & Von Korff, M. (2004). Telephone psychotherapy and telephone care management for primary care patients starting antidepressant treatment: A randomized controlled trial. *Journal of the American Medical Association, 292*, 935–942.

Spitzer, R. L., Kroenke, K., & Williams, J. B. (1999). Validation and utility of a self-report version of PRIME-MD: The PHQ primary care study. Primary Care Evaluation of Mental Disorders. Patient Health Questionnaire. *Journal of the American Medical Association, 282*, 1737–1744.

Spitzer, R. L., Williams, J. B., Kroenke, K., Linzer, M., DeGruy, F. V., III, Hahn, S. R., et al. (1994). Utility of a new procedure for diagnosing mental disorders in primary care: The PRIME-MD 1000 study. *Journal of the American Medical Association, 272*, 1749–1756.

Stanley, M. A., Hopko, D. R., Diefenbach, G. J., Bourland, S. L., Rodriguez, H., & Wagener, P. (2003). Cognitive-behavior therapy for late-life generalized anxiety disorder in primary care: preliminary findings. *American Journal of Geriatric Psychiatry, 11*, 92–96.

Stanley, M. A., Novy, D. M., Kunik, M. E., Wilson, N. L., Overall, J., & Rhoades, H. M. (2005). Preliminary outcomes: Late-life GAD in primary care. In M. S. Stanley (Chair), *Expanding the boundaries for late-life anxiety: Symptoms and treatment in primary care*. Symposium, Association for the Advancement of Behavior Therapy, Washington, DC, November.

Stanley, M. A., Roberts, R. E., Bourland, S. L., & Novy, D. M. (2001). Anxiety disorders among older primary care patients. *Journal of Clinical Geropsychology, 7*, 105–116.

Tolin, D. F., Robison, J. T., Gaztambide, S., & Blank, K. (2005). Anxiety disorders in older Puerto Rican primary care patients. *American Journal of Geriatric Psychiatry, 13*, 150–156.

Unutzer, J., Katon, K., Callahan, C. M., Williams, J. W., Jr., Hunkeler, E., Harpole, L., et al. (2002). Collaborative care management of late-life depression in the primary care setting: A randomized controlled trial. *Journal of the American Medical Association, 288*, 2836–2845.

Veazey, C., Stanley, M. A., Hopko, D., Densmore, D., & Kunik, M. (2003). *Randomized controlled trial of a cognitive behavioral intervention for anxiety and depression in Chronic Obstructive Pulmonary Disease*. Presented at the 37th Annual Convention of the Association for Advancement of Behavior Therapy, Boston, November.

Webb, S. A., Diefenbach, G., Novy, D. M., Kunik, M., Rhoades, H. M., Weiss, B. J., et al. (2005). *Utility of GAD-Q-IV and PSWQ to identify late-life GAD in a referred primary care sample*. Presented at the 39th Annual Convention of the Association for Advancement of Behavior Therapy, Washington, DC, November.

Weissman, M. M., Myers, J. K., Tischler, G. L., Holzer, C. E., III, Leaf, P. J., Orvaschel, H., et al. (1985). Psychiatric disorders (DSM-III) and cognitive impairment among the elderly in a U.S. urban community. *Acta Psychiatrica Scandinavica, 71*, 366–379.

Wetherell, J. L., Birchler, G. D., Ramsdell, J., & Unutzer, J. (2007). Screening for generalized anxiety disorder in geriatric primary care patients. *International Journal of Geriatric Psychiatry, 22*, 115–123.

Wetherell, J. L., Gatz, M., & Craske, M. G. (2003). Treatment of generalized anxiety disorder in older adults. *Journal of Consulting & Clinical Psychology, 71*, 31–40.

Wetherell, J. L., Kaplan, R. M., Kallenberg, G., Dresselhaus, T. R., Sieber, W. J., & Lang, A. J. (2004). Mental health treatment preferences of older and younger primary care patients. *International Journal of Psychiatry and Medicine, 34*, 219–233.

Wetherell, J. L., Lenze, E. J., & Stanley, M. A. (2005). Evidence-based treatment of geriatric anxiety disorders. *Psychiatric Clinics of North America, 28*, 871–896.

Wetherell, J. L., Sorrell, J. T., Thorp, S. R., & Patterson, T. L. (2005). Psychological interventions for late-life anxiety: A review and early lessons from the CALM study. *Journal of Geriatric Psychiatry and Neurology, 18*, 72–82.

Wittchen, H. U., Kessler, R. C., Beesdo, K., Krause, P., Hofler, M., & Hoyer, J. (2002). Generalized anxiety and depression in primary care: Prevalence, recognition, and management. *Journal of Clinical Psychiatry, 63*, 24–34.

Zinbarg, R. E., Barlow, D. H., Liebowitz, M., Street, L., Broadhead, E., Katon, W., et al. (1994). The DSM-IV field trial for mixed anxiety-depression. *American Journal of Psychiatry, 151*, 1153–1162.

4
Cognitive-Behavior Therapy for Late-Life Insomnia

Kristen C. Stone, Andrea K. Booth, and Kenneth L. Lichstein

Practitioners knowledgeable in psychologically treating late-life insomnia improve the overall care of patients by (1) using insomnia complaints and symptoms to screen for medical, mental, and organic sleep disorders and (2) treating the insomnia instead of referring the patient to a sleep specialist, thereby ensuring efficient and effective individual tailoring of the treatment through high familiarity with the patient. Cognitive-behavior treatment for insomnia is viable with older adults and may be preferred as an alternative to sleep medication.

Insomnia refers to the complaint of difficulty initiating or maintaining sleep or of nonrestorative sleep accompanied by significant daytime distress or impairment (American Psychiatric Association, 1994). Insomnia is usually predominantly characterized by the inability to fall asleep, stay asleep, fall back asleep after early morning waking, or some combination of these difficulties (Ancoli-Israel, Pat-Horenczyk, & Martin, 2001) and is further differentiated by its duration. Most individuals have experienced transient insomnia, which results from an acute stressor (e.g., a deadline) and last for a period of days. Chronic insomnia occurs when the insomnia does not relent due to maladaptive coping skills (e.g., trying harder to get sleep by going to bed earlier) or is caused by physical, psychosocial, environmental, and/or behavioral factors (Ancoli-Israel et al.). Older adults experience an increase in acute stressors, environmental changes, physical and mental difficulties, and detrimental psychosocial factors and are, thus, more predisposed to insomnia than the general population. Also, insomnia is more prevalent (20–40%; e.g., Foley, Monjan, Brown, Simonsick, Wallace, & Blazer, 1995; Mellinger, Balter, & Uhlenhuth, 1985; Miles & Dement, 1980) and more severe in older adults than in younger adults (Lichstein, Durrence, Riedel, Taylor, & Bush, 2004), and older adults report suffering from insomnia more often than any other sleep disorder (Ancoli-Israel et al.).

The empirical literature on late-life insomnia suggests that problems with sleep must be interpreted within the context of aging (Nau, McCrae, Cook, & Lichstein, 2005). Aging has been associated with decreased sleep efficiency, variable and declining sleep time, increased light sleep (stages 1–2), increased frequency of arousals, and declining deep sleep (stages 3–4; cited in Nau et al.). Individuals experiencing such changes may present with an insomnia complaint despite an age-normative sleep pattern. Distinguishing between insomnia and adjustment problems associated with developmental sleep changes proves challenging for the clinician and makes thorough assessment of sleep difficulties and related daytime functioning necessary for accurate diagnosing and corresponding treatment planning.

Assessment Considerations Linked to Treatment Planning

Investigating the sleep problem's origin, type, frequency, history, severity, and effects on normal functioning allows for both accurate diagnosis and treatment that is tailored, as much as possible, to the unique constellation of causes of the patient's sleep problems (Cook, Nau, & Lichstein, 2005). The multifaceted assessment also serves to rule out organic sleep disorders, such as sleep apnea, that require different treatments from those of insomnia. In fact, some insomnia medications are contraindicated for people with sleep apnea. Tools effectively utilized to assess insomnia include sleep diaries, objective measurements, and a clinical interview.

Sleep diaries are used to collect daily information on the sleep patterns of the patient (Cook et al., 2005). The patient should be asked to provide information over a period of 2 weeks prior to the initial interview. The diary should include questions assessing amount of time spent napping, time of entering the bed to sleep, time to fall asleep, number of awakenings during the night, time spent awake during the night, final time awake, time of exiting the bed, subjective satisfaction with quality of sleep, and details on medications or alcohol used (see Cook et al. for an example of a sleep diary). The patient should estimate these variables instead of watching the clock so as not to exacerbate anxiety and compound difficulty falling asleep. Due to their subjective nature, sleep diaries are likely to overestimate sleep onset latency (SOL) and wake time after sleep onset (WASO), but because treatment outcomes are evaluated by the subjective satisfaction of the individual patient, sleep diaries correspond well as an assessment tool. The patient's satisfaction will be reflected in their perceptions and subsequent recordings of sleep (Ancoli-Israel et al., 2001).

Valid and reliable sleep questionnaires, such as the Pittsburgh Sleep Quality Index (PSQI; developed by Buysse, Reynolds, Monk, Berman, & Kupfer, 1989), are often brief, can detail sleep patterns, and aid in differential diagnosis. Furthermore, daytime functioning questionnaires, listed in Table 1 with corresponding cutoff scores, can rule out other disorders (e.g., severe apnea in the absence of self-reported daytime sleepiness) and assess subjective distress in older adults with insomnia (OAWI). When the cutoff score for at least one measure is exceeded, significant daytime impairment is indicated (Cook et al., 2005).

Objective measures can also be used to assess sleep and daytime functioning of OAWI, namely, polysomnography (PSG) and actigraphy. PSG is the standard objective sleep assessment and consists of electro-oculography (EOG),

TABLE 1. Daytime functioning questionnaires.

Measure	Cutoff score	Source of measure
Epworth Sleepiness Scale	≥7.4	Johns (1991)
Fatigue Severity Scale	≥5.5	Krupp, LaRocca, Muir-Nash, and Steinberg (1989)
State-Trait Anxiety Inventory	≥37	Spielberger, Gorsuch, Lushene, Vagg, and Jacobs (1983)
Geriatric Depression Scale	≥15	Yesavage et al. (1983)

electroencephalography (EEG), chin and anterior tibialis, electromyography (EMG), electrocardiography (ECG), oximetry and measures of respiratory effort and airflow (Cook et al., 2005). Due to the overlapping clinical presentation of many sleep, mental, and physical disorders, PSG is often useful in determining if the presenting symptoms are manifestations of organic sleep disorders (Ancoli-Israel et al., 2001). PSG is indicated when symptoms of periodic leg movement disorder (PLMD; e.g., bed partner reports leg twitches of patient during the night), sleep-related breathing disorder (SBD; e.g., severe daytime sleepiness), or narcolepsy (e.g., cataplexy) are revealed in the subjective assessment. PSG may also be warranted when treatments of insomnia have not been effective or when diagnosis is uncertain (Lichstein & Riedel, 1994).

An actigraph is a device, typically worn on the wrist, that measures movement to infer sleep and wakefulness. Actigraphy appears valid in normal sleepers, but in people with insomnia (PWI) validity is tenuous. Although some actigraphs have shown adequate validity at detecting sleep and wake (with the exception of SOL) in PWI (e.g., Lichstein, Stone, et al., 2006), the sensitivity thresholds have not been standardized and validity studies have not been conducted with most instruments. The reliability of actigraphy is not in question, having been demonstrated in numerous studies; thus actigraphy is accurate at detecting sleep changes over time. In older adults it can also be used to monitor activity and napping during the day.

Finally, a clinical interview is necessary to clarify the current pattern of sleep, the quality of sleep, and the resulting daytime effects (Cook et al., 2005). Information about nighttime activity should be corroborated with the patient's bed partner, if possible. The clinician should also inquire about the development and history of the sleep complaint, current and previous treatment, health status and medical history, and psychological functioning and history. To rule out alternative sleep disorders, including primary hypersomnia and circadian rhythm sleep disorder, it is important to investigate snoring, twitching legs, uncontrollable sleepiness, and medication usage. Examples of extensive clinical interviews are found elsewhere (e.g., Cook et al.; Moller, Barbera, Kayumov, & Shapiro, 2004).

An important assessment factor with OAWI is the strong possibility of secondary insomnia (SI). Primary insomnia is indicated only when the problem is not caused by another disorder; SI is indicated when a physical or psychological condition or substance causes and maintains insomnia. Attention to the presence of possible SI is always warranted with OAWI given that preexisting physical conditions place older adults at increased risk (Lichstein, Gellis, Stone, & Nau, 2006). However, assessing the secondary nature of the insomnia is complex because temporal sequence is difficult to establish (e.g., whether the other condition preexisted the insomnia and whether or not the insomnia subsides when the other condition subsides and returns in the same corresponding fashion). Comorbidity is common – in which each condition exists independent of the other or inconsistently exacerbates the other. In other words, even if the insomnia is secondary initially, it easily transforms into an independent condition (i.e., primary) capable of causing or exacerbating other conditions. The *International Statistical Classification of Diseases and Related Health Problems: 10th Revision* (ICD-10, 2003) and the *International Classification of Sleep Disorders* (American Sleep Disorders Association, 2005) provide additional conceptualizations of insomnia, insomnia types based on etiologies, and criteria that may help with challenging diagnostic assessments. The body of literature investigating SI treatment response reveals it to be virtually indistinguishable from the primary insomnia treatment response (Nau et al., 2005). In light of this amenability to treatment combined with the difficulty of diagnosis, SI should always be treated.

Treatment

PWI can be effectively treated by integrating behavioral, cognitive, and educational approaches (Cook et al., 2005); older adults are no exception. Individuals do not need to have a diagnosis of primary insomnia for successful treatment to ensue. Sleep problems caused by or exacerbated by other conditions can be treated concurrently with these other conditions. Treatment for older adults include such modifications as compulsory education about elderly sleep norms, passive relaxation techniques preferable for those with arthritis, stimulus control instructions for those with limited mobility, and flexibility with napping tolerance.

Behavioral Interventions

Behavioral interventions demonstrating effectiveness include sleep hygiene, sleep restriction therapy (SRT)/sleep compression, stimulus control, and relaxation (Cook et al., 2005). These interventions should occur in combination in a sequence that best suits the patient's need (Lichstein & Morin, 2000). For example, if a patient is highly anxious, starting with relaxation may help ease anxiety, make the patient feel more comfortable, and increase coping skills for the future. Six to ten sessions may be needed but variability in treatment length results from the severity of the insomnia, treatment adherence, and patient motivation. Less frequent sessions for maintenance are common.

Sleep hygiene. Sleep hygiene education is geared toward discouraging behaviors that are sleep inhibiting and encouraging behaviors that are sleep promoting (Cook et al., 2005). OAWI may need to make adjustments to the frequency, duration, and timing of activities in their lives in order to foster a more sleep-conducive environment. During the day, OAWI should minimize naps and caffeine consumption. Activities that should be avoided before bedtime include exercising, consuming alcohol and nicotine, and eating large meals. Individual differences in how behaviors affect sleep account for varying degrees of importance for adopting certain behaviors and letting go of others. OAWI should be encouraged to eliminate or limit afternoon naps due to the effects on circadian cycles (Cook et al.).

Sleep restriction therapy/sleep compression. In OAWI there is often a discrepancy between *perception* of sleep need and *actual* sleep need (Cook et al., 2005). SRT addresses this discrepancy by underscoring the importance of sleeping according to need, rather than desire. Increasing the urge to sleep (homeostatic drive) and maintaining circadian rhythmicity are key components of this intervention.

SRT increases homeostatic drive and consolidates sleep by decreasing unnecessary time in bed (TIB) and sleep duration. First, the therapist uses the patient's sleep diaries to estimate sleep need (Cook et al., 2005). The actual sleep need is usually derived by obtaining an average total sleep time (TST) over a 2-week period. TST is computed by subtracting the individual's total wake time from their total TIB. Based on an average of the patient's TST, a TIB allotment for the first week of treatment is determined. Because a TIB allotment is derived from patient's sleep reports, which are usually underestimates of actual time slept, the initial TIB allotment may not meet the patient's sleep requirement. Such sleep deprivation, however, only serves to increase homeostatic drive and consolidate sleep.

After a TIB allotment is determined, consistent bed times and wake times should be established (Cook et al., 2005). After the patient sleeps for a week on this new schedule, the therapist calculates the patient's sleep efficiency (SE) for that week (SE = TST/TIB). If the patient's mean SE ≥ 90% with reports of significantly impaired daytime functioning (i.e., the patient is sleeping well but may not have met his/her sleep need), the TIB allotment is increased 15–30 minutes. If the patient's mean SE drops below 85% the TIB allotment is decreased by 15–30 minutes. This pattern continues until there are no reports of significant daytime impairment and the patient's SE is consistently ≥85%. Ultimate success, however, is determined by the patient's subjective satisfaction.

Sleep compression differs from SRT because it incrementally reduces TIB (Lichstein, Riedel, Wilson, Lester, & Aguillard, 2001). The amount of the reduction is determined by averaging the total wake time from the patient's sleep diaries and dividing by the number of compression sessions (approximately six). This gradual reduction of TIB allotment often reduces resistance, increases adherence, and aids in adaptation to treatment for OAWI. Successful outcomes of SRT/sleep compression are maintained by varying degrees of continued compliance. Some older adults may have to restrict TIB indefinitely; others may be flexible - for instance, on the weekends - with the allotted TIB and established bed/wake times.

Stimulus control. Stimulus control therapy is based on conditioning principles and operates under the assumption that for many PWI the bedroom and bed have become sleep-preventing stimuli (Ancoli-Israel et al., 2001). The bed and bedroom *can* be strong cues for sleep. However, repeatedly engaging in behaviors that interfere with sleep *in the bed/bedroom* – such as tossing and turning, worrying, reading, watching TV, or exercising – creates a strong psychological association between bed and wake that must be unlearned and replaced with a strong association between bed and sleep. Bootzin and Epstein (2000) offer the following instructions for stimulus control therapy:

1. *Only use the bedroom and bed for sleep (or sex).* Engaging in behaviors that are not sleep promoting within the bed/bedroom violates the sanctity of the bed/bedroom. The bedroom should be reprioritized as a special room that is used only for the activity of sleep.
2. *Go to bed only when sleepy.* This instruction is useful for many OAWI who go to bed for a variety of reasons other than feeling sleepy, such as following a bed partner's routine. Instead, somatic feelings of sleepiness should be the only prompt for bed.
3. *Leave the bed and bedroom if you have not fallen asleep within approximately 20 minutes . In another room, engage yourself in a relaxing activity. Go back to bed only when you feel sleepy.* Breaking the habit of lying in bed struggling to sleep eliminates sleep-preventative associations between the bed and wakefulness. Instead of struggling to sleep, wakefulness should be a cue to leave the bedroom. Engaging in a quiet activity (e.g., reading) in another room redirects one's attention from frustration and reserves the bedroom for sleep.
4. *If you have difficulty falling asleep after returning to bed, once again, remove yourself from the bed and bedroom and return to bed only when sleepy. Repeat this procedure as often as needed. Also, follow this instruction if you have difficulty falling asleep after an awakening in the night.* Emphasizing leaving the bedroom when having trouble sleeping prepares the patient for the first few difficult nights of conditioning, forming new habits, and breaking old ones. Offering support and validating frustration is important for motivation and for treatment adherence of OAWI.
5. *Regardless of the quality or quantity of sleep obtained, arise every morning at the same time.* Maintaining consistency will promote a constant sleep rhythm; compensating for lost sleep by staying in bed will alter the sleeping cycle and may delay the onset of sleep that night. In addition, lingering in bed hoping for more sleep can promote feelings of frustration and reinforce negative associations.
6. *Limit or avoid napping.* Napping has the potential to deconsolidate sleep and negatively affect circadian rhythmicity. However, for older adults it may also provide a boost of energy for the remainder of the day. If the nap is short, this benefit may be produced without interfering with homeostatic drive at night. If a nap is necessary, it is recommended that it be taken before 2:00 p.m. and last less than 1 hour. Daily naps should occur at the same time everyday and should occur only in the bedroom. Furthermore, the stimulus control instructions should be applied during nap time (i.e., being awake longer than 20 minutes should signal leaving the bedroom).

Individually tailoring these stimulus control procedures is often necessary for older adults with special circumstances (Cook et al., 2005). For instance, individuals with physical disabilities may find it difficult to leave the bedroom repeatedly during the night. Instead, they may want to sit up in bed and engage in a relaxing activity until feeling sleepy as a way to break the habit of tossing and turning.

Relaxation. Often it is difficult for OAWI to relax due to physical tension and mental arousals that result from worry. Relaxation serves to decrease physical and mental arousals that pose as sleep barriers. This technique encompasses comfortable positioning, a calming and repetitive stimulus, and an environment that is quiet and serene (Cook et al., 2005). When teaching relaxation, the clinician usually asks patients to close their eyes for a period of 10–20 minutes while engaging in either progressive or passive relaxation techniques (see Lichstein, 2000, for more details on conducting relaxation procedures). Progressive muscle relaxation consists of alternating between contrasting states of muscle tension and relaxation. Tensing or tightening different muscle groups and then releasing the tension leads to the ability to become fully aware of what relaxed muscles feel like and to reproduce the sensation (Lichstein). Older adults with muscle pain or arthritis may have difficulty tensing muscles and may prefer passive relaxation, which focuses on breathing and soothing bodily sensations guided by slow, repetitive phases. As with the other behavioral treatments, relaxation should be individually tailored to the client's needs. Lengthening or removing parts of the procedure as desired by the patient will increase effectiveness and compliance. In addition to deep breathing and muscle relaxation, the procedure can include autogenic phrases such as, "I am at peace. My arms and legs feel heavy and warm" on which the patient can focus. Encouraging the patient to self-administer the procedure in session can allow the therapist an opportunity to observe execution and make suggestions to enhance relaxation, such as decreasing the pace, or increasing attention to areas of the body that appear particularly tense.

Cognitive Interventions

Cognitive therapy for insomnia is designed to restructure the attitudes and beliefs held by patients regarding sleep and to teach the patient how to interpret situations in adaptive ways (Cook et al., 2005). To achieve these goals the therapist must first elicit the automatic thoughts about the causes and effects of the patient's sleep problems through verbal questioning. This inquiry provides an opportunity to determine if the patients' cognitions perpetuate the cycle of sleep loss. The expectation that napping will help catch up on "lost" sleep and that anything less than 8 hours of sleep a night is unhealthy are examples of maladaptive cognitions. OAWI may also incorrectly attribute the solitary cause of their sleep problems to age, pain, or something else beyond their control (Cook et al.). Other types of misconceptions include dichotomous thinking (e.g., I'll either sleep or die trying), overgeneralization (e.g., I have not slept in a month), fortune-telling (e.g., I know I'll never be able to sleep), and selective recall (e.g., I only slept for about one hour last night). Faulty cognitions about sleep lead to heightened arousal and feelings of anxiety that prevent the onset of sleep.

Once the faulty beliefs and automatic thoughts have been identified in session, the therapist must help the patient identify these faulty perceptions as they occur daily, thus, increasing opportunities to challenge the validity of the automatic thoughts. The therapist encourages the patient to search for more adaptive and rational thoughts. This process is often difficult because many of the distorted beliefs have been held for a long time. Educating the OAWI about age-appropriate norms, wide variability within the norms, and reduced sleep needs can help dispel maladaptive expectations. Explaining that insomnia can be caused by a variety of factors and almost always successfully treated allows the patient to regain control. The patient should also be encouraged to explore the feelings that arise when a particular thought emerges (Cook et al., 2005) to underscore the connection between how certain thoughts affect emotions which have subsequent effects on behavior. In conclusion, the goals of cognitive therapy are to increase the patient's awareness of automatic, dysfunctional thoughts and to teach the patient to self-sufficiently recognize these thoughts, challenge their validity, and replace them with more adaptive thoughts.

Case Example

Mrs. S was a 66-year old retiree, who reported a reoccurrence of insomnia despite having taken a benzodiazipine nightly for the past 6 years. Mrs. S's physical health was good with the exception of migraine headaches. Initial 2-week sleep diaries revealed that Mrs. S was experiencing prolonged SOL and WASO 4–5 nights a week. Mrs. S reported long hours of tossing and turning throughout the night and low energy throughout the day. Mrs. S's therapist helped her gradually withdraw from her sleeping medication before introducing a cognitive-behavior therapy for insomnia (CBTI) treatment package consisting of sleep hygiene, stimulus control, and relaxation. After 1 week of treatment, Mrs. S admitted to drinking coffee and napping and to failing to adhere to her instruction to only go to bed when sleepy. Thus, treatment goals were tailored to better fit Mrs. S's lifestyle. For instance, it was decided that Mrs. S would allow herself to lie down when having a migraine. Furthermore, with help from her therapist, she began a gradual withdrawal from caffeine after a 40-year "two-cup-a-day" habit. Finally, Mrs. S agreed to discuss with her husband the idea of her not going to bed when he did, which was their usual routine.

In the ensuing month of CBTI, Mrs. S noticed moderate sleep gains but remained dissatisfied with her sleep. She collaborated with her psychologist and primary care doctor on a treatment plan that involved 20 mg of Ambien along with continued CBTI. From that point forward, treatment focused primarily on monitoring Mrs. S's adherence to previous instructions and on Mrs. S's faulty cognitions about sleep such as "I cannot sleep because my brain is broken." Her reluctance not to accompany her husband to bed and strategies for coping with her husband's snoring were also addressed. Mrs. S's compliance to treatment and her satisfaction with her sleep increased during this treatment phase, and she reduced her Ambien intake to 5–10 mg per night. At termination, Mrs. S. stated "I still wish I slept as well as my husband, but my sleep is not making me miserable anymore. And I am proud not to be so dependent on medicine."

As with this case, many OAWI are already on a sleeping medication when they seek cognitive-behavioral intervention for their insomnia and will require assistance from their prescribing physician and education from their sleep therapist to allay their fears regarding discontinuing this medication. Furthermore, the instruction to only go to bed when sleepy is frequently difficult for OAWI who enjoy earlier bedtimes and who know of many others their age able to maintain early bedtimes. It is important to offer empathy and support for OAWI as they form new habits that are sleep promoting. Older adults benefit from time spent assessing barriers to compliance in order to discover realistic goals that can be gradually attained.

Combining Pharmacological and CBT Interventions

Rarely should sleep medication be the first or only line of defense for treating chronic insomnia. One important reason is that older adults typically take multiple medications and, thus, are at increased risk for drug interactions. Furthermore, in older adults, benzodiazapines have been linked with slow-wave sleep suppression (Poyares, Guilleminault, Ohayon, & Tufik, 2004), increased hip fractures (e.g., Ray, Griffin, & Downey, 1989), and increased car accidents (e.g., Hemmelgarn, Suissa, & Huang, 1997). Finally, sleep medications are primarily intended for short-term use, and chronic use can lead to hypnotic-dependent insomnia (Nau et al., 2005). Hypnotic-dependent insomnia is characterized by increased bouts of sleep problems during periods of high tolerance to sleep medication. Withdrawing sleep medication (in hopes of at least restoring sleep patterns to baseline levels) is usually problematic because of psychological or physiological dependence or both. Fortunately, cognitive-behavior sleep therapy – combined with gradual sleep medication withdrawal or used alone – is generally effective in treating hypnotic-dependent insomnia (Nau et al.).

Despite their flaws, sleep medications usually improve sleep relatively quickly and are readily available. In fact, not only is sleep medication the most common intervention for sleep problems across ages, rates of usage are higher among the elderly than among younger adults (Nau et al., 2005). Furthermore, there is evidence that combining psychological and pharmacological interventions for insomnia is effective. Research (e.g., Hauri, 1997; Morin, Colecchi, Stone, Sood, & Brink, 1999) suggests that short term, there seems to be no incremental effectiveness of combining psychological and pharmacological treatments; they are equally effective either alone or in combination. However, for a sustained treatment effect, CBTI or a combination of CBTI and medication may be more effective than medication alone.

The majority of these studies on combining sleep medication with CBTI were designed so that both treatments began at the same time. However, for most OAWI medication exists as the only form of treatment for a period of time before CBTI is employed. There is no empirical basis for this sequencing trend. In fact, Vallières, Morin, and Guay (2005) demonstrated that beginning treatment with medication alone, followed by CBTI, leads to a period of sleep loss that is regained months after treatment. This period of sleep loss was not experienced in the CBTI-alone condition or in the condition that began with CBTI and medication followed by CBTI alone.

Issues of Diversity

Unique aspects of sleep associated with diverse groups of OAWI may be concurrent with factors attributed to the aging process among elderly people presenting for treatment. Being knowledgeable about differing sleep patterns across diverse groups of older adults expedites assessment, enriches treatment planning, and aids in setting realistic treatment goals.

SES and Ethnicity/Culture

Research on ethnic diversity with regard to insomnia is scarce and inconclusive (e.g., Durrence & Lichstein, 2006; Jean-Louis et al., 2001; Robert, Roberts, & Chan, 2006). Whether or not discrepancies in research findings can be accounted for by factors such as age is unknown. Gellis et al. (2005) investigated differences of insomnia patterns across socioeconomic status (SES) and found that lower levels of education increases the likelihood of experiencing insomnia and daytime complaints attributed to insomnia and that these patterns are consistent across ethnicity, gender, and age. Furthermore, Gellis et al. points to the confounding influence of SES on ethnicity in most studies and, therefore, denies the existence of a founded relationship between ethnicity and insomnia. Findings reveal no SES differences on alcohol or caffeine intake (Gellis et al.), suggesting that insomnia in individuals with lower SES may not be stemming from poor sleep hygiene. Differences in access or response to insomnia treatment according to SES, ethnicity, or culture are largely unknown. Among approximately 200 studies of CBTI, there is not one mention of treating African-Americans with insomnia. Obviously, ethnicity should be of greater focus in insomnia treatment research.

Cognitive Impairment

The associations between mental functioning and sleep are proncunced among elderly people suffering from cognitive impairments. Poor sleep may exacerbate their cognitive problems, and their sleep may be disturbed by the medications taken to stabilize their mental faculties. Sleep disturbance and nighttime wakefulness are clinical features of dementia (Moller et al., 2004). Interestingly, Piani, Brotini, Dolso, Budai, & Gigli (2004) found that cognitively impaired older adults had less trouble falling asleep and woke less frequently during the night; they also snored less. They presented with enuresis and fell off the bed significantly more often, however, than healthy older adults and had higher levels of daytime sleepiness.

Most studies demonstrate *increased* nighttime wakefulness among older adults with dementia compared with controls (Moller et al.). In patients with dementia, sleep disturbances should be carefully and frequently assessed due to the overlap of symptoms between cognitive deficits and insomnia (e.g., diminished alertness and concentration, disturbed mood; Moller et al.). With regard to assessing and treating sleeping problems in older adults with cognitive decline, PSG can be helpful to rule out sleep apnea and PLMD, which – if diagnosed and treated – can greatly increase daytime alertness and correspondingly improve cognition and sleep continuity and regularity. Older medications, such as tricyclic antidepressants, should be avoided due to higher risk of anticholinergic and cardiovascular effects among older adults. Finally, factors such as the environment and sleep hygiene play a major role in regulating these individuals' sleeping patterns. Optimal levels of light and mobility should be sought during the day with corresponding dim light and relaxation in the evenings.

Conclusion

Older adults are at increased risk for insomnia due to a normal decline in sleep need that is handled with varying degrees of adjustment and due to factors such as retirement, health conditions, and death of spouses. Improvements in sleep can occur and with them come improved quality of life and decreased risk for anxiety, depression, substance abuse, and suicide (Taylor, Lichstein, & Durrence, 2003).

The goal of cognitive-behavior sleep therapy is not for the OAWI to return to previous sleep patterns of younger years (Cook et al., 2005). Older adults should not aim for deeper or longer sleep but for more consolidated, continuous, and restorative sleep. Objective and subjective assessment of daytime functioning, sleep parameters, and levels of distress should be used to evaluate treatment outcome. A mean SE% of 85 or greater, high levels of daytime functioning, and a reduction in sleep complaints are all indicative of successful treatment. Ultimate satisfaction can be determined by the older adult's reported improvement in quality of life.

References

American Psychiatric Association. (1994). *Diagnostic and statistical manual of mental disorders* (4th ed.). Washington DC: American Psychiatric Association.

American Sleep Disorders Association. (2005). *International classification of sleep disorders, revised: Diagnostic and coding manual.* Rochester, MN: Author.

Ancoli-Israel, S., Pat-Horenczyk, R., & Martin, J. (2001). Sleep disorders. In A. S. Bellack & M. Hersen (Series Eds.) & B. Edelstein (Vol. Ed.), *Comprehensive clinical psychology: Clinical geropsychology* (Vol. 7, pp. 307–326). Oxford, England: Elsevier Science Ltd.

Bootzin, R. R., & Epstein, D. R. (2000). Stimulus control. In K. L. Lichstein & C. M. Morin (Eds.), *Treatment of late-life insomnia* (pp. 167–184). Thousand Oaks, CA: Sage.

Buysse, D. J., Reynolds, C. F., Monk, T. H., Berman, S. R., & Kupfer, D. J. (1989). The Pittsburgh Sleep Quality Index: A new psychiatric practice and research. *Psychiatry Research, 28*, 193–213.

Cook, K. G., Nau, S. D., & Lichstein, K. L. (2005). Behavioral treatment of late-life insomnia. In L. VandeCreek (Ed.), *Innovations in clinical practice: Focus on adults* (pp. 65–81). Sarasota, FL: Professional Resource Press.

Durrence, H. H., & Lichstein, K. L. (2006). The sleep of African Americans: A comparative review. *Behavioral Sleep Medicine, 4*, 29–44.

Foley, D. J., Monjan, A. A., Brown, S. L., Simonsick, E. M., Wallace, R. B., & Blazer, D. G. (1995). Sleep complaints among elderly persons: An epidemiologic study of three communities. *Sleep, 18*, 425–432.

Gellis, L. A., Lichstein, K. L., Scarinci, I. C., Durrence, H. H., Taylor, D. J., Bush, A. J., et al. (2005). Socioeconomic status and insomnia. *Journal of Abnormal Psychology, 114*, 111–118.

Hauri, P. J. (1997). Can we mix behavioral therapy with hypnotics when treating insomniacs? *Sleep, 20*, 1111–1118.

Hemmelgarn, B., Suissa, S., & Huang, A. (1997). Benzodiazepine use and the risk of motor vehicle crash in the elderly. *Journal of the American Medical Association, 278*, 27–31.

International Statistical Classification of Diseases and Related Health Problems: 10th Revision (2003). http://www3.who.int/icd/vol1htm2003/fr-icd.htm.

Jean-Louis, G., Magai, C. M., Cohen, C. I., Zizi, F., von Gizycki, H., DiPalma, J., et al. (2001). Ethnic differences in self-reported sleep problems in older adults. *Sleep, 24*, 926–933.

Johns, M. W. (1991). A new method for measuring daytime sleepiness: The Epworth Sleepiness Scale. *Sleep, 14*, 540–545.

Krupp, L. B., LaRocca, N. G., Muir-Nash, J., & Steinberg, A. D. (1989). The Fatigue Severity Scale: Application to patients with multiple sclerosis and systemic lupus erythematosus. *Archives of Neurology, 46*, 1121–1123.

Lichstein, K. L. (2000). Relaxation. In K. L. Lichstein & C. M. Morin (Eds.), *Treatment of late-life insomnia* (pp. 185–206). Thousand Oaks, CA: Sage.

Lichstein, K. L., Durrence, H. H., Riedel, B. W., Taylor, D. J., & Bush, A. J. (2004). *Epidemiology of sleep: Age, gender, and ethnicity*. Mahwah, NJ: Erlbaum

Lichstein, K. L., Gellis, L. A., Stone, K. C., & Nau, S. D. (2006). Primary and secondary insomnia. In S. R. Pandi-Perumal & J. M. Monti (Eds.), *Clinical pharmacology of sleep* (pp. 1–9). Basel, Switzerland: Birkhauser Verlag.

Lichstein, K. L., & Morin, C. M. (2000). Treatment overview. In K. L. Lishstein & C. M. Morin (Eds.), *Treatment of late-life insomnia* (pp. 111–124). Thousand Oaks, CA: Sage.

Lichstein, K. L., & Riedel, B. W. (1994). Behavioral assessment and treatment of insomnia: A review with an emphasis on clinical application. *Behavior Therapy, 25*, 659–688.

Lichstein, K. L., Riedel, B. W., Wilson, N. M., Lester, K. W., & Aguillard, R. N. (2001). Relaxation and sleep compression for late-life insomnia: A placebo-controlled trial. *Journal of Consulting and Clinical Psychology, 69*, 227–239.

Lichstein, K. L., Stone, K. C., Donaldson, J., Nau, S. D., Soeffing, J. P., Murray, D., et al. (2006). Actigraphy validation with insomnia. *Sleep, 29*, 232–239.

Mellinger, G. D., Balter, M. B., & Uhlenhuth, E. H. (1985). Insomnia and its treatment. Prevalence and correlates. *Archives of General Psychiatry, 42*, 225–232.

Miles, L., & Dement, W. C. (1980). Sleep and aging. *Sleep, 3*, 119–220.

Moller, H. J., Barbera, J., Kayumov, L., & Shapiro, C. M. (2004). Psychiatric aspects of late-life insomnia. *Sleep Medicine Reviews, 8*, 31–45.

Morin, C. M., Colecchi, C. A., Stone, J., Sood, R. K., & Brink, D. (1999). Behavioral and pharmacological therapies for late-life insomnia. *Journal of the American Medical Association, 281*, 991–999.

Nau, S. D., McCrae, C. S., Cook, K. G., & Lichstein, K. L. (2005). Treatment of insomnia in older adults. *Clinical Psychology Review, 25,* 645–672.

Piani, A., Brotini, S., Dolso, P., Budai, R., & Gigli, G. L. (2004). Sleep disturbances in elderly: A subjective evaluation over 65. *Archives of Gerontology and Geriatrics, 9*, 325–331.

Poyares, D., Guilleminault, C., Ohayon, M. M., & Tufik, S. (2004). Chronic benzodiazepine usage and withdrawal in insomnia patients. *Journal of Psychiatric Research, 38*, 327–334.

Ray, W. A., Griffin, M. R., & Downey, W. (1989). Benzodiazepines of long and short elimination half-life and the risk of hip fracture. *Journal of the American Medical Association, 262*, 3303–3307.

Robert, R. E., Roberts, C. R., & Chan, W. (2006). Ethnic differences in symptoms of insomnia among adolescents. *Sleep, 29*, 359–365.

Spielberger, C. D., Gorsuch, R. L., Lushene, R., Vagg, P. R., & Jacobs, G. A. (1983). *State-Trait Anxiety Inventory (Form Y)*. Palo Alto, CA: Consulting Psychologists Press.

Taylor, D. J., Lichstein, K. L., & Durrence, H. H. (2003). Insomnia as a health risk factor. *Behavioral Sleep Medicine, 1*, 227–247.

Vallières, A., Morin, C. M., & Guay, B. (2005). Sequential combinations of drug and cognitive behavioral therapy for chronic insomnia: An exploratory study. *Behaviour Research and Therapy*, 43, 1611–1630.

Yesavage, J. A., Brink, T. L., Rose, T. L., Lum, O., Huang, V., Adey, M., et al. (1983). Development and validation of a geriatric depression screening scale: A preliminary report. *Journal of Psychiatric Research, 17*, 37–49.

5
A Relapse Prevention Model for Older Alcohol Abusers

Larry W. Dupree, Lawrence Schonfeld,
Kristina O. Dearborn-Harshman, and Nancy Lynn

With the aging of America, interest on improving behavioral health services for older adults has increased, including attention to those with alcohol abuse problems. Early studies focused on the possible benefits of elder-specific, group treatment programs (Kofoed, Tolson, Atkinson, Toth, & Turner, 1987; Schonfeld, Dupree, & Rohrer, 1995) or evaluating the benefits of group treatment approaches for older adults (Carstensen, Rychtarik, & Prue, 1985; Dupree, Broskowski, & Schonfeld, 1984; Graham, Brett, & Baron, 1997; Schonfeld et al., 2000).

Prevalence estimates of alcohol problems vary according to the criteria used to define older adults and the methodology used to survey the problem. Early estimates suggested that between 2 and 10% of adults ages 60 and older (60+) were problem drinkers (Gomberg, 1980). According to the National Household Survey on Drug Abuse (NHSDA) 9.4% of adults ages 55+ were heavy drinkers and 2.3% were binge drinkers. The findings from NHSDA also indicated that males were 3–4 times more likely to exhibit heavy or binge drinking than females, and that there were no statistically significant differences among whites, blacks, and Hispanics with respect to rates of binge drinking or heavy drinking.

In medical settings, estimates vary as well. Some have suggested that from 4 to 10% of older primary care patients and up to 14% among older emergency room patients exhibit alcohol problems (Adams & Smith-Cox, 1997). In one study of primary care patients ages 65+ who consumed alcohol, 21% were found to be "harmful drinkers" (drinking that results in adverse events) and 26% were found to be hazardous drinkers (consumption with such risks) (Fink, Elliott, Tsai, & Beck, 2005). In veterans hospitals, where most patients are males, one study found that 23% of patients with substance abuse problems were aged 55+ (Moos, Mertens, & Brennan, 1993).

The costs of alcohol problems are high, but usually reflect the medical care required to address the consequences of abusive drinking rather than actual substance abuse treatment. A 1992 Congressional report from the Select Committee on Aging indicated that 21% of hospital patients age 60 or older in the US had a diagnosis of alcoholism and in 1990 alone, estimated costs were as high as $60 billion (HR Report No. 102–852, 1992).

Issues of Diversity

Some suggest that women need specialized treatment programs, and have not done well in traditional treatment (Marsh, Thomas, D'Aunno, & Smith, 2000). Others suggest that women can be treated alongside men if certain components of the program are designed to meet their needs (Johnson, Wiechelt, Ahmed, & Schwartz, 2003; Satre, Mertens, & Weisner, 2004; Uziel-Miller & Lyons, 2000). Johnson et al. (2003) found that women in traditional addiction treatment showed reduced substance use and other problems. Satre et al. treated men and women 55–77 years old via supportive group therapy, education, relapse prevention (RP), and family oriented therapy. The study indicated that older women had better drinking outcomes than men.

Arroyo, Miller, and Tonigan (2003) evaluated the effectiveness of AA 12-step facilitation therapy (TSF) versus CBT and MET for Hispanics. There were no major differences in the response from the Hispanics to the three treatments; however, non-Hispanic whites responded better to TSF than Hispanics with less frequent drinking. Tonigan (2003) attempted to compare and contrast Hispanic, black, and white participants' engagement and outcome in treatment, and sought to determine if culturally incongruent treatment resulted in different outcomes for Hispanics and blacks. The participants were randomly assigned to either CBT, TSF, or MET. Hispanics and blacks did as well during the 12-month follow-up as whites on at least two measures of drinking. No evidence was found for the assumption that culturally incongruent manual-guided therapy has an adverse impact on the treatment outcome of minority groups.

Screening and Assessment

Screening and assessment of older adults creates a challenge for service providers. Unlike their younger counterparts, many older adults with alcohol problems are not identified by law enforcement or health care systems. Since many older problem drinkers are widowed or divorced, retired, and not driving as frequently, the probability of being detected by significant others is reduced. It is important that the helping professions rely on different strategies to identify such problems. For example, aging services case managers who provide in-home services can be a valuable source for identifying the person who drinks at home and alone, one of the most common antecedents to drinking in this age-group (Dupree et al., 1984; Schonfeld et al., 2000).

SAMHSA's Center for Substance Abuse Treatment (CSAT) included screening guidelines in Treatment Improvement Protocol (TIP) # 26 on "Substance Abuse Among Older Adults" (CSAT, 1998). According to the TIP #26, older adults should be screened for substance use problems when experiencing transitions or major life changes such as menopause, a newly "empty nest," retirement, or providing care for loved ones. Physical conditions such as sleep complaints, cognitive impairments, liver function abnormalities, and slurred speech are also considered a trigger for alcohol screening. Such screening requires a sensitive and nonjudgmental approach, involving active listening and motivational interviewing, recognizing that some older adults have a stigma about alcohol and drug use problems. Interviewing friends and family may be necessary in developing a full picture of a person's risk for substance abuse.

Several brief screens have been recommended by the Center for Substance Abuse Treatment (CSAT, 1998, 2006). The Short-Michigan Alcoholism Screening Test-Geriatric version is an elder appropriate and valid screen for identifying alcohol problems. CSAT also recommends two instruments that are not elder specific. These are the Alcohol Use Disorders Identification Test (AUDIT; Babor, de la Fuenta, Saunders, & Grant, 1992) and the CAGE (Mayfield, McLeod, & Hall, 1974).

O'Connell et al. (2004) reported the sensitivities and specificities for these instruments with respect to their use with older adults. For the CAGE, sensitivity ranged from 38 to 92%, and specificity from 48 to 99%. Studies of the MAST-G indicate a range in sensitivity from 50 to 100%, and specificity from 27 to 93%. Studies examining the AUDIT revealed sensitivity from 28 to 81% and specificity from 86 to 100%.

Treatment Approaches

Two publications by SAMHSA's Center for Substance Abuse Treatment (CSAT) have promoted CBT models for elders. The first publication is the Treatment Improvement Protocol (TIP) #26 on "Substance Abuse Among Older Adults" (CSAT, 1998). It represented the consensus of an expert panel on best practices for identifying and treating substance abuse problems in older people. The second publication is a model and curriculum describing a CBT and self-management treatment approach: Substance Abuse Relapse Prevention for Older Adults: A Group Treatment Approach (Dupree & Schonfeld, 2005). In this chapter, we focus on the TIP panel's recommendations and the treatment approach presented in the relapse prevention curriculum.

For more than 30 years, numerous studies have supported the effectiveness of cognitive-behavioral treatment (CBT) for alcohol problems. Holder, Longabaugh, Miller, and Rubonis (1991) demonstrated that treatment programs focusing on social skills training, self-control training, and other behavioral interventions have higher rates of success and cost effectiveness. Programs employing CBT teach clients the skills necessary to recognize high-risk situations for drinking, interrupt or restructure the thoughts and situations that are likely to lead to drinking, with the ultimate goal of preventing relapse. As Fleming (2002) noted, the alcohol treatment field is moving away from a focus on abstinence and toward a harm reduction public health paradigm.

The Relapse Prevention Model

The RP model was based on research from the 1970s and 1980s involving substance abusers who relapsed following completion of treatment. Interviewing these individuals, Marlatt and Gordon (1980, 1985) found that most relapsed in response to negative emotional states such as anger, depression, anxiety, or guilt; conflicts with other people; or pressure from others to use alcohol or drugs. Few relapses occurred in response to physical withdrawal, cravings, and urges. The RP model not only explains why relapses occur, but also suggests treatment strategies for teaching skills necessary to deal with high-risk situations. In this model, if the individual has adequate coping skills, it enhances self-efficacy, or the confidence to cope with the situation. As a result, the likelihood of alcohol or drug use is reduced. In contrast, without adequate coping skills, there is decreased self-efficacy, and subsequent increased expectancy that a drink or use of a drug would help. This increases the probability for a lapse and a possible "Abstinence Violation Effect" (or AVE). The AVE is a belief that if the goal is violated, the person is a failure. That in turn may lead to continued use and a full relapse.

To address the treatment needs within the RP model, a cognitive-behavioral model is employed. Identification of high-risk situations that occurred prior to treatment are viewed as potential determinants of relapse during or after treatment. CBT approaches rely on behavior analysis to identify each person's drinking behavior chain leading to and following substance use, and use of behavioral principles to teach the appropriate coping skills necessary to prevent relapse.

Aspects of CBT have been employed in varying forms. Certain aspects of CBT are used in brief advice or brief intervention sessions, conducted in one or two outpatient meetings. More typically, CBT is employed in outpatient, individual, intensive outpatient treatment, and residential programs (Benshoff, Harrawood, & Koch, 2003).

Both the TIP and previous reviews (Dupree & Schonfeld, 1998; Schonfeld & Dupree, 1997) offered several recommendations for treatment of older adults: (1) emphasize age-specific, group treatment with supportive approaches, avoiding confrontation; (2) focus on negative emotional states such as depression, loneliness, overcoming losses (e.g., death of a loved one, retirement); (3) rebuild the social support network (i.e., teach the older person the skills necessary to rebuild the network); (4) employ staff members who are experienced in working with older persons, and have interest in doing so; (5) develop linkages with aging services, medical services, and institutional settings, for both referral into treatment and referral out, as well as "case management"; and (6) develop the pace and content of treatment for the older person.

Our experience with CBT in this area began in 1979 with the development, implementation, and evaluation of the Gerontology Alcohol Project ("GAP") at the Florida Mental Health Institute (Dupree et al., 1984). The treatment program focused on late-onset alcohol abusers, admitting those age 55 or older with an age of onset of their alcohol problems occurring after age 50. GAP was a day treatment program, with group treatment utilizing cognitive-behavioral approaches such as cognitive restructuring, thought-stopping and covert assertion, problem solving, and assertiveness training. These CBT approaches were combined with self-management techniques such as self-monitoring and self-reinforcement. One year outcomes indicated that 75% of program graduates maintained their drinking goals, either abstinence or limited drinking goals, depending on the individual and the original negotiation between client and counselor. In addition, GAP clients demonstrated improved psychosocial measures such as increased social support and decreased depression, and fewer relapses. The GAP treatment model was later applied to medication misuse, elders with early onset alcohol abuse (Schonfeld & Dupree, 1991), and older veterans (Schonfeld et al., 2000).

The CBT/Self-Management Model

The treatment model is now available from SAMHSA's CSAT in a format involving 16 weekly group treatment sessions (Dupree & Schonfeld, 2005). It involves three sequential stages: behavioral analysis of each client's drinking behavior (resulting in the development of personal drinking behavior chains), teaching the individual to identify his/her personal high-risk situations for drinking, and finally, teaching the individual the skills necessary to prevent relapse.

The first stage, implemented before any treatment occurs, involves behavior analysis of the antecedents and consequences of drinking or drug use behavior, using a structured interview. The original version known as the Gerontology Alcohol Project Drinking Profile (GAP-DP) was an elder-appropriate modification of the original Drinking Profile by Marlatt (1976). The GAP-DP was used by counselors to identify antecedents and consequences of drinking on a typical day of alcohol use. As shown in the recent CSAT (Dupree & Schonfeld, 2005) relapse prevention curriculum, the instrument was further modified to include issues related to any substance use problem and is now referred to as the Substance Abuse Profile for the Elderly (SAPE). This structured interview is a mandatory first step because it helps the counselor identify an alcohol abuser's drinking behavior chain (antecedents to, and consequences following, a drink) in addition to describing the person's history and characteristics related to alcohol problems. The interview is to guide the counselor in aiding the client to identify his/her situations or thoughts, feelings, cues, and urges that precede the first drink on a typical day of drinking; and the-short term consequences (positive and negative ones) and long-term consequences that follow that first drink. The client's answers to the questions from the SAPE are recorded (written down) by the counselor and later typed up according to the categories of the behavior chain. These are considered the client's high-risk situation for alcohol use. As reproduced from the CSAT relapse prevention manual (Dupree & Schonfeld), Fig. 1 illustrates the behavior chain and its flow from antecedents to behavior to consequences.

Drinking Behavior Chain				
Antecedents →	**Behavior** →		**Consequences**	
Situations/Thoughts + Feelings + Cues + Urges →	first drink on a typical day of drinking	→	Immediate/ short term (negative or positive)	Long term (negative)

FIGURE 1. Drinking Behavior Chain.

In the second stage, the information recorded is provided to the client and also used as a teaching aid within the first required group treatment session. In a module described as the "A–B–Cs group" (antecedents–behavior–consequences), clients are taught how to recognize the A–B–Cs using simple, but general examples first, then their own personal "high-risk situations" for substance use, using the answers provided on the SAPE. The group members can modify or add to their own A–B–C list as the group discussions prompt. Not only do clients learn to break down the components of their own behaviors while in the modules, each learns to complete a weekly "alcohol self-monitor log" for each week. A log consists of a table printed on a single sheet of paper with columns prompting the client to describe how he/she responded to the urge to drink. Under the columns related to "Antecedents" are the headings or prompts: day/date, where the person was prior to the first drink, who was with the person, what happened, any particular feelings noted, and time of the urge. Under the column labeled "Behavior," the client is asked to record if the urge to drink was resisted or not. Under the consequences column, the client is asked to describe the events that occurred after either resisting the urge, or drinking; and how much they drank in total. Logging urges and drinking behavior is an important step in teaching clients to understand their behaviors away from the program.

In the third stage, the counselor teaches clients the skills necessary to address their individual high-risk situations (the links in their personal drinking chains) and prevent relapse. In this stage, clients enter a series of treatment modules or classes taught by a trained counselor. Each module is a scripted lesson plan, containing lecture content, instructions for conducting behavior rehearsals, pretest and posttest quizzes and ratings of cognitive and behavioral rehearsals, and other exercises.

As illustrated in Table 1, the CSAT Relapse Prevention model covers nine group treatment topics beginning with the A–B–Cs group (second stage noted above) and proceeding to the self-management skills groups (third stage). Each topic is presented as a group treatment module, complete with lesson plan. Each topic begins with a listing of goals for the session, required visual materials (e.g., flipcharts, diagrams), and contains a complete script for the session, exercises and quizzes to test the client's knowledge of what is being taught. The table also categorizes the CBT and self-management skills taught within those modules.

CBT methods include techniques such as thought-stopping, covert assertion, cognitive restructuring, and problem-solving skills. To teach these skills, sessions include didactic and experiential exercises: overt and covert behavior rehearsals. The term behavior rehearsal refers to a structured, scripted role play. In rehearsals, the clients practice skills in a nonthreatening, simulated situation that resembles a real situation likely to occur outside of the treatment setting, and that has led to drinking in the past, i.e., the high-risk situation. The instructor selects a sample situation from the curriculum that best resembles the situation the client indicates is a personal high-risk situation for drinking. In covert behavior rehearsal, the client is asked to vocalize (say aloud) in front of the others in the group, what he or she would do in the sample situation. In overt behavior rehearsal, the client role plays with the instructor. After a rehearsal,

TABLE 1. Description of the CBT/self-management modules and skills.

Module	Content	Methods used
A–B–Cs of substance use	• Identify personal high-risk situations • Identify antecedents and consequences for high-risk situations • Diagram personal drinking behavior chain	• Functional analysis of behavior using general examples • Self-monitor urges by completing a weekly log
Social pressure	• How to refuse an offer of a drink • How to say no to a persistent "pusher" without being aggressive	• Behavior rehearsal/role plays • Basic assertiveness and communication skills
At home and alone	• How to avoid being lonely and becoming sad • Identify pleasant events and alternative behaviors	• Thought-stopping • Cognitive-restructuring • Covert behavior rehearsal
Thoughts and negative emotions	• Identify negative thoughts and feelings preceding drinking • Identify positive self-statements	• Thought-stopping • Cognitive-restructuring • Covert assertion
Anxiety and tension	• Identify signs of tension building • Interrupt negative, ruminative thoughts • Begin to relax • Generate alternative solutions and try them out	• Brief relaxation • Thought-stopping • Cognitive-restructuring • Covert assertion • Problem-solving skills
Anger and frustration	• Appropriate assertiveness skills rather than aggressive or passive behaviors • DESC Model (describe objectionable behavior to the other person, express feelings, specify changes, and indicate beneficial consequences)	• Thought-stopping • Cognitive-restructuring • Covert assertion
Controlling substance use cues	• CARD model: identify behaviors that compete with drinking, avoid cues, rearrange your life, dispose of cues)	• Covert behavior rehearsal
Coping with urges	• Urges are considered self-statements that give one permission to drink • CRASH Model – (remember consequences, remove from situation, find incompatible activity, use skills learned, call for help)	• Covert behavior rehearsal
Preventing a slip from becoming a relapse	• Make a positive statement about stopping • Get out of the situation or get rid of the alcohol • Call for help	• Covert behavior rehearsal

the client receives positive encouragement and feedback from the instructor, as well as other clients in the group.

Within the Dupree and Schonfeld (2005) relapse prevention model for older adults, there is major emphasis on teaching clients to deal with negative emotional states that typically lead to drinking. As detailed in the CSAT RP curriculum manual, clients learn to recognize and deal with negative and typically inaccurate ruminative thoughts. CBT skills taught include thought-stopping to interrupt ruminations, followed by covert assertion, i.e., replacing those thoughts with positive and more accurate self-statements. In the session "lecture," they are taught how to recognize such self-statements and how to interrupt the thought via subvocalization. This is then practiced in a matter of minutes under supervision of the group leader. Once the unwanted thought is interrupted, clients learn to relax and implement "covert assertion." For example, a person may indicate that his ruminative thought is a repetitive, negative self-statement such as: "I'm no good. No one likes me." The instructor might query the client to prompt a more accurate and positive self-statement through discussions with the client and the other members of the group. An example might be: "I can count on my good friends, Joe and Sylvia, if I need help." This is facilitated using a blackboard and a "double-ledger" approach, in which a negative self-statement is written on the left side of a black board, with the requirement that a more accurate and positive self-statement be written on the right side of the blackboard. For each negative statement a client might identify, another positive statement will be generated.

For modules that involve private behaviors known only to the client, but not observable to others, the client is taught how to use the covert behavior rehearsal method. This is used when addressing high-risk situations associated with depression, grief, being at home and alone, anxiety and tension, cues for drinking, and the urge to drink, as well as steps he or she might take if experiencing a slip. In other modules, such as the social pressure module or the module on how to manage anger and frustration (where other people are likely to be antecedents for drinking), overt behavior rehearsal is useful.

General problem-solving skills are taught within several of the modules. Clients are taught how to state the problem that leads to drinking in a manner that guides the person toward a solution. For example, if a Ms. Smith reports she drinks out of loneliness and boredom she might define the goal for solving the problem as "how to make new friends," rather than how to stop feeling lonely. Once the goal has been identified, the individual is taught to generate possible solutions, prioritize them, and try out the solution with the highest priority.

Finally, to extend the CBT skills taught, clients are taught to use self-management (SM) techniques to ensure their use in future high-risk situations. SM involves the client applying behavioral techniques rather than a therapist doing so, as is the case with basic behavior modification. Kanfer (1975) described three categories of self-management that are now incorporated into our RP model. The first, self-monitoring, increases the client's awareness that a problem behavior is occurring. The client is taught to collect data about the behavior and its consequences. In the RP model, older adults are taught to use alcohol self-monitoring logs of their urges to drink or actual drinking episodes. The completed logs are turned in and reviewed by the counselor, who in turn provides feedback to the client aimed at preventing future slips or relapses from occurring. A second self-management method is the completion of homework or assignments. For example, clients might be assigned to call a local alcohol hotline to learn what the hotline counselor might say, or how he/she would help them. Such assignments can be used to diminish the stress of what might occur outside of treatment, when calling for help might be necessary.

The third category is a behavior contract, an agreement between the client and the therapist, often written down and signed by both parties. The contract spells out the expected behavioral goal and the consequence the client will experience if the behavior does occur or the consequence if the behavior does not occur.

In addition to the above three self-management techniques, clients learn self-reinforcement techniques. For example, the man who has successfully avoided taking a certain route home that includes his favorite tavern, might reward himself by engaging in a preferred nondrinking activity.

Determining Discharge Readiness

A client may be discharged from the active phase of treatment when several criteria are met. First, clients must demonstrate that they have successfully attained their negotiated drinking goals. Should it be evident that this goal cannot be attained or maintained, it might be renegotiated.

A second discharge criterion is that the individual must receive passing scores or ratings on quizzes or behavior rehearsals included in the curriculum. Such performance indicates that knowledge and skills as taught in the groups were learned successfully. Individuals who do not reach criteria or passing scores may require remediation or one-on-one tutoring until they do so.

Finally the individual must demonstrate the ability to diagram his or her personal drinking behavior chain. The individual should be able to recall the high-risk situations and behavior chain that are pertinent to him or her. The importance of diagramming and understanding the behavior chain cannot be overemphasized. In outpatient treatment, a slip can occur during the active stages of treatment as well as during follow-up or aftercare. A prerequisite is that clients should feel comfortable enough to report the occurrence of a slip or a relapse without fear of being labeled or accused of failure. When the client reports the slip, the client and the therapist review the events and diagram the components of the behavior chain preceding the slip. Once diagrammed, they rehearse the skills to be used the next time such circumstances occur. Also, the slip may be used in group as a "tutorial" for others.

Thus, discharge from treatment is more than simply completing a certain number of sessions. Rather, it is based on demonstration that the skills taught in treatment are actually learned.

Follow-Up and Aftercare

Follow-up may refer to a schedule of appointments for visits or phone contacts that permit the counselor to assess the client's success or progress over time. In GAP, we scheduled both in-person and telephone follow-ups on schedule that gradually "weaned" the person from the treatment program. GAP clients returned for visits to the program at 2 weeks, then at 1, 2, 3, 6, and 12 months, with assessments being administered at 1, 3, 6, and 12 months. The term aftercare suggests that the individual may also receive some form of treatment after completion of the active phase of treatment. In GAP, individuals could attend a session with active clients if the client had experienced urges to drink, or even a slip. Individuals who slipped would have an individual session with a counselor to diagram the events leading to and following the slip, followed by a review of the skills that would be used the next time such events occurred again.

Each program will decide the length of follow-up. The critical time as noted by research on determinants of relapse, may be the first 90 days after completing the active stage of treatment. If the program's labor force and budget permit, 1-year follow-up may be the appropriate time to formally discharge the person from the follow-up status and the program. However, if feasible, a client should be allowed telephone contact with a counselor on an as needed basis.

Suggestions for Counselors

The continuum of approaches for treating older adults with alcohol problems as described in this chapter is shown in Table 2. It begins with identifying suspected cases through outreach efforts, followed by screening, in-depth analysis of

TABLE 2. Recommendations for CBT/self-management treatment of older adults.

Strategy	Recommendations for providers
Outreach and casefinding	1. Educate and develop a network of community services/agencies that have frequent contact with older adults 2. Develop agreements such that the network members "pre-screen" their clients using brief screens 3. Maintain a regular schedule of telephone and face-to-face contact with network agencies' staff
Screening and admission to treatment	1. Use age-appropriate and valid screening instruments such as the Short-Michigan Alcoholism Screening Test-Geriatric version 2. Use the principles of Motivational Interviewing to place the weight of evidence on the client and to avoid labeling and confrontations 3. Involve the client by negotiating the drinking goal
Analysis of drinking behavior	1. Use the SAPE interview to identify and record the components of the drinking behavior chain 2. Record the client's answers and provide the client with a copy to use throughout the treatment program 3. Teach each client to recognize his/her personal antecedents and consequences 4. Teach clients to record urges and slips, antecedents and consequences on the alcohol self-monitor log
Self-management skills	1. Teach relapse prevention skills using the curriculum 2. Rely on behavior rehearsals, covert behavior rehearsal, and written quizzes to encourage learning
Discharge readiness	1. Based on absence of problematic or risky drinking 2. Based on abilities and learning as demonstrated in group performance, rating scales, and quizzes 3. Based on ability to diagram and understand personal drinking behavior chain
Follow-up and aftercare	1. Negotiate a schedule for follow-up contact and visits. Time between visits or contacts should gradually increase until discharge from follow-up 2. Provide "touch-up" sessions when slips occur or the urge to drink becomes problematic 3. Base length of follow-up on the program's resources and the client's progress in treatment

drinking behavior, CBT/self-management skills training, discharge from active treatment, and aftercare.

Using the principles mentioned thus far, we illustrate the utility of the CBT/self-management approach with a case example.

Case Example: The Widow Who Drank Alone

Mrs. S was a 65-year-old widow living alone. At age 42, she began drinking liquor after work on a daily basis "to ease the discomfort of menopause." Beginning at age 55 she twice entered treatment for alcohol problems, and periodically attended AA from ages 60 to 64. She remained abstinent until a close friend died. About the same time, her daughter began a new job, working long hours, and pressuring Mrs. S to babysit her two young children. Mrs. S also wanted to work, but gave in to her daughter. She reported falling into a deep depression about her losses, pressure from her daughter, and her life in general; and began drinking at least a pint of bourbon each day. A subsequent hospitalization prompted her referral to the outpatient treatment program. Upon admission, Mrs. S scored very high on formal anxiety and depression measures. Also, using behavior rating scales, Mrs. S's daughter described her mother as anxious, confused and depressed.

Drinking Pattern

Using the SAPE assessment and treatment planning instrument (Substance Abuse Profile for the Elderly) to assess Mrs. S's drinking behavior in the 30 days preceding her last drink prior to admission, she reported drinking up to a fifth of bourbon or vodka on a typical day. Yet, she rated this as being less than the amount she consumed in previous 30-day periods.

Mrs. S usually felt depressed and lonely, or angry with her daughter for pressuring her to babysit for the grandchildren. Instead of being active, Mrs. S usually drank at home and alone, hoping that she would feel less lonely or more relaxed. Instead of expressing to her daughter that she did not wish to babysit as frequently as she did, she became angry, depressed and drank with the immediate consequences of the first drink being positive (e.g., felt relaxed, less lonely, less angry).

Intervention

While in treatment, Mrs. S attended modules that taught her via cognitive and behavior rehearsal how to relax without drinking; and how to manage depression and loneliness, anxiety and tension, and anger and frustration.

The latter by learning how to respond more assertively to her daughter without alienating her. She was also taught how to manage drinking cues and how to prevent a slip/lapse from becoming a relapse. In addition, she was encouraged to participate in social activities. Mrs. S successfully completed treatment less than 3 months after admission and was followed up at 1-, 3-, 6-, and 12-month appointments.

During follow-ups, there were occasional slips, none of which resulted in relapse. She never returned to her pretreatment pattern of steady drinking, quantity consumed, or drunken behavior. More importantly, Mrs. S filled out alcohol self-monitoring logs during the follow-up phase. On several occasions during the follow-ups, she reported feeling the urge to drink, but used the methods learned in group to manage her feelings, control alcohol cues and urges, prevent relapse, and avoid drinking.

Looking at three major points in time (admission, discharge, and final 12 month follow-up) Mrs. S's self-administered depression scores went from very high to no indication of depression. Her anxiety scores also dropped markedly. Concomitant behavior ratings by her daughter also indicated great improvement in the areas previously noted (anxiety, depression, confusion).

By the 12-month follow-up, Mrs. S began part-time work at a gift shop, went out regularly with nondrinking friends, and reported maintaining telephone contact with several other friends. Upon discharge from the 12-month follow-up program, she reported no drinking, no urges to drink, and the relationship with her daughter markedly improved. The daughter confirmed those outcomes. Thus, both Mrs. S's and her daughter's ratings indicated that Mrs. S had improved significantly, and was using the skills taught.

Pharmacological Approaches

There are currently only three medications approved by FDA for treatment of alcohol dependence: disulfiram, naltrexone, and acamprosate. Older adults with cardiovascular disease and other health issues are prescribed disulfiram with caution (Barrick & Connors, 2003).

Naltrexone acts as an opioid receptor antagonist and prevents the "high" that alcohol brings to the individual by blocking the release of dopamine (Oslin, 2005) and thus, inhibits the drug "award" effect. Studies of the use of naltrexone have primarily been conducted with younger populations, and been inconsistent, with some suggesting negative or no effect of the drug (Oslin). Oslin reported that naltrexone has shown improved treatment outcomes when used in combination with psychosocial treatment approaches.

Acamprosate has been shown to reduce the craving for alcohol (Mann, 2004) once the patient stops drinking, therefore helping to reduce relapse episodes. Use of acamprosate with nonpharmacological therapies has produced mixed results (Fenney, Young, Connor, Tucker, & McPherson, 2002; De Wildt et al., 2002).

Naltrexone in combination with acamprosate has also shown mixed results (COMBINE, 2006; Kiefer et al., 2003).

One of the few studies directed at older adults tested the efficacy of combining naltrexone with sertraline, an SSRI, to treat older adults with major depression and alcohol dependence (Oslin, 2005). There were no significant outcomes in the improvement of either depression or drinking response in the naltrexone–sertraline–psychosocial support group as compared to the placebo–sertraline–psychosocial group. Oslin speculated that the efficacy of naltrexone may be reduced as age increases; and may be associated with the different reasons older adults turn to drinking or their motivation for quitting.

Summary

The CBT/relapse prevention model described in this chapter offers a structured approach for counselors that not only provides didactic content and measures of knowledge acquisition, but also serves to maintain fidelity of the model. Although designed as a group treatment approach, the curriculum has been used to guide individual (one-on-one) treatment.

The model has advantages over "traditional" approaches to substance abuse treatment. It avoids stigma associated with substance abuse, as well as age (i.e., "ageism"). It has been used with different racial and ethnic groups, and men and women. It is relatively "culture-free" in that each person's individual drinking behavior chain serves as the basis for developing the treatment plan to help the client generate solutions appropriate to his or her life and high-risk situations. Most importantly, the CBT and self-management methods provide clients with the skills necessary to help them prevent future relapse.

References

Adams, W. L., & Smith-Cox, N. (1997). Epidemiology of problem drinking among elderly people. In A. M. Gurnack (Ed.), *Older adults' misuse of alcohol, medicines, and other drugs* (pp. 1–23). New York: Springer Publishing.

Arroyo, J. A., Miller, W. R., & Tonigan, J. S. (2003). The influence of Hispanic ethnicity on long-term outcomes in three alcohol-treatment modalities. *Journal of Studies on Alcohol, 64*, 98–104. Retrieved February 27th, 2006, from Health & Wellness Resource Center.

Babor, T. F., de la Fuenta, J. R., Saunders, J., & Grant, M. (1992). *AUDIT: The Alcohol Use Disorders Identification Test: Guidelines for its use in primary health care.* Geneva, Switzerland: World Health Organization.

Barrick, C., & Connors, G. J. (2003, October/November). Relapse prevention and maintaining abstinence in older adults with alcohol use disorders. *Drugs & Aging, 19*, 583–594, 2003.

Benshoff, J. J., Harrawood, L. K., & Koch, D. S. (2003). Substance abuse and the elderly: Unique issues and concerns. *Journal of Rehabilitation, 69*, 43–48.

Carstensen, L. L., Rychtarik, R. G., & Prue, D. M. (1985). Behavioral treatment of the geriatric alcohol abuser: A long term follow-up study. *Addictive Behaviors, 10*, 307–311.

Center for Substance Abuse Treatment. (1998). *Substance abuse among older adults* (TIP Series, No. 26). DHHS Publication No. (SMA): 02–3688. Rockville, MD: Substance Abuse and Mental Health Services Administration.

Center for Substance Abuse Treatment Alcohol Use Among Older Adults. 2006. Pocket Screening Instruments for Health Care and Social Service Providers. DHHS Publication accessed August 4, 2006 http://kap.samhsa.gov/products/brochures/pdfs/Pocket_2.pdf.

COMBINE Study Research Group. (2006). Combined pharmacotherapies and behavioral interventions for alcohol dependence: The COMBINE Study: A Randomized Controlled Trial. *Journal of the American Medical Association*, *295*, 2003–2017.

De Wildt, W. A. J. M., Schippers, G. M., Van Den Brink, W., Potgieter, A. S., Deckers, F., & Bets, D. (2002). Does psychosocial treatment enhance the efficacy of acamprosate in patients with alcohol problems? *Alcohol and Alcoholism*, *37*, 375–382.

Dupree, L. W., Broskowski, H., & Schonfeld, L. (1984). The Gerontology Alcohol Project: A behavioral treatment program for elderly alcohol abusers. *The Gerontologist*, *24*, 510–516.

Dupree, L. W., & Schonfeld, L. (1998). Older alcohol abusers: Recurring treatment issues. In E. Gomberg, A. Hegedus, & R. Zucker (Eds.), *Alcohol problems and aging*. (pp. 339–357). Research Monograph No. 3 Bethesda, MD: National Institute on Alchol Abuse and Alcholism.

Dupree, L. W., & Schonfeld, L. (2005). *Substance abuse relapse prevention for older adults: A group treatment approach*. Center for substance abuse treatment, DHHS Publication No. (SMA) 05-4053. Rockville, MD: Substance Abuse and Mental Health Services Administration.

Fenney, G. F. X., Young, R. M. D., Connor, J. P., Tucker, J., & McPherson, A. (2002). Cognitive behavioural therapy combined with the relapse-prevention medication acamprosate: Are short-term treatment outcomes for alcohol dependence improved? *Australian and New Zealand Journal of Psychiatry*, *36*, 622–628.

Fink, A., Elliott, M. N., Tsai, M., & Beck, J. C. (2005). An evaluation of an intervention to assist primary care physicians in screening and educating older patients who use alcohol. *Journal of the American Geriatrics Society*, *53*, 1937–1943.

Fleming, M. (2002). Identification and treatment of alcohol use disorders in the elderly. In A. M. Gurnack, R. Atkinson, & N. J. Osgood (Eds.), *Treating alcohol and drug abuse in the elderly* (pp. 85–108). New York: Springer.

Gomberg, E. S. L. (1980). *Drinking and problem drinking among the elderly*. Ann Arbor: Institute of Gerontology, University of Michigan.

Graham, K., Brett, P. J., & Baron, J. (1997). A harm reduction approach to treating older adults: The clients speak. In P. G. Erickson, D. M. Riley, Y. W. Cheung, & P. A. O'Hare (Eds.). *Harm reduction: A new direction for drug policies and programs* (pp. 428–452). Toronto: University of Toronto Press.

Holder, H., Longabaugh, R., Miller, W. R., & Rubonis, A. V. (1991). The cost effectiveness of treatment for alcoholism: A first approximation. *Journal of Studies on Alcohol*, *52*, 517–540.

HR Report 102–852. (1992). Alcohol abuse and misuse among the elderly. A report by the Chairman of the Subcommittee on Health and Long-Term Care of the Select Committee on Aging, House of Representatives.

Johnson, J. L., Wiechelt, S. A., Ahmed, A. U., & Schwartz, R. P. (2003). Outcomes for substance user treatment in women: Results from the Baltimore Drug and Alcohol Treatment Outcomes Study. *Substance Use and Misuse, 38*(11–13), 1807–1829.

Kanfer, F. (1975). Self-management methods. In F. H. Kanfer & A. P. Goldstein (Eds.), *Helping people change: A textbook of methods* (pp. 309–355). New York: Pergamon Press, Inc.

Kiefer, F., Jahn, H., Tarnaske, T., Helwig, H., Briken, P., Holzbach, R., et al. (2003). Comparing and combing naltrexone and acamprosate in relapse prevention of alcoholism. *Archives of General Psychiatry, 60*, 92–99.

Kofoed, L., Tolson, R., Atkinson, R. M., Toth, R., & Turner, J. (1987). Treatment compliance of older alcoholics: An elder-specific approach is superior to "mainstreaming." *Journal of Studies on Alcohol, 48,* 47–51.

Mann, K. (2004). Pharmacotherapy of alcohol dependence: A review of the clinical data. *CNS Drugs, 18*, 485–504.

Marlatt, G. A. (1976). The Drinking Profile: A questionnaire for the behavioral assessment of alcoholism. In E. J. Mash & L. G. Terdal (Eds.), *Behavior therapy assessment: Diagnosis, design, and evaluation* (pp. 121–137). New York, NY: Springer.

Marlatt, G. A., & Gordon, J. R. (1980). Determinants of relapse: Implications for the maintenance of behavior change. In P. O. Davidson & E. O. Davidson (Eds.), *Behavior therapy assessment: Diagnosis, design, and evaluation* (pp. 410–452). New York: Springer.

Marlatt, G. A., & Gordon, J. R. (1985). *Relapse prevention: Maintenance strategies in the treatment of addictive behaviors*. New York: Guilford Press.

Marsh, J. C., Thomas, J., D'Aunno, T. A., & Smith, B. D. (2000). Increasing access and providing social services to improve drug abuse treatment for women with children. *Addiction, 95*(8), 1237–1247.

Mayfield, D., McLeod, G., & Hall, P. (1974). The CAGE questionnaire: Validation of a new alcoholism screening instrument. *American Journal of Psychiatry, 131*, 1121–1128.

Moos, R. M., Mertens, J. R., & Brennan, P. L. (1993). Patterns of diagnosis and treatment among late-middle-aged and older substance abuse patients. *Journal of Studies on Alcohol, 54*, 479–487.

O'Connell, H., Chin, A., Hamilton, F., Cunningham, C., Walsh, J. B., Coakley, D., et al. (2004). A systematic review of the utility of self-report alcohol screening instruments in the elderly. *International Journal of Geriatric Psychiatry, 19*, 1074–1086.

Oslin, D. W. (2005). Treatment of late-life depression complicated by alcohol dependence. *American Journal of Geriatric Psychiatry, 13*, 491–500.

Satre, D. D., Mertens, J. R., & Weisner, C. (2004). Gender differences in treatment outcomes for alcohol dependence among older adults. *Journal of Studies on Alcohol, 65*(5), 638.

Schonfeld, L., & Dupree, L. W. (1991). Antecedents of drinking for early- and late-onset elderly alcohol abusers. *Journal of Studies on Alcohol, 52*, 587–591.

Schonfeld, L., & Dupree, L. W. (1997). Treatment alternatives for older alcohol abusers. In A. Gurnack (Ed.), *Older adults' misuse of alcohol, medicines, and other drugs* (pp. 113–131). New York: Springer.

Schonfeld, L., Dupree, L. W., Dickson-Fuhrmann, E., Royer, C. M., McDermott, C. H., Rosansky, J. S., et al. (2000). Cognitive-behavioral treatment of older veterans with substance abuse problems. *Journal of Geriatric Psychiatry and Neurology, 13*, 124–129.

Schonfeld, L., Dupree, L. W., & Rohrer, G. E. (1995). Age-related differences in antecedents to substance abuse. *Journal of Clinical Geropsychology, 1*, 219–227.

Tonigan, J. S. (2003). Project MATCH treatment participation and outcome by self-reported ethnicity. *Alcoholism: Clinical and Experimental Research, 27*, 1340–1344.

Uziel-Miller, N. D., & Lyons, J. S. (2000). Specialized substance abuse treatment for women and their children: An analysis of program design. *Journal of Substance Abuse Treatment, 19*, 355–367.

6 Cognitive-Behavioral Pain Management Interventions for Long-Term Care Residents with Physical and Cognitive Disabilities

P. Andrew Clifford, Daisha J. Cipher, Kristi D. Roper,
A. Lynn Snow, and Victor Molinari

Fifty to 72% of long-term care (LTC) residents suffer from chronic or intermittent pain that limits activities of daily living (ADL) (Bressler, Keyes, Rochon, & Badley, 1999; Cipher & Clifford, 2004). The efficacy of nonpharmacological treatment for chronic pain management (psychotherapy, biofeedback, relaxation training, hypnosis, physical therapy, exercise, and behavioral interventions) is well established in the clinical psychology literature for both older and younger adults (Cipher, Fernandez, & Clifford, 2001, 2002; Simmons, Ferrell, Schnelle, 2002). Specifically, cognitive-behavioral therapy (CBT) has been an effective modality within an interdisciplinary care team approach focusing on improving the quality of life (QOL) in chronic pain patients (Morley, Eccleston, & Williams, 1999). Sorkin (1990) suggests that older and younger patients with chronic pain are more alike than different in their response to CBT, particularly if the patients are cognitively intact and only suffering from mild levels of physical disability. But when a LTC resident suffers from both cognitive and physical impairments, the assessment and treatment process must be modified considerably (Cipher & Clifford; Cipher, Clifford, & Roper, 2007; Snow et al., 2004). This chapter focuses on CBT for LTC residents with pain and comorbid physical or cognitive impairments, aiming to help clinicians meet the challenges of modifying CBT for this special population. While this chapter highlights the kinds of modifications required to accommodate the physical and cognitive impairments common in LTC settings, the approaches presented herein are to a large extent also applicable to community-dwelling older adults across the spectrum of ability and disability.

Pain in Long-Term Care

At any given time, 45–80% of NH residents are in some amount of pain (Ferrell, Ferrell, & Rivera, 1995; Weiner, Peterson, & Keefe, 1999; Weiner, Peterson, Ladd, McConnell, Keefe, 1999, Werner, Cohen-Mansfield, Watson, & Pasis, 1998).

Furthermore, many of these residents suffer from persistent pain at high levels. An analysis of the national minimum data set (MDS) found that 15% of NH residents were reported to be in pain for two consecutive quarterly MDS assessments, and that 41% of these residents were in severe pain at the second assessment (Teno, Bird, & Mor, 2001). More than half of the residents of LTC or skilled nursing facilities (SNFs) report persistent or recurrent pain at an average of "moderate" severity or worse (Clifford, Cipher, & Roper, 2005b). Cipher, Clifford, and Roper (2006) observed that the most common behavioral disturbances of chronic and acute pain patients in LTC were anhedonia, depressed mood, withdrawal, low activity levels, low appetite, and weight loss. This finding suggests that pain may contribute to the development of these problematic QOL indicators, highlighting the importance of appropriate pain management for such individuals. Therefore, when residents report that they "can't enjoy life," that they "don't feel like doing anything", and resist getting out of bed, or request to be in bed most of the time, or report that they "don't feel like eating," psychologists are prudent to assess for the presence of chronic pain conditions, even if the resident denies pain initially.

Special Considerations Regarding Pain in Older Persons with Dementia

Persons with dementia are at particularly high risk for the under-recognition and under-treatment of pain (Scherder & Bouma, 1998). Rates of pain, as documented in charts, are one-third to one-half of the rates found when verbal NH residents with mild dementia are interviewed by researchers (Ferrell et al., 1995). In a study of persons with advanced dementia admitted to a tertiary care setting, 13 of 19 had previously undiagnosed or unstable fractures (Morrison & Siu, 2000). Finally, several studies report that only a quarter or less of the demented individuals identified as in pain received analgesics (Bernabei et al., 1998; Won et al., 1999). Such findings are particularly disturbing given the evidence for the effectiveness of pain treatment in this population (Douzijan et al., 1998; Simons & Malabar, 1999).

Problems of under-recognition and under-treatment are due to a variety of factors, including pain attitudes of both patients and staff (such as fear of addiction and belief in the value of stoicism), lack of pain knowledge in both patients and staff, and lack of systemic pain management approaches in many facilities (Snow, et al., 2004; Weiner, Peterson, Ladd, et al., 1999; Won et al., 1999). Another important contributor to under-recognition and under-treatment is the high prevalence of cognitive impairment in LTC facilities (Morrison & Siu, 2000). Residents in LTC who are suffering from cognitive impairment, such as dementia, stroke, or brain injury, often have difficulty communicating about their pain experience (Cipher, Clifford, & Roper, 2006; Scherder et al., 2005). When there are communication barriers, behavioral observation is the gold standard for pain assessment (AGS Panel on Persistent Pain in Older Persons, 2002; American Medical Directors Association, 1999). Pain is often expressed in the form of behavioral disturbances, and may include agitation and other observable behaviors associated with discomfort such as facial grimaces, vocalizations, and guarding, immobilization, or rubbing of body parts (Buffum, Miaskowski, Sands, & Brod, 2001; Cipher, Clifford, & Roper, 2007). Unfortunately, agitated, disruptive, and withdrawn behaviors are not unique to pain, and are also observed in the presence of a wide variety of other problems, such as nausea, thirst, depression, and under- or over-stimulation. This is another reason why healthcare providers have trouble identifying LTC residents who suffer from pain (Snow et al., 2004).

Evidence of the Effectiveness of CBT for Older Adults with Chronic Pain

With regard to the psychotherapeutic treatment of chronic pain, there has been much research demonstrating its efficacy conducted with the comparatively younger community-dwelling employed populations (Loeser, Butler, Chapman, & Turk, 2001). However, the transition of psychotherapeutic treatment outcome research from the working population to the older adult, particularly the frail older adult population, has not yet been fully made. A few preliminary effectiveness studies have investigated the impact of CBT interventions for the treatment of pain in older adults, with promising results. An uncontrolled pilot study of CBT administered in individual sessions to senior housing center residents with chronic low back pain resulted in significant reductions in self-reported pain intensity and pain-related disability scores (Reid, Otis, Barry, & Kerns, 2003). A controlled study of CBT administered in group sessions to nursing home residents with chronic pain reported that those in the experimental group reported less pain and pain-related disability at the end of the intervention (Cook, 1998). Although both of these studies were restricted to cognitively intact persons, they do demonstrate that CBT treatment effects can be achieved with older adults in LTC settings.

Two studies have evaluated a more complex, real-world approach to treatment of pain in LTC that promotes multidisciplinary treatment of persons with cognitive impairment. One randomized controlled trial evaluated a nursing home behavioral consultation service in which a four-member team (psychiatrist, psychologist, two nurses) developed behavioral care plans for residents with advanced dementia and severe behavioral disturbances. Some of the care plans addressed pain management. The experimental group demonstrated a significantly greater decrease in challenging behaviors (Opie, Doyle, & O'Connor, 2002). In a preliminary study of the effectiveness of geriatric multimodal cognitive-behavioral therapy (GMCBT) among cognitively impaired residents who were suffering from persistent pain, patients exhibited significant reductions in pain severity, behavioral disturbances, activity interference due to pain, emotional distress due to pain, and depression after receiving eight twice biweekly sessions of GMCBT. These residents also improved significantly in their cooperation with healthcare staff during ADL care (Cipher, Clifford, & Roper, 2007). Below we delineate the GMCBT approach of providing behavioral and cognitive interventions to LTC residents suffering from chronic pain.

Psychological Assessment for Pain Management

A comprehensive psychological evaluation is necessary for effective psychological treatment of chronic pain. Components of the evaluation should include: (1) psychosocial history that includes screening for historical and recent trauma and stressors; (2) assessment of personality and psychophysiological styles affecting pain reporting, emotional suffering, and the development of dysfunctional pain and illness behaviors; (3) cognitive status or impairment that affects pain experience, reporting, and presentation; (4) psychiatric history and current psychiatric symptoms; (5) review of a recent medical history and physical examination, which may delineate medical conditions that are associated with specific types of pain and noxious sequela; (6) practical effects of pain and suffering on the patient's ADL; (7) dysfunctional behaviors; and finally (8) cooperation and compliance with previous and current medical care (and necessary assistance with ADL). Data from each of these assessment areas should be utilized in the development of a comprehensive case conceptualization that will be used to formulate treatment goals and specific treatment options.[1]

Psychosocial History

A general psychosocial history includes: place of birth, mother's and father's education and occupation, childhood traumas, family of origin dysfunction, resident's education, interpersonal and marital history, adult family life, occupational history, adult relationships with children, lifelong friendships, current social support, reasons for admission to LTC, and recent stressors or life changes. Motivational themes, values, and traumatic experiences over the resident's life history need to be considered in the case conceptualization and treatment planning process. There is also growing empirical evidence showing an association between lifetime trauma (childhood and adult) with the development of chronic pain syndromes and poor medical outcomes (e.g., Goldberg, 1994; Schofferman, Anderson, Hines, Smith, & Keane, 1993). Whenever appropriate and with cultural sensitivity, the occurrence of childhood and adult trauma needs to be assessed and considered when treating chronic pain syndromes that involve coexisting mood and behavioral disorders. Posttraumatic stress disorder symptoms are often triggered by late life chronic pain and disability. Along with possible trauma, the psychosocial history will also be the basis for formulating the resident's lifetime motivational themes

---

[1] Roper, K. D., Reid, T., & Clifford, P. A. *The Mind Body Wellness psychological assessment and treatment handbook*. Unpublished manuscript available at www.mindbodywellnesspc.com.

and current developmental issues. The implementation of GMCBT techniques will rely heavily on historical motivational themes and values, particularly for residents who suffer from mild to moderate dementia.

Cultural, Personality, and Psychophysiological Styles Affecting Pain Experience and Expression

Cultural, personality, and psychophysiological styles greatly affect pain presentation and experience (Clifford, Cipher, & Schumaker, 2003; Cipher, Clifford, & Schumaker, 2002; Cipher, Fernandez, & Clifford, 2002; Wickramasekera, 1993). In general, these psychosocial factors affect how an individual will report the experience of pain. Social/cultural roles and individual differences influence the way a person experiences pain by filtering the perception of pain stimuli, which in turn shapes the meaning or appraisal of that pain perception, which leads eventually to socially reinforced expressions of pain and illness behaviors (emotional, verbal, and nonverbal). In general, patients can either minimize or intensify pain experiences which then may be reflected in idiosyncratic pain behaviors (Dworkin & Sherman, 2001).

One way of delineating these important individual differences involves the construct of psychophysiological styles (Cipher, Fernandez, & Clifford, 2002). From this framework, some people are viewed as being emotionally reserved, stoic, socially conventional, psychologically naïve, and practically minded. People reporting or displaying behaviors consistent with this style are classified as Repressors. They tend to underreport all negative emotions including pain levels and tend to withdraw from others when in moderate to severe pain. They increase personal activity levels with mild pain in order to distract themselves; they suffer silently and often delay or postpone seeking professional care. Another group of people classified as Amplifiers tend to be emotionally hypersensitive and catastrophic in their perceptual style. They will overreport mild levels of pain, and forcefully express their distress or influence others for emotional comfort or secondary gains, and often act out dysfunctionally in their exaggerated illness role. Assessing differences regarding individual mind–body self-report bias (or psychophysiological style) is an important skill that psychologists bring to the interdisciplinary pain management team, and this assessment will help determine both appropriate pharmacological and nonpharmacological interventions (Cipher, Fernandez, & Clifford, 2002; Cipher & Clifford, 2003).

Traditional assessment of personality can be helpful in determining psychophysiological styles. Personality testing is recommended in the treatment of ambulatory older adults living in the community or LTC facilities if cognitive impairment is minimal. When the LTC resident suffers from significant cognitive impairment and physical disabilities, personality assessment can be based on clinical interview, observations, family recollections, and healthcare provider reports regarding historical and current behaviors and attitudes. The clinicians' understanding of how different personality styles affect clinical rapport and compliance to treatment is important in order to maximize treatment effectiveness (Clifford, Cipher, & Schumaker, 2003; Cipher & Clifford, 2003; Cipher, Clifford, & Schumaker, 2002).

As the LTC resident progresses in dementia, understanding of the resident's historical personality traits can be helpful in formulating motivational themes and values used in therapeutic coaching and behavioral therapy.

Cognitive Assessment

For purposes of case conceptualization and treatment planning, brief cognitive assessment should be conducted to determine the resident's cognitive strengths and weaknesses, which will guide CBT modifications. The Neurobehavioral Cognitive Status Examination (Kiernan, Mueller, Langston, & Van Dyke, 1987) and the Dementia Rating Scale-2 (Jurica, Leitten, & Mattis, 2004) are sufficient testing tools in most cases. A dementia staging assessment, such as the Global Deterioration Scale (Reisberg, Ferris, de Leon, & Crook, 1982) or the Clinical Dementia Rating scale (Hughes, Berg, Danziger, Coben, & Martin, 1982) can be helpful for understanding the level of cognitive impairment in terms of multidimensional effects on functioning. Table 1 briefly discusses how the assessment and treatment processes might be adapted to the resident's level of cognitive impairment.

As dementia progresses, verbal assessment of the patient becomes increasingly problematic because of the language and higher order processing deficits that accompany even mild dementia (White & Cummings, 1999). In order to validly self-report about one's internal state, one must be able to understand the rater's questions, recall how one has been feeling, and accurately interpret one's internal experience – e.g., depression, anxiety, pain (Snow, Graham et al., 2005; Snow, Kunik et al., 2005). Although no cognitive assessment instrument score cutoffs have been established to strictly guide the clinician regarding when self-reports are no longer accurate, there is evidence that as cognitive impairment increases, the ability of the patient to provide information about more abstract and complex concepts decreases. The clinician is advised to modify assessment approaches for persons with dementia by interpreting responses to standardized questions with caution, using questions with simpler response formats (e.g., yes/no), and gathering data from behavioral observations and collaterals as well as from interviews of the patient (see Table 1 below).

Along with establishing an understanding of the patient's cognitive abilities and level of impairment, it is helpful to consider the possible type or etiology of the dementia, because specific types of dementia may affect the patient's perception of pain and depression. Scherder et al. (2003) report preliminary data suggesting that people with Alzheimer's dementia present with decreased affective components of pain and have a higher pain tolerance than vascular dementia patients who tend to have increased affective components of pain and less pain tolerance. Frontal–temporal dementias show decreased cognitive-evaluative and sensory-discriminative components of pain perception. Patients with multiple sclerosis or Parkinson's disease without dementia present with increased affective suffering secondary to pain and may be less pain tolerant (Scherder et al., 2005).

TABLE 1. CBT Assessment and treatment considerations across progressive levels of dementia.

Reisberg's global deterioration scale (RGDS)	FAST	Acquired or lost ability	Dev. age	Assessment and treatment considerations
Stage 1: *Cognitively normal.* No complaints or indications of memory problems	1	Holds a job	Adult	COGNISTAT: usually within normal limits. At these levels of functioning, delirium (due to medications, anesthesia, TIA, traumatic brain injury, etc.), depression, motivation, education, cultural, and severe life changes – stress, need to be assessed and their possible effects on test performance considered in interpretation Normal or Very Mildly Impaired adults benefit from traditional developmentally appropriate CBT. Assessment focuses on inter-personal styles, and adapting behavioral goals and CBT interventions to conform to the outpatient's personal goals for quality of life
Stage 2: *Very mild deficits. Forgetfulness*, people continue to function competently but complain of forgetting once familiar names and where they put things like keys	2	Holds a job	Adult	The GMPI, GLDS, and PRADLI are appropriate for these levels of cognitive impairment, but are only in LTC settings. The clinician uses the GMPI 10-point Likert scale, and uses both behavioral and verbal anchors. Clients who are psychophysiological amplifiers will tend to over report pain and discomfort levels, often saying that their pain is 15 on 10-point scale, yet they interact with only moderate levels of pain behaviors or inconsistent pain behaviors. At these levels of cognitive impairment traditional CBT techniques are utilized
Stage 3. *Mild cognitive decline. Early confusion.* People have trouble remembering the names of objects, recent names, material just read. They lose keys, wallet, or checkbook. They get lost easily in unfamiliar surroundings	3	Holds a nondemanding job	12+	Stage 3: COGNISTAT is likely to show only mild to moderate impairment in visual spatial construction, and short term memory. Starting at this stage verbal assessments should be supplemented with behavioral observations and collateral interview. Behavioral and collateral information becomes increasingly integral as cognitive impairment worsens. Interpersonal and psychophysiological styles are still important, but now CBT becomes more behavioral, and cognitive interventions are more concrete and rule oriented. Some clients may find keeping behavioral records to be particularly helpful and comforting
Stage 4. *Moderate cognitive decline. Late confusion.* People get lost going to familiar destinations and have difficulty handling finances and recognizing familiar individuals. They forget recent events and are uncertain both of the time and date	4	Handles simple finances	8–12	Stage 4: COGNISTAT is likely to show moderate to severe short-term memory impairment, moderate to severe impairment in visual spatial construction ability, and normal to mild impairment in Naming, Similarities, Orientation, and Social Judgment. Interpersonal and psychophysiological styles are still important for accurate assessment of mood, but family/collateral input is even more important. Positive behavioral outcomes will be dependent on the quality of social support the client has at his or her residence. Interventions are increasingly

				more behavioral, and cognitive interventions are more concrete-rule oriented and motivational using the client's longstanding values to motivate compliance to structured behavioral interventions. Daily supervision or assistance is needed
		Needs some assistance on a daily basis, and needs daily or 24-h supervision because of poor safety awareness and inability to function in stressful situations		At these levels of cognitive impairment the GMPI 10-point scale with verbal and behavioral anchors are used to socialize the client to behavioral goals and to the idea of active coping with chronic illness involving pain and/or other debilitating symptoms. Clinical ratings are more dependent on staff and family input. Mild to moderate cognitive impairment often requires the therapist to be more directive and behavioral in the assessment and treatment process. If the client is severely distressed by symptoms, verbal and behavioral anchors help the client give input in assessing level of suffering and interference associated with the illness and symptoms. At these levels, cognitive functioning is often sensitive to exacerbations in physical chronic conditions (e.g., CHF, diabetes, chronic renal failure, vascular disease, repeated urinary and respiratory infections, COPD)
Stage 5. *Severe decline. Early dementia*. Individuals cannot survive without ongoing assistance, make significant errors in dressing, and get confused about the season and year. They forget their address, phone number, high school, and where they are	5	Selects proper clothing	5–7	Stage 5: COGNISTAT is likely to show severe impairment in short-term memory, visual spatial construction ability and Orientation; mild to moderate impairment in Naming, Similarities, and Social Judgment. Normal to mild impairment in Calculation. At this stage the person is now very egocentric and emotional in his or her thinking and reasoning process. While they recognize social rules, as part of long-term memory, their cognitive decline is so great that they cannot employ abstract rules or interpersonal empathy in order to govern their mood swings and emotional behaviors. Therefore, successful behavioral interventions and monitoring are meditated through 24-h caregivers who collaborate with the psychologist and family members. At this level of dementia the person can still communicate. Sensations fairly well intact, but psychophysiological styles may become more exaggerated or diverge from historical tendencies

(*Continued*)

TABLE 1. CBT Assessment and treatment considerations across progressive levels of dementia—*cont'd.*

Reisberg's global deterioration scale (RGDS)	FAST	Acquired or lost ability	Dev. age	Assessment and treatment considerations
Stage 6. *Severe cognitive decline. Middle dementia.* People forget close relatives' names but still know their own. They have little if any idea where they live, what day and year it is, and what they ate at their last meal. Their behavior changes, they seem nervous, tense, and agitated. They may wander aimlessly or lose interest in doing what they used to enjoy. They may become suspicious and paranoid, and may lash out verbally or physically. They may talk to imaginary companions. They have bladder and bowel accidents	6a b c d e	Dresses unaided Showers unaided Toilets unaided Controls urine Controls bowels	5 4 4 3–4.5 2–3	Stage 6: COGNISTAT shows severe impairment in short-term memory, visual spatial construction ability, and Orientation; moderate to severe impairment in Naming, Calculation and Social Judgment. As dementia progresses, thinking and communication becomes severely impaired. Families will report that historical personality traits, interpersonal and psychophysiological styles are either exaggerated or significantly altered. Psychological assessment becomes increasingly behavioral and systems oriented. The psychologist establishes therapeutic relationship with resident, works behaviorally using cognitive themes important to the resident, then coordinates structure environmental and behavioral interventions involving family and LTC staff/caregivers
Stage 7. *Very severe cognitive decline. Late dementia.* They become incontinent, cannot feed themselves, they lose the ability to walk, to talk, to sit up, and eventually they become vegetative	7a b c d e f	Speaks w/5–6 words Speaks w/1 word Walks Sits up Smiles Holds head up	15 months 1 year 1 year 6–10 months 2–4 months 1–3 months	Stage 7: COGNISTAT becomes less useful because of "floor effects." As the person loses verbal abilities, the multidisciplinary team must rely on nonverbal communication and behavioral analysis of dysfunctional behaviors and mood shifts, which may signal pain or other discomfort associated with chronic illness. Assessment and treatment utilize validation and behavioral techniques in collaboration with other healthcare providers. Assessment and treatment are often nonverbal, but very dependent on social facilitation via tone of voice, behavioral modeling, facial expressions, therapeutic touch, and body language

At this level the clinician's ratings on the GMPI, GLDS, and PRADLI are increasingly dependent on clinical observations and the input of staff and family, sitters, and friends familiar with the client's behaviors. Behavioral anchors on the GMPI are used almost exclusively in making clinical ratings along the 10-point scale. Ratings are best done during afternoon nurse shift changes where both morning and swing shifts are available for input. Family members are also interviewed before making behavioral ratings. As the client approaches the sixth stage, PRN analgesics are less appropriate, and clients typically refuse or resist ADL care and social activities secondary to pain or discomforting symptoms. Behavioral apathy or aggressive acting out are the most common responses to ADL care when suffering chronic discomforting symptoms, thus routine analgesics with behavioral interventions that increase activities is the indicated treatment in nonhospice residents. As the patient becomes nonverbal consider approaches presented in Herr, Bjoro and Decker (2006)

The table is a brief summary/adaptation of Reisberg's Global Deterioration Scale, his Functional Assessment Staging Tool (FAST), and his delineation of developmental milestones or "inabilities" associated with the "retrogenesis" process in Alzheimer's disease. This table also includes general information designed to help clinical psychologists become familiar with the use of the COGNISTAT in the staging of Alzheimer's dementias in order to make assessment and treatment formulations. Residents and patients with vascular dementias involving aphasia, will often perform significantly worse on the COGNISTAT (Reisberg, Auer, Franssen, Monteiro, & Kenowsky, 1999; Reisberg et al. 1982).

Psychiatric History and Current Medical Symptoms

As with all psychological evaluations, a thorough psychiatric history should be conducted in preparation for treatment. Although 15–20% of the general population at any given time will meet criteria for a psychiatric diagnosis, people suffering from acute and chronic medical conditions have significantly higher rates of psychiatric disorders. The general medical patient has a psychiatric prevalence rate of 25–30% (Ford, 1992, 1995); the medical patient with chronic medical conditions shows a rate of 25–40% (Wells, Golding, & Burnam, 1988). When chronic medical conditions involve chronic pain, the psychiatric comorbidity rate increases to 50–75% (Fishbain, Goldberg, Meagher, Steele, & Rosomoff, 1988). Finally, psychiatric disorders are more common in LTC settings than in the general medical population. The American Association of Geriatric Psychiatry reports that 80–90% of LTC residents suffer from psychiatric disorders (Rovner et al., 1990). About 10–50% of LTC residents suffer from depression/anxiety disorders, 50–90% of LTC residents suffer from dementia (Burgio, 1997; Davis, Buckwalter, & Burgio, 1997), and 64–83% will exhibit behavioral disturbances associated with the above conditions (Allen-Burge, Stevens, & Burgio, 1999; Cipher & Clifford, 2004).

Medical Conditions Associated with Acute and Chronic Pain

Psychologists providing CBT for pain management need to be aware of medical conditions associated with pain. There are several mechanisms of pain – nociceptive, deafferentation, sympathetic mediated, myofascial, and visceral, which all have different characteristics of presentation, etiology, and response to treatment. Common medical conditions in LTC associated with pain include osteoporosis, sites of recent and old bone fractures, arthritis, degenerative joint disease, fibromyalgia, vascular disease, cancer, claudication, diabetes, amputations, oral and dental pathology, renal conditions, neuropathies, Parkinson's disease, multiple sclerosis, gastrointestinal conditions, diverticulitis, and connective tissue disease (American Medical Director's Association, 2003; Teno, Bird, & Mor, 2001). Therefore, the consulting psychologist needs to be familiar with the sequelae of any medical condition diagnosed by the medical staff, particularly when working with dementia patients who may not be able to appropriately communicate the nature of their pain or symptoms.

Pain Assessment

Pain is defined by the Taxonomy Commission for the International Association for the Study of Pain as "an unpleasant sensory and emotional experience associated with actual or potential tissue damage, or described in terms of such damage." Pain is generally acknowledged to be a complex, subjective phenomenon that encompasses nociceptive/perceptual, cognitive, emotional, and behavioral/functional factors (Dworkin & Sherman, 2001; Turk, Meichenbaum, & Genest, 1983;

Snow et al., 2004). Therefore, an assessment of a resident's pain should also examine these factors (Gagliese & Melzack, 1997). There is a wealth of pain assessment instruments available; the following discussion provides a general overview and then more specifically focuses on those instruments used in the GMCBT approach of assessment and treatment.

Nociceptive/Perceptual Assessments

One-Item Pain Rating Scales

Two of the most commonly used measures of pain intensity in LTC facilities are the visual analogue scale (VAS) and the numeric rating scale (NRS). The VAS is a self-report scale that is typically 6-cm wide with labels at each endpoint, such as "no pain" on one end and "worst possible pain" on the other. Respondents put a mark on the scale, and the mark is assigned a number. The numeric rating scale (NRS) is similar to the VAS, but is a 10-point Likert scale, with labels associated with certain points on the scale. Because a large body of research supports the reliability and validity of the NRS in a wide variety of types of pain and clinical settings, the Veterans Healthcare Administration has adopted it nation-wide (Department of Veterans Affairs, 2000).

Minimum Data Set 2.0

Another common pain assessment method in LTC is the MDS three-point pain intensity scale, with points labeled as "mild pain" (1 point) to "moderate pain" (2 points) to "horrible or excruciating pain" (3 points; Briggs Corporation, 1997). This scale is a part of the MDS and also assesses cognitive abilities, functional capacity, communication abilities, mood, vision, hearing, psychosocial issues, and general health condition. The MDS is typically used in SNFs, and is verbally administered and rated by LTC staff on a quarterly basis. Fries and colleagues demonstrated that the MDS pain scale is highly consistent with VAS ratings, and can be used in place of the VAS in LTC settings (Fries, Simon, Morris, Feldstrom, & Bookstein, 2001). Unfortunately, as with the VAS, the scaling of the MDS pain assessment scale can be inaccurate, since there may be a vast difference between "moderate" and "horrible and excruciating" pain. Moreover, there are no explanatory statements on the scale itself that detail the criteria for the pain ratings, thus resulting in potential inconsistency among raters. A word of caution when viewing pain ratings done as part of the routine quarterly MDS assessments done by nursing staff – although such ratings might provide some historical context regarding pain experiences, one will not be sure under what conditions the data were collected.

Behavioral Observational Pain Severity Scales

Verbal pain assessment should always be supplemented with behavioral observational assessment in persons with cognitive impairment. Ability to respond to a verbal pain assessment is generally associated with level of cognitive impairment (e.g., Chibnall & Tait, 2001; Cohen-Mansfield, 2002; Wynne, Ling, & Remsburg, 2000). For example, one study reported that, of five simple pain intensity scales administered to a group of cognitively impaired nursing home residents, 83% of the residents could complete at least one of the instruments; on the other hand, only 17% could complete all five instruments (Ferrell et al., 1995). There are currently at least seven published behavioral observational scales with some reliability and validity evidence for persons with dementia. All require the rater to observe the patient and rate presence/absence, intensity, or frequency of behaviors. A recent review evaluates the available scales and their psychometric evidence (Herr, Bjoro, & Decker, 2006). Snow et al. (2004) provide a conceptual model for understanding the factors that affect assessment of pain in persons with cognitive impairment, and guidelines for how to develop a multidimensional assessment approach for this population.

Assessment of ADL and Behavioral Dysfunction Associated with Pain

Psychosocial Resistance to Activities of Daily Living Index (PRADLI) (Clifford, Cipher, & Roper, 2003). The PRADLI is an 8-item clinically rated instrument that assesses the resident's level of functional independence and cooperation with eight psychosocially related ADL. The eight domains are out of bed time, eating habits, dressing, toileting, bathing, medical compliance, restorative care, and social/recreational activities. These items are rated on a 7-pronged scale, with one representing the lowest levels of independence and cooperation, and 7 representing the highest levels of independence and cooperation. The PRADLI has been evidenced to have high internal consistency ($\alpha = 0.88$) and high convergent validity (from 0.95 to 0.98; Clifford, Cipher, & Roper, 2003). Cipher and Clifford (2004) found that the PRADLI was sensitive to the effects of pain mediated by both dysfunctional behaviors (measured by the GLDS) and depression (measured by the Geriatric Depression Scale – GDS).

Geriatric Level of Dysfunction Scale (GLDS) (Clifford, Cipher, & Roper, 2005a). The GLDS is a clinician-rated instrument that assesses the average intensity, frequency, duration, and number (count) of each of 19 possible dysfunctional behaviors, including agitation, verbal aggression, withdrawal, and physical aggression. All ratings are made on a 7-pronged scale. For the rating of intensity, lower numbers represent lower intensity of the behavioral dysfunctions. For the rating of frequency, lower numbers represent the lower frequencies of behavior and for the rating of duration, lower numbers represent less total time per day of behavioral dysfunction. These ratings have excellent internal consistency ($\alpha = 0.96$). Test–retest coefficients have ranged between 0.86 and 0.94 among three independent raters. Cipher and Clifford (2004) found that pain measured by the GMPI predicted dysfunctional behaviors better than dementia levels. Cipher et al. (2006, 2007) also found that different levels of dementia predicted different types of dysfunctional behaviors associated with chronic and acute pain.

Multidimensional Assessment Batteries

Several multidimensional assessments have been developed and validated for younger populations with chronic pain disorders (Kerns, Turk, & Rudy, 1985; Turk & Melzack, 2001), as well as for the ambulatory elderly (Ferrell, Stein, & Beck, 2000). Clifford, Cipher & Roper (2003) and Clifford et al. (2005a, 2005b) developed a battery of assessment instruments that can be used to assess residents in various levels of LTC who are often immobile and suffering from pain associated with numerous acute and chronic medical conditions. The following batteries assess a combination of the four factors (nociceptive/perceptual, cognitive, emotional and behavioral/functional) although none comprehensively covers all four factors, and thus all need supplementation.

The Geriatric Pain Measure (GPM) (Ferrell et al., 2000). GPM was designed to assess pain and associated limitations in the ambulatory elderly who may be experiencing multiple medical problems. The GPM is an "interviewer-assisted" instrument, meaning that the items are read to the patient and/or clarified in order to ease completion. The GPM is a vast improvement upon the VAS, Numeric Rating Scale, and MDS in its assessment of both the severity of pain and functional limitations due to pain.

The Geriatric Multidimensional Pain and Illness Inventory (GMPI) (Clifford et al., 2005b). The GMPI is a 12-item instrument designed to assess pain and its functional, social, and emotional consequences in LTC. In the absence of pain, the GMPI assesses the severity of the resident's primary chronic medical condition and its functional and emotional consequences. The first item is, "How bad is your pain or illness right now?" Other items include, "How much have you suffered because of your pain or illness this last week?", and "How irritable have you been this last week because of your pain or illness?" All items are rated on a 10-point scale, with each point associated with specific behavioral criteria. The scaling of the items is behaviorally oriented because GMPI ratings are based only on what the clinician and the staff members can *observe*. The GMPI has high convergent validity ($r = 0.88$), high internal consistency ($\alpha = 0.88$), and test–retest reliabilities for the three subscales range from 0.62 to 0.96. Higher values are indicative of higher levels of pain and/or higher levels of functional/social/emotional difficulties (also see Table 1).

GMCBT

Case Conceptualization and Psychological Care Plans

The initial psychological evaluation concludes with an individualized case conceptualization that summarizes the findings of the psychological evaluation, delineating treatment goals and intervention options that accommodate to the resident's personality and psychophysiological style, dementia level, and historical and recent traumatic or stressful life experiences.

The standardized GMCBT evaluation utilizes the GMPI, PRADLI, GLDS, COGNISTAT, the GDS (Yesavage, Brink, & Rose, 1983), and/or the Cornell Scale for Depression in Dementia (Alexopoulos, Abrams, Young, & Shamoian, 1988). The GMCBT assessment and treatment model is illustrated in Fig. 1, which is adapted from Cipher and Clifford's (2004) mediational model of QOL for resident's suffering from psychiatric and medical comorbidities involving dementia, chronic pain, and behavioral disturbance. These assessment instruments are also employed on a weekly or monthly basis to assess the clinical effectiveness of GMCBT interventions and/or interdisciplinary care.

In formulating a care plan, the psychologist consults with the resident, family members, and relevant medical, nursing, rehabilitation and dietitian staff in order to prioritize treatment goals. Common treatment goals include improved pain tolerance, increased pleasurable and meaningful activity levels, behavioral pain management (e.g., distraction, ice and heat, relaxation, postural adaptations in wheel chairs), improved depressed/anxious/irritable moods, increased compliance with medical/rehabilitation care plans, improved ADL compliance, increased out of bed and room time, weight maintenance (motivation to eat), recreational/social/spiritual functional restoration/maintenance, decreased family conflict/dysfunction/distress,

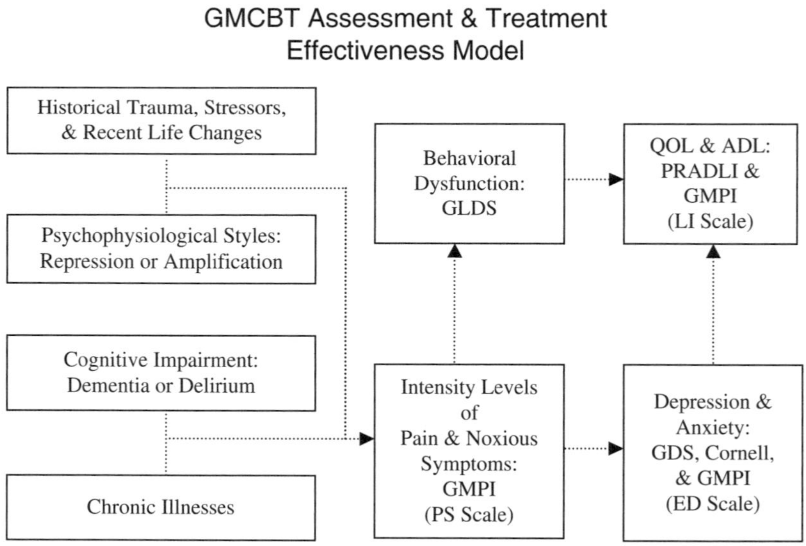

FIGURE 1. Mind Body Wellness Assessment and Treatment Effectiveness Model for Quality of Life in Long Term Care. The mediational model suggests that dysfunctional behaviors measured by the GLDS are best predicted by noxious or illness symptoms, cognitive impairment, and medical conditions. The GMPI is a mediation variable for both depression and dysfunctional behaviors as they relate to Quality of Life and PRADLI measures. This model suggest that pain and noxious medical symptoms are important intervention areas in effecting behavioral, mood, and functional impairment levels in residents with dementia living in LTC (Clifford et al., in press).

adjustment to recent physical disabilities or limitations, adjustment to need for LTC placement, improved cognitive functioning associated with reduced analgesic and psychotropic side effects, and decreased frequency of inappropriate behaviors. After the psychological care plan is completed and agreed upon by the resident, involved family members and other healthcare providers, GMCBT interventions are initiated. The vectors illustrated in Fig. 1 suggest causation and points of intervention when formulating psychological interventions and interdisciplinary care plans.

The behavioral anchors used in the PRADLI (item 1 – out of bed time, and item 8 – time participating in personal hygiene, social and recreational activities) and the GMPI Life Interference Scale were partly designed to correspond to common behavioral benchmarks signifying either dysfunction or relative levels of QOL. Degrees of change within each of the items delineate specific behavioral goals that are relevant to and consistent with Federal LTC Quality Indicators, which are based on MDS and maintained by all LTC facilities. Effective GMCBT must be integrated into the assisted living and NH facilities' system of care and their MDS assessment system.

GMCBT: A Comprehensive Approach to Pain Management

GMCBT is inclusive of any psychotherapeutic technique that can be operationalized into cognitive-behavioral terms. While standardized, GMCBT continually incorporates empirically demonstrated techniques into its intervention armamentarium. Like other healthcare treatments (psychopharmacology, physical, and occupational therapy), GMCBT is self-correcting – if one technique does not work, another is tried; it focuses on multiple problems and attempts to relieve parallel and interacting difficulties. It always seeks improvement in the general functioning of the LTC resident and the amelioration of the disorder or presenting symptoms (Seligman, 1995).

Dementia patients in rehabilitation care will need multiple sessions per week if pain and dysfunctional behaviors are interfering with medical and physical progress. In general, the more cognitively impaired the resident, the more behavioral the psychological interventions should be. Cognitive interventions are always employed, but the therapist also takes the role of an empathic listener, who validates (Feil, 1993) the resident's experience, and then within the resident's frame of reference, suggests alternative appraisals of the situation using perceptual themes consistent with the resident's lifelong values and personality style. Essentially, the GMCBT therapist is a therapeutic coach who facilitates behavioral change in the resident during the therapeutic 20–45 min session once or twice a week. With family and resident consent, individual therapeutic sessions are conducted with other healthcare providers present, watching and then participating in the cognitive-behavioral facilitation of desired behaviors and assessment processes. As with all CBT, assessment and treatment are continuous. Effective techniques are developed and continuously adapted through the treatment process, until daily healthcare providers and family can facilitate and maintain (through appropriate reinforcement and motivational techniques) the desired behavioral and emotional goals.

Residents with milder levels of dementia often receive more Socratic and interpersonal exploration of psychosocial issues, particularly if they have a history of being socially skilled and psychologically minded. Conventional Repressors and distressed

Amplifiers, however, tend to respond better to practical behavioral coaching that relieves symptoms and improves pain tolerance and QOL. If the resident is "time-oriented" (with only mild–moderate short-term memory and orientation impairment), CBT interventions are more client-directed and incorporate more interpersonally collaborative cognitive restructuring techniques. Cognitive restructuring generally facilitates residents' reappraisal of their medical conditions, which may be contributing to problematic pain behaviors, mood disturbances, or interpersonal difficulties with caregivers or family members. With empathic understanding and reappraisal of the situation, the resident is then therapeutically coached to cooperate with behavioral techniques that improve pain tolerance and increase enjoyable activities. It is important to produce an emotionally pleasant rapport, and some type of symptom relief or behavioral success early in the GMCBT process.

Case Study

Assessment. Patricia (we will call her) was an 84-year-old white widowed female who was admitted to an SNF after suffering a left hip fracture 2 weeks prior. She was living alone before her fall, and she lay injured in her bathroom for 36 h before she was discovered by a neighbor and taken to the hospital. The hip was surgically repaired and Patricia spent 10 days in the acute care hospital before being transferred to the SNF. Past medical history included osteoporosis with compression fractures in her lumbar and thoracic spine, hypertension, rheumatoid arthritis – hands were significantly deformed and impaired in functioning, osteoarthritis, multiple degenerative joints, noninsulin dependent diabetes (for the last 6 years), coronary artery disease with CABG × 2 four years ago. After 2 weeks in the SNF, Patricia was referred by her internist at the request of the rehab therapist because of poor motivation and refusal to participate consistently in physical and occupational therapies. She was 5′ 6″ tall and weighed 110 lbs, with weight loss of over 10 lbs in the last 6 weeks. Patricia's internist had placed her on an antidepressant 4 years ago. The social worker had noted that Patricia appeared to be depressed, but denied "feeling anything." Patricia consented to psychological evaluation, but seemed annoyed at the idea.

Patricia reported that she was born in Kansas. Her father was a farmer and her mother a homemaker of four children, Patricia being the youngest by 8 years. Her childhood was unremarkable. She graduated high school, and attended 1 year of college before returning home to care for her mother who died of cancer at age 62. Two years later her father died after suffering two strokes in 1 year. A year later, Patricia married at age 24. She followed her husband John, to Dallas, Texas where he became a successful insurance executive. She had one son who died in the Vietnam War. While mostly a homemaker, Patricia worked as a church secretary for 27 years before retiring at age 74. She and her husband remained active in a local Church until John was diagnosed with Alzheimer's dementia 7 years ago. She cared for him at their home until one violent episode, which precipitated placement in LTC. He died 18 months ago. Patricia has one brother still living in Oklahoma, and a grandniece who lives within an hour's drive.

Patricia answered the psychologist's questions without emotion. When empathy was attempted, she became annoyed and indignant, and protested "you're wasting yours and my time.... Why do you care anyway?" Neuropsychological testing (COGNISTAT) indicated she was disoriented to place, date, and day of week. She knew the year and her age and that she was in a nursing home in Dallas. Attention, Calculation, Verbal Comprehension, Repetition, and Naming were all within normal limits. Construction and Delayed Memory were moderately impaired as was Orientation. Abstract Reasoning and Judgment were both mildly impaired, but her frustration tolerance was very low and after 20 min of neuropsychological testing, she became increasingly irritable with the psychologist's questioning. Her GDS score was 12/15. Dysfunctional behaviors (GLDS) included (1) verbal aggression, which was distressing to staff and occurred several times a day; (2) depression, withdrawal, low motivation, denial of suicidal intentions but verbalized the wish "that I would just die"; (3) low activity level; (4) loss of appetite and weight; and (5) noncompliance with restorative care. Each of the last four behaviors was judged to be interfering with Patricia's medical care and status. Her GMPI scales indicated moderately severe Pain and Suffering 6.3/10; severe Life Interference due to pain and illnesses 9/10; severe Emotional Distress – depression 7/10, frustration 10/10, anxiety and tension 6/10; and she felt unable to cope with her condition 10/10. Her PRADLI scores indicated she was resisting assistance for dressing, bathing, toileting, and rehabilitation, and she refused to be fed solid food but was drinking a high caloric shake several times a day. She refused to get out of bed, and would not assist in any personal hygiene activities even though she was capable.

Diagnostic impressions were major depression-severe with mild cognitive impairment (Reisberg's Global Deterioration Scale score of 3/7 – early confusion). Patricia's premorbid personality style was likely cooperative, conventional, with considerable social skills involving etiquette and assertiveness, suggesting probable obsessive–compulsive personality traits. Her psychophysiological style was repressor, but with noted social skills and a potential for some psychological processing. Both historical and recent stressors were intense and her current defensiveness was probably a longstanding coping style considering the extent of interpersonal losses she had experienced over her lifetime.

Interventions. Patricia was surprised to see the psychologist a second time, and again said, "you're wasting yours and my time." She was tired of living and wished she would "just die." She was not actively suicidal, but also was not interested in rehabilitation. She reported that her pain levels were bearable when in bed and that she just wanted to be left alone. Early sessions explored her cognitive appraisal of her now worthless life. The following issues were explored: palliative care, end of life concerns, loss of her husband and son, and the isolation she experienced in the last 3 years of caring for her husband. She expressed disgust with herself and her dependence, and the insistence of PT/OT that she do more for herself. Her lack of cooperation with rehabilitation was putting her at risk of being discharged from the SNF to LTC. She was also showing signs of a stage one decubitis ulcer on the sacral area. The psychologist reviewed his role as a rehabilitation psychologist during the first month of therapy, emphasizing the necessity that Patricia increase her time out of bed, but the therapist only facilitated "uptime" after first validating Patricia's sense of worthlessness associated with her profound disability. The psychologist offered counter evidence and/or reappraisals that suggested she was not worthless and that her life did have value, often leading to discussions about the limitations in her QOL. During the first month of therapy, Patricia needed time to vent her frustration. After validation of her experience, some time was spent sensitizing her to the disadvantage of her current behaviors and the advantages of her compliance with rehabilitation. Behavioral progress was slow at first. She agreed to "off loading" of her sacral area and sitting up for meals in bed, which caused increased pain. Eventually she agreed to routine acetaminophen, which made her compliance with uptime and PT/OT easier.

Prior to her husband's death, one of Patricia's core values was to do her duty without complaint. She resented having to put her husband in LTC and pitied "the poor (pathetic) people" in LTC. Now she was one of them. Using the "double standard technique," (Burns, 1989) and her Christian faith, the psychologist and Patricia developed a new motivational theme, "love yourself as you have loved others." Patricia's existential crisis was partly resolved with a new philosophical orientation to physical disability and LTC. After family consultation with her brother who was also her Power of Attorney, he visited Patricia and reinforced this new perception of herself and her situation within LTC.

During the second and third months of GMCBT, Patricia was increasingly compliant with behavioral interventions that increased out of bed time to over 9 h a day. Gradual increases in "uptime" started in her room, and then involved activities out of her room. She learned to feed herself with adaptive silverware. Cognitive interventions helped lessen depression and anger that was associated with social isolation, self-pity, and shame at needing assistance in ADL care. When sitting intervals increased to over 2 h, opiate analgesics were necessary. In the third month of psychotherapy after PT/OT discharge, Patricia continued to make progress in enjoyable activities out of her room. These changes were facilitated by collaborative activity scheduling, pleasure/mastery lists, and repeated cognitive restructuring techniques using motivational themes, advantages vs. disadvantages, reappraisal of the situation, situational analysis, rational/realistic thinking. After 6 weeks of twice weekly sessions, the psychologist met with Patricia weekly for three more months.

Outcomes. After 24 sessions of GMCBT, Patricia's GDS score was 3/15. Dysfunctional behaviors (GLDS) included (1) occasional depressed mood involving withdrawal or low motivation; and (2) low activity levels. Other behavioral problems were considered to be resolved by staff. Her GMPI scales indicated mild Pain and Suffering; mild Life Interference due to pain and illnesses. Her PRADLI scores were out of bed at least 7–9 h a day; eating with setup and prep; cooperating with moderate assistance for dressing, bathing, toileting, and restorative care; cooperating with minimal assistance for medications; and participating 7–9 h a day in hygiene, personal, social, or recreational activities per day.

Relapse and Interventions. Three years later, Patricia suffered a serious upper respiratory infection and was hospitalized for 30 days in a long-term acute care (LTAC) hospital. After being discharged back to her LTC facility, the same psychologist was consulted while she was receiving rehabilitation. Staff reported more confused since the hospitalization, sleeping at least 15h a day. When awake she was cooperative, but constantly requested to be put back in bed if up for more than 30min. Valid neuropsychological testing indicated cognitive impairment was significantly more impaired than 3 years ago. Most days, staff complained of "Sundowning," when they tried to facilitate rehabilitation activities in the later afternoon. Reisberg's Global Deterioration Scale was judged to be 5/7, with noted intermittent delirium mostly in the afternoons. Psychiatry Service had been consulted at the LTAC, and was following Patricia at the LTC SNF. All physicians and other healthcare providers complained of poor motivation, and indicated that they thought Patricia could do more in rehabilitation.

Her GMCBT assessment battery reflected moderate depression – a GDS score of 8/15. While her GDS was not as high as her previous score, her Cornell Depression Scale was 26 confirming the psychologist's clinical impression that Patricia was more depressed now than 3 years ago. Her dysfunctional behaviors (GLDS) included (1) depression, withdrawal, low motivation, and intense complaints of fatigue and weakness; (2) low activity levels; (3) loss of appetite and weight; and (4) noncompliance with restorative care. Each of these four behaviors was judged to be interfering with Patricia's medical care and status. Her GMPI scales indicated moderate Pain and Suffering 5.3/10; severe Life Interference due to pain and illnesses 9.4/10; moderate Emotional Distress. Her PRADLI scores were 1/7 for six of the eight items, indicating she was resisting total assistance for dressing, bathing, toileting, and rehabilitation, and she refused solid food and had to be encouraged to drink a high caloric shake several times a day. She cooperated and got out of bed but would cry and request to be put back into bed within 10–15min. While facial expressions indicated distress and possibly pain, she tended to deny pain and complained of fatigue.

The psychologist then met with Patricia in the Rehabilitation area; within 10–15min of her being up she repeatedly asked to be put into bed. She denied pain. Prior to the next session, the psychologist asked that PRN analgesics be given before the session. While she complained of fatigue, Patricia was able to complete the session and go on to work with the OT after her session. Analgesics were then made routine. One way of conceptualizing Patricia's current psychological status, is that as Patricia's dementia progressed her personality changed from one who was an angry perfectionist when confronted with severe disability and pain, to one who was passive and now responded with a "learned-helplessness" style. While compliant with healthcare staff, she could not or would not admit to pain or ask for any PRN medications. She attempted to comfort herself with excessive rest. GMCBT sessions continued to make use of motivational themes first developed 3 years prior. The psychologist was supportive, validating, and socially reinforcing of Patricia's historical values. The psychologist worked along side PT/OT and coached Patricia to put forth effort.

Patricia needed step by step cueing for all behaviors and almost constant reinforcement. Complicating Patricia's depression and cognitive status, was the loss of her brother who had died 4 months ago. Recent radiology studies also indicated new stress fractures and possible spinal stenosis. Patricia had both biological and psychological reasons for pain, but she experienced it all as distressing fatigue. The need for constant encouragement and social reinforcement tapered quickly over the first 4 weeks of two per week sessions, and this trend generalized across other healthcare providers. The psychologist worked 2 months with Patricia, until her discharge back to LTC. As her activities improved so did her depression.

Upon discharge to LTC, Patricia's GDS score was 4/15; her Cornell Depression Scale was 6; her dysfunctional behaviors (GLDS) and GMPI ratings were not significantly different from those prior to her LTAC hospitalization. Her PRADLI scores were, however, lower due to permanent loss of functional capacity, but she was cooperative with all ADLs. She was out of bed 6–9 h a day, attending 1–2 activities a day, but was "Max Assist" for all ADLs except for Eating.

Pharmacological Interventions

Pharmacological interventions for chronic and acute pain include systemic nonopioid analgesics, systemic opioid analgesics, antidepressants, muscle relaxants, *N*-methyl-D-aspartate receptor antagonists, anticonvulsants, local anesthetic drugs, and topical medications (Loeser et al., 2001). The American Geriatrics Society (1998) and the American Medical Director's Association (2003) have both published guidelines for pharmacological pain management.

Diversity Issues

Cultural, ethnic, psychosocial, and individual differences involving personality and psychophysiological style were discussed above regarding the effects these factors have on pain experience and expression (Cipher & Clifford, 2003; Cipher, Fernandez, & Clifford, 2002; Cipher, Clifford, & Schumaker, 2002; Clifford, Cipher, & Roper, 2003; Wickramasekera, 1993). Other diversity issues affecting psychological treatment include religious beliefs and practice (Brennan & Heiser, 2004; Richards & Bergin, 2000). Pain, illness, and disability often trigger existential issues and spiritual questing in elderly patients (Kriseman, 2005; McCord et al., 2004). Families and friends often struggle to make sense of a situation where a loved one suffers with chronic pain and cognitive/physical impairment. Integrating the patient's spirituality, religious beliefs, and practices into a comprehensive treatment plan can be very helpful in improving the patient's ability to cope with pain and disability (Frankl, 1997; Kriseman; Miller, 1999). There is great need for further research in this area as the United States aging population becomes more and more diverse.

References

AGS Panel on Persistent Pain in Older Persons. (2002). The management of persistent pain in older persons. *Journal of the American Geriatrics Society, 50*(Suppl. 6), S205–S224.

Alexopoulos, G. S., Abrams R. C., Young R. C., & Shamoian, C. A. (1988). Cornell scale for depression in dementia. *Biological Psychiatry, 23*(3), 271–284.

Allen-Burge, R., Stevens, A. B., & Burgio, L. D. (1999). Effective behavioral interventions for decreasing dementia-related challenging behavior in nursing homes. *International Journal of Geriatric Psychiatry, 14*, 213–232.

American Geriatric Society Panel on Chronic Pain in Older Persons. (1998). The management of chronic pain in older persons. *Journal of the American Geriatrics Society, 46*, 635–651.

American Medical Director's Association. (2003). *Chronic pain management in the long term care setting clinical practice guideline*. Columbia, MD: American Medical Directors Association. Available online: http://www.amda.com

Bernabei R., Gambassi G., Lapane K., et al., for the SAGE Study Group. (1998). Management of pain in elderly patients with cancer. SAGE Study Group. Systematic assessment of geriatric drug use via epidemiology. *JAMA, 279*, 1877–1882.

Brennan, M., & Heiser, D. (Eds.). (2004). *Spiritual assessment and interventions with older adults*. Binghamton, NY: Haworth Pastoral Press.

Bressler, H. B., Keyes, W. J., Rochon, P. A., & Badley, E. (1999) The prevalence of low back pain in the elderly: A systematic review of the literature. *Spine, 24*(17), 1813–1819.

Briggs Corporation. (1997). *Minimum data set Version 2.0*. Des Moines: Briggs.

Buffum, M. D., Miaskowski, C., Sands, L., & Brod, M. (2001). A pilot study of relationship between discomfort and agitation in patients with dementia. *Geriatric Nursing, 22*(2), 80–85.

Burgio, L. (1997). Direct observation of behavioral disturbances of dementia and their environmental context. *International Psychogeriatrics, 8*(3), 343–346.

Burns, D. D. (1989). *The feeling good handbook*. New York: Plume.

Chibnall, J. T., & Tait, R. C. (2001). Pain assessment in cognitively impaired and unimpaired older adults: A comparison of four scales. *Pain, 92*, 173–186.

Cipher, D. J., & Clifford, P. A. (2003). Treatment outcome varies with coping style in chronic pain management. *The Pain Clinic, 15*(1), 35–44.

Cipher, D. J., & Clifford, P. A. (2004). Dementia, pain, behavioral disturbances, and ADLs: Toward a comprehensive conceptualization of quality of life in long-term care. *International Journal of Geriatric Psychiatry, 19*(8), 741–748.

Cipher, D. J., Clifford, P. A., & Roper, K. D. (2006). Behavioral manifestations of pain in the demented elderly. *Journal of the American Medical Directors Association, 7*, 355–365.

Cipher, D. J., Clifford, P. A., & Roper, K. D. (2007). Efficacy of geropsychological treatment in long-term care. *Clinical Gerontologist, 30*(3), 23–40.

Cipher, D. J., Clifford, P. A., & Schumacker, R. E. (2002). The heterogeneous chronic pain personality: Diverse coping styles among sufferers of chronic pain. *Alternative Therapies in Health and Medicine, 8*(6), 93–102.

Cipher, D. J., Fernandez, E., & Clifford, P. A. (2001). Cost effectiveness and health care utilization in a multidisciplinary pain center: Comparison of three treatment groups. *Journal of Clinical Psychology in Medical Settings, 4*(8), 437–444.

Cipher, D. J., Fernandez, E., & Clifford, P. A. (2002). Coping style influences compliance with multidisciplinary pain management. *Journal of Health Psychology, 7*(5), 665–673.

Clifford, P. A., Cipher, D. J., & Roper, K. D. (2003). Assessing resistance to activities of daily living in long-term care. *Journal of American Medical Directors Association, 4*, 313–319.

Clifford, P. A., Cipher, D. J., & Roper, K. D. (2005a). Assessing dysfunctional behaviors in long-term care. *Journal of American Medical Directors Association, 6*(5), 300–309.

Clifford, P. A., Cipher, D. J., & Roper, K. D. (2005b). The Geriatric Multidimensional Pain and Illness Inventory: A new instrument measuring pain and illness in long-term care. *Clinical Gerontologist, 28*(3), 45–62.

Clifford, P. A., Cipher. D. J., & Schumacker, R. E. (2003). Compliance and stages of change in multidisciplinary pain centers. *The Pain Clinic, 15*(4), 335–367.

Cohen-Mansfield, J. (2002). Relatives' assessment of pain in cognitively impaired nursing home residents. *Journal of Pain Symptom Management, 24*, 562–571.

Cook, A. J. (1998). Cognitive-behavioral pain management for elderly nursing home residents. *Journal of Gerontology B Psychological Sciences and Social Sciences, 53*, P51–P59.

Davis, L. L., Buckwalter, K., & Burgio, L. D. (1997). Measuring problems in dementia: Developing a methodological agenda. *Advanced Nursing Science, 20*(1), 40–55.

Department of Veterans Affairs Geriatrics and Extended Care Strategic Healthcare Group, National Pain Management Coordinating Committee. (2000). *Pain as the 5th vital sign toolkit*. Washington, DC: Veterans Health Administration.

Douzijan, M., Wilson, C., Shultz, M., Berger, J., Tapnio, J., & Blanton, V. (1998). A program to use pain control medication to reduce psychotropic drug use in residents with difficult behavior. *Annals of Long Term Care, 6*, 174–179.

Dworkin, S. F., & Sherman, J. J. (2001). Relying on objective and subjective measures of chronic pain: Guidelines for use and interpretation. In D. C. Turk & R. Melzack, (Eds.), *Handbook of pain assessment* (pp. 619–638). New York: Guilford Press.

Feil, N. (1993). *The validation breakthrough*. Baltimore: Health Professionals Press.

Ferrell, B. A., Ferrell, B. R., & Rivera, L. (1995). Pain in cognitively impaired nursing home patients. *Journal of Pain and Symptom Management*, 10(8), 591–598.

Ferrell, B. A., Stein, W. M., & Beck, J. C. (2000). The Geriatric Pain Measure: Validity, reliability and factor analysis. *Journal of the American Geriatrics Society, 48*, 1669–1673.

Fishbain, D., Goldberg, M., Meagher R., Steele, R., & Rosomoff, H. (1986). Male and female chronic pain patients categorized by DSM-III psychiatric diagnostic criteria. *Pain, 26*, 181–197.

Ford, C. V. (1992). Illness as a lifestyle: The role of somatization in medical practice. *Spine, 22*, S338–S343.

Ford, C. V. (1995). Dimensions of somatization and hypochondriasis. *Neurologic Clinics, 13*, 241–253.

Frankl, V. E. (1997). *Man's search for ultimate meaning*. New York: Plenum Press.

Fries, B. E., Simon, S. E., Morris, J. N., Feldstrom, C., & Bookstein, F. L. (2001). Pain in U.S. nursing homes: Validating a pain scale for the Minimum Data Set. *The Gerontologist, 41*(2), 173–179.

Gagliese, L., & Melzack, R. (1997). Chronic pain in elderly people. *Pain, 70*, 3–14.

Goldberg, R. T. (1994). Childhood abuse, depression, and chronic pain. *Clinical Journal of Pain, 10*, 277–281.

Herr, K., Bjoro, K., & Decker, S. (2006). Tools for assessment of pain in nonverbal older adults with dementia: A state-of-the-science review. *Journal of Pain and Symptom Management, 31*, 170–192.

Hughes, C. P., Berg, L., Danziger, W. L., Coben, L. A., & Martin, R. L. (1982). A new clinical scale for the staging of dementia. *British Journal of Psychiatry, 140*, 566–572.

Jurica, P. J., Leitten, C. L., & Mattis, S. (2001). Dementia Rating Scale - 2 Lutz, FL: Psychological Assessment Resource, Inc.

Kerns, R. D., Turk, D. C., & Rudy, T. E. (1985). The West-Haven Yale Multidimensional Pain Inventory (WHYMPI). *Pain, 23*, 345–356.

Kiernan, R., Mueller, J., Langston, J., & Van Dyke, C. (1987). The Neurobehavioral Cognitive Status Examination: A brief but differentiated approach to cognitive assessment. *Annals of Internal Medicine, 107*, 481–485.

Kriseman, N. L. (2005). *The caring spirit approach to eldercare.* Baltimore: Health Professionals Press.

Loeser, J. D., Butler, S. H., Chapman, C. R., & Turk, D. C. (2001). *Bonica s management of pain.* Philadelphia: Lippincott Williams & Wilkins.

McCord, G, Gilchrist, V. J., Grossman, S. D., King, B. D., McCormick, K. E., Operandi, A. M., et al. (2004). Discussing spirituality with patients: A rational and ethical approach. *Annals of Family Medicine, 2*, 356–361.

Miller, W. R. (1999). *Integrating spirituality into treatment.* Washington DC: American Psychological Association.

Morley, S., Eccleston, C., & Williams, A. (1999). Systematic review and meta-analysis of randomized controlled trials of cognitive behaviour therapy and behaviour therapy for chronic pain in adults, excluding headache. *Pain, 80*, 1–13.

Morrison, R. S., & Siu, A. L. (2000). A comparison of pain and its treatment in advanced dementia and cognitively intact patients with hip fracture. *Journal of Pain Symptom Management, 19*, 240–248.

Opie, J., Doyle, C., & O'Connor, D. W. (2002). Challenging behaviours in nursing home residents with dementia: A randomized controlled trial of multidisciplinary interventions. *International Journal of Geriatric Psychiatry, 17*, 6–13.

Reid, M. C., Otis, J., Barry, L. C., & Kerns, R. D. (2003). Cognitive-behavioral therapy for chronic back pain in older persons: A preliminary study. *Pain Medicine, 4* 223–230.

Reisberg, B., Auer, S. R., Franssen, E., Monteiro, I., & Kenowsky, S. (1999). A rational psychological approach to the treatment of behavioral disturbances and symptomatology in Alzheimer's disease based upon recognition of developmental age. In N. Brunello, S. Z. Langer, & G. Racagni (Eds.), *Mental disorders in the elderly: New therapeutic approaches* (Vol. 13, pp. 102–109). International Academy for Biomedical Drug Research, Basel: Karger.

Reisberg, B., Ferris, S. H., de Leon, M. J., & Crook, T. (1982). The Global Deterioration Scale for assessment of primary degenerative dementia. *American Journal of Psychiatry, 139*, 1136–1139.

Richards, P. S., & Bergin, A. E. (2000). *Handbook of psychotherapy and religious diversity.* Washington DC: American Psychological Association.

Roper, K. D., Reid, T., & Clifford, P. A. (2007). The Mind Body Wellness Geriatric Psychological assessment and treatment handbook. Unpublished manuscript available at http://www.mindbodywellnesspc.com.

Rovner, B. W., German, P. S., Broadhead, J., Morriss, R. K., Brant, L. J., Blaustein, J., & Folstein, M. F. (1990). The prevalence and management of dementia and other psychiatric disorders in nursing homes. *International Psychogeriatrics 2*, 13–24.

Scherder, J. A., & Bouma, A. (1998). Visual analogue scales for pain assessment in Alzheimer's disease. *Gerontology, 46*, 47–53.

Scherder, E., Oosterman, J., Swaab, D, Herr, K., Ooms, M., Ribbe, M., et al. (2005). Recent developments in pain in dementia. *British Medical Journal*, *330*, 461–463.

Scherder, E. J., Sergeant, J. A., & Swaab, D. F. (2003). Pain Processing in dementia and its relation to neuropathology. *Lancet Neurology*, November *2*(11), 677–686.

Schofferman, J., Anderson, D., Hines, R., Smith. G., & Keane, G. (1993). Childhood psychological trauma and chronic refractory low-back pain. *Clinical Journal of Pain*, *9*, 160–165.

Seligman, E. P. (1995). The effectiveness of psychotherapy: The consumer reports study. *American Psychologist*, *50*(12), 965–974.

Simmons, S. F., Ferrell, B. A., & Schnelle, J. F. (2002). Effects of a controlled exercise trial on pain in nursing home residents. *Clinical Journal of Pain*, *18*, 380–385.

Simons, W., & Malabar, R. (1995). Assessing pain in elderly patients who cannot respond verbally. *Journal of Advanced Nursing*, *22*, 663–669.

Snow, A. L., Graham, D., Molinari, V., Orengo, C. A., Doody, R., Norris, M., et al. (2005). Factors affecting deficit awareness in persons with dementia. *Dementia and Geriatric Cognitive Disorders*, *20*, 133–139.

Snow, A. L., Kunik, M., Molinari, V., Orengo, C. A., Doody, R., Graham, D., et al. (2005). Accuracy of self-reported depression in persons with dementia. *Journal of the American Geriatrics Society*, *53*(3), 389–396.

Snow A. L., O'Malley, K. J., Cody, M., Kunik, M. E., Ashton, C. M., Beck, C., et al. (2004). A conceptual model of pain assessment for noncommunicative persons with dementia. *The Gerontologist*, *44*(6), 807–817.

Sorkin, B. A., Rudy, T. E., Hanlon, R. B., Turk, D. C., & Stieg, R. L. (1990). Chronic pain in old and young patient: Differences appear less important than similarities. *Journal of Gerontology*, *45*(3), 64–68.

Teno, J., Bird, C., & Mor, V. (2001). *The prevalence and treatment of pain in US nursing homes*. Providence, Rhode Island: Center for Gerontology and Healthcare Research, Brown University.

Teno, J. M., Weitzen, S., Wetle, T., & Mor, V. (2001). Persistent pain in nursing home residents. *JAMA*, *285*, 2081.

Turk, D. C., Meichenbaum, D., & Genest, M. (1983). *Pain and behavioral medicine: A cognitive-behavioral perspective*. New York: Guilford Press.

Turk, D. C., & Melzack, R. (Eds.). (2001). *Handbook of pain assessment*. New York: Guilford Press.

Weiner, D., Peterson, B., & Keefe, F. (1999). Chronic pain-associated behaviors in the nursing home: Resident versus caregiver perceptions. *Pain*, *80*, 577–588.

Weiner, D., Peterson, B. L., Ladd, K., McConnell, E., Keefe, F. J. (1999). Pain in nursing home residents: An exploration of prevalence, staff perspectives, and practical aspects of measurement. *Clinical Journal of Pain*, *15*, 92–101.

Wells, K. B., Golding, J. M., & Burnam, M. A. (1988). Psychiatric disorder In a sample of the general population with and without chronic medical conditions. *American Journal of Psychiatry*, *145*, 976–981.

Werner, P., Cohen-Mansfield, J., Watson, V., & Pasis, S. (1998). Pain in participants of adult day care centers: Assessment by different raters. *Journal of Pain Symptom Management*, *15*, 8–17.

White, K. E., & Cummings, J. E. (1999). Neuropsychiatric aspects of Alzheimer's disease and other dementing illnesses. In S. C. Yudofsky & R. E. Hale (Eds.), *The American psychiatric press textbook of neuropsychiatry* (pp. 823–854). Washington, DC: American Psychiatric Press.

Wickramasekera, I. (1993). Assessment and treatment of somatization disorders: The high risk model of threat perception. In J. W. Rhue, S. J. Lynn, & I. Kirsch (Eds.), *Handbook of clinical hypnosis* (pp. 587–621). Washington, DC: American Psychological Association.

Won, A., Lapane, K., Gambassi, G., Bernabei, R., Mor, V., & Lipsity, L. A. (1999). Correlates and management of nonmalignant pain in the nursing home. *Journal of the American Geriatrics Society*, *47*, 936–942.

Wynne, C. F., Ling, S. M., & Remsburg, R. (2000). Comparison of pain assessment instruments in cognitively intact and cognitively impaired nursing home residents. *Geriatric Nursing, 21*, 20–23.

Yesavage, J., Brink, T., & Rose, T. (1983). Development and validation of a geriatric depression scale: A preliminary report. *Journal of Psychiatric Research*, *17*, 37–49.

7 Reducing Psychosocial Distress in Family Caregivers

Ann M. Steffen, Judith R. Gant, and Dolores Gallagher-Thompson

Background

Families remain the most common source of assistance for older adults with physical and/or cognitive limitations (National Alliance for Caregiving and the American Association of Retired Persons, 1997; Schulz & Martire, 2004). A national phone survey found that nearly one in four US households included at least one self-identified caregiver who gave unpaid assistance to an impaired or physically frail relative over the age of 50 (NAC/AARP, 1997). The term "caregiving" encompasses a wide range of activities, from management of medications and appointments to bathing, dressing, and toileting the individual. Typically, tasks change over time as the conditions in question either become more stable, deteriorate, or in some instances, improve. Family and friends often experience considerable stress when trying to provide extensive caregiving in addition to the other demands of their everyday lives – they are often referred to as the "sandwich generation" since many are caught in the middle between employment and the responsibilities of caring for their husbands and children (and perhaps grand-children), as well as the responsibilities of caring for an impaired parent or parent-in-law. Spouses also experience considerable distress, typically focusing on their sense of loss as changes occur in the marital relationship (Davidson, 2006). Although it is true that caregiving can occur in the context of *any* significant physical and/or emotional disorder, most of the intervention research has been conducted with relatives of older adults with Alzheimer's disease or another form of dementia. In addition, studies have found that these dementia family caregivers are generally more distressed than caregivers of physically impaired elders (Ory, Yee, Tennstedt, & Schulz, 2000). Given space limitations, this chapter will focus on interventions with dementia family caregivers.

Overview of Caregiver Distress: Why Do We Need Interventions for Caregivers?

Multiple studies over the past 20 years have reported significant negative psychosocial impacts from long-term dementia caregiving. These include high rates of clinical depression and/or depressive symptoms (Gallagher, Rose, Rivera, Lovett, & Thompson, 1989; Schulz & Martire, 2004; Schulz, O'Brien, Bookwala, & Fleissner, 1995), high rates of other negative emotions such as anger, frustration, burden, and fear (Ory et al., 2000), and other indices of distress, such as family conflict over caregiving, significant emotional strain, financial hardship, and reduced time for leisure pursuits (Ory et al.; Pinquart & Sorensen, 2003). In addition, studies have reported negative effects of caregiving on the physical health of the primary caregiver. For example, Vitaliano et al. (2002) and Vitaliano, Young, and Zhang (2004) found that male caregivers had a greater prevalence of heart disease than noncaregiving men of the same age. Greater risk for high blood pressure and the metabolic syndrome were also reported. Others have found complaints of bodily aches and pains and greater prevalence of diabetes, allergies, and use of nonprescription pain medication to be common in dementia caregivers (Coon, et al., 2004; Pinquart & Sorenson, 2003).

Numerous interventions have been developed to address these problems, and new psychosocial interventions are continuing to be tested (Steffen, 2006). In the past decade, there have been multiple reviews published as to their efficacy (Bourgeois, Schulz, & Burgio, 1996; Brodaty, Green, & Koschera, 2003; Schulz, Martire, & Klinger, 2005; Sorensen, Pinquart, & Duberstein, 2002). It is difficult to compare results across reviews, however, because different criteria have been used at different times to establish treatments as "evidence-based." A recent review by Gallagher-Thompson and Coon (2007) that used strict criteria for identifying interventions as evidence-based found that three categories could be so considered at the present time; psychoeducational skill-building programs (e.g., Coon, Thompson, Steffen, Sorocco, & Gallagher-Thompson, 2003); psychotherapy (e.g., Gallagher-Thompson & Steffen, 1994); and multicomponent interventions (those using several distinct types of treatment such as support groups plus family meetings and case management; e.g., Mittleman, Roth, Coon, & Haley, 2004; Zarit & Zarit, 2007). Psychoeducational interventions, which derive from behavioral and cognitive theories and therapies (cf. Beck, Rush, Shaw, & Emery, 1979; D'Zurilla, 1986; Lewinsohn, 1974; Lewinsohn, Munoz, Youngren, & Zeiss, 1986) are often preferred by family members. The term is used to distinguish this kind of intervention from both traditional psychotherapy and from community-based support or educational programs. Most family caregivers, including the majority of those with diagnosable mental health conditions, do not identify themselves as candidates for psychotherapy or counseling. To appeal to these individuals, treatments are typically not labeled as individual or group psychotherapy; instead, such programs are often presented to caregivers with titles such as "skill training sessions" or "stress management classes" (Burgio, Hardin, Sinnott, Janosky, & Hohman, 1995).

That said, it would also be well to add a note of caution here: most clinical researchers in this field recognize that there is no "one size fits all" intervention that will be successful. What type of intervention to use in a clinical setting, with a given caregiver, depends on the presenting problem, the care-recipient's stage of dementia, caregivers' coping strategies and general psychological and physical health, and available resources in the community. On a national level, attempts have been made to delineate different types of services that may be useful to caregivers at different points in the process – Coon, Ory, and Schulz (2004) describe five basic services included in the National Family Caregiver Support Program, a recently enacted piece of legislation that provides funding to the states for these services. They are information and referral, assistance with access to services, caregiver support, respite care, and supplemental services as needed. However, these services are not theory-driven; they are not grounded in the psychological or psychiatric literature, nor do they try to specify mechanisms of change.

In addition, most of the published reviews that have focused on developing an evidence base for X, Y, or Z treatment have failed to identify any *one consistently effective method* to reduce psychological distress, or improve well-being, in dementia family caregivers. It appears that types of distress and strain change considerably over time during the course of one's caregiving "career" and therefore the kinds of interventions that would successfully treat them also must change over time (Gallagher-Thompson & Steffen, 1994; Gottlieb & Wolfe, 2002; Zarit & Zarit, 2007). For example, family therapy may be very helpful for certain problems, such as resolving interpersonal conflicts about what the diagnosis means (in early stages) or about placement decision-making (in later stages) (Arguelles, Klausner, Arguelles, & Coon, 2004). Home-based behavioral and skill-building interventions may be effective for caregivers who are concerned about such middle stage issues as safety in the home and who need help improving the physical and social components of the home environment (Gitlin & Gwyther, 2004). Finally, few studies have included the care-recipient in the design; most focus solely on the primary caregiver, yet it is reasonable to assume that working with the dyad could significantly improve quality of life for both individuals. A notable exception to this can be found in the work of Linda Teri and colleagues (Teri, Logsdon, Uomoto, & McCurry, 1997) in which caregivers were taught to decrease depression in the care-recipient by developing a program of shared pleasant events. A later study by this group showed the positive effects of regular mild exercise to improve mood of both caregiver and care-recipient (Teri et al., 2003). In addition, a later chapter in this volume (Fisher, Drossel, Ferguson, Cherup, & Sylvester, 2007) addresses this issue in depth. Readers interested in the application of behavioral and cognitive interventions to persons with dementia who reside in nursing homes (where paid, not family, caregivers are typically involved) are referred to a recent book on issues in long-term care consultation and intervention (Hyer & Intrieri, 2006). For the purposes of this chapter, we will focus on description of behavioral and cognitive psychoeducational approaches, where the evidence base is strong for their application to community-dwelling family caregivers.

Diversity Issues in Intervention Research

Ethnic and cultural differences are also extremely important in the choice of intervention. There are two large-scale multisite studies called the Resources for Enhancing Alzheimer's Caregivers Health (REACH) projects that can be used to guide the choice of effective interventions. These studies were conducted with significant numbers of Black/African American and Hispanic/Latino caregivers (in addition to Caucasian/Anglo caregivers) to evaluate what kind of intervention program(s) would be effective with these groups (see Schulz et al., 2003, for overall description of the first REACH project). Several publications from these studies may be of interest to the reader; for example, in the first REACH study, two sites enrolled Hispanic/Latino caregivers and four sites enrolled African Americans – in addition to Caucasians. The Palo Alto CA site enrolled primarily Mexican Americans; at the Miami FL site, Cuban Americans were the focus. Each site found that the interventions they chose to use (psychoeducational skill building at Palo Alto and family therapy plus technological support at Miami) were very effective with the non-Caucasians to reduce depression and increase use of positive coping strategies (Eisdorfer et al., 2003; Gallagher-Thompson, Coon, Solano, Ambler, Rabinowitz, & Thompson, 2003). At the Alabama site, a psychoeducational program that focused on teaching skills for managing problematic behaviors of the care-recipient was successful with African Americans (Burgio, Stevens, Guy, Roth & Haley, 2003), and at the Philadelphia site, an occupational-therapy-based in-home intervention program that taught home safety and environmental modification was also successful with African Americans (Gitlin et al., 2003). Note that these programs were "tailored" to be ethnically and culturally appropriate (Gallagher-Thompson et al., 2003) and at the sites serving Hispanic/Latino caregivers, interventions were conducted in Spanish as well as English. In the second REACH study, the most successful interventions from the first project were used to create a single, multicomponent intervention that was done in the same way across sites. It included teaching skills for stress management and management of problematic care-recipient behaviors, modifying negative thoughts about caregiving, and increasing everyday pleasant events. Over 600 new ethnically diverse caregivers participated; results indicated that, compared to the control condition, those in the active intervention were significantly more improved in several quality of life indicators. This was true in both the Hispanic/Latino and Caucasian groups, regardless of relationship. In the African Americans, spouse caregivers improved more than daughters or other relatives (REACH II investigators, 2006).

Taken together, results of the second REACH project and the Gallagher-Thompson and Coon (2007) review strongly support the conclusion that both psychoeducational skill-building interventions (such as those focused on behavior management, depression management, etc.,) and some specific multicomponent interventions that include several distinct approaches can currently be considered evidence based. This evidence base extends beyond Caucasian caregivers (to Hispanic/Latinos and Black/African Americans), although empirical studies with other ethnic groups (e.g., Asian groups such as Chinese, Filipino, Japanese, and Asian Indian caregivers) remain to be conducted. Individuals interested in clinically oriented information about working with diverse family caregivers are referred to Yeo and Gallagher-Thompson (2006).

Assessment Issues and Recommendations

Clinicians who work with family caregivers of cognitively impaired older adults would do well to conduct an assessment of mental health symptoms, such as depression and anxiety levels, before initiating treatment. Such assessments can be done again, during treatment, to measure change. Scales such as the Center for Epidemiological Studies – Depression self-report measure of symptoms (Radloff, 1977) have been used in much of the clinical research discussed here, as well as in practice. It has clear cutoff scores to help determine if depression is in a clinical range, which can be helpful to indicate if a referral for more intensive psychiatric or psychological treatment is needed. In addition, the Revised Memory and Behavior Problem Checklist (Teri et al., 1992) asks caregivers to report whether the care-recipient displayed a number of specific problem behaviors over the past week (i.e., memory related behaviors such as repeating questions, or disruptive behaviors such as yelling, and depressive behaviors, such as being tearful). If the caregiver endorses the item, then he/she is asked to rate level of distress or "bother" it caused on a Likert scale of 0, not at all, to 4, extremely. Some studies have also found that self-efficacy (at baseline) is an important moderator of treatment effects (Rabinowitz et al., 2006). It can be assessed readily with the Revised Caregiver Self-Efficacy Scale (Steffen, McKibbin, Zeiss, Gallagher-Thompson, & Bandura, 1997) which asks caregivers to rate their confidence that they can ask for respite, respond effectively to problem situations with the care-recipient, and control upsetting thoughts about caregiving. Changes in self-efficacy may also serve to mediate change in outcome measures and treatment responsiveness; this remains to be explored. Other measures to consider include: positive aspects of caregiving (Tarlow et al., 2004); religious coping (which may be particularly relevant for ethnically diverse caregivers; Pargament, Koenig, & Perez, 2000), and overall quality of life. The interested reader is referred to the review chapter by Gottlieb, Thompson, and Bourgeois (2004) for discussion of these and other assessment tools.

We would like to recommend that in addition to assessing the caregiver, it would be wise to also evaluate cognitive and behavioral functioning in the care-recipient. Many times, however, professionals do not have direct access to the care-recipient, and will need to rely on caregivers' reports to understand care-related challenges and demands. The Index of Activities of Daily Living Scale (Katz, Ford, Moskowitz, Jackson, & Jaffe, 1963) provides information concerning care-recipients' levels of functional impairments, with each item reflecting a type of more severe disability common in late-stage dementia (e.g., inability to bathe or toilet oneself). There is also a measure of the Instrumental Activities of Daily Living (Lawton & Brody, 1969) tapping into such domains as medication management, ability to remember and keep appointments, and to find one's way outside the home. It is useful with early stage patients. As noted above, the checklist inquiring about memory and behavior problems can also be helpful to determine the level of "objective" burden of the caregiver (Teri et al., 1992).

Besides the use of these kinds of psychometric scales or questionnaires, behavioral and cognitive methods emphasize the role of specificity in identifying goals for intervention. With that in mind, we also suggest use of an idiographic assessment tool that allows for the measurement of problem areas specific to individual caregivers. For example, many of our studies have used some variant of a target complaints interview where caregivers are asked "As you think about your caregiving situation, what are the three things that are most stressful for you to deal with?" Then, for each of the three situations described, the caregiver is asked to rate the degree to which the situation was accompanied by negative feelings (e.g., upset, sadness, frustration, or irritation, on a Likert scale of 1, not at all, to 5, extremely). These ratings can then be averaged across the three situations to yield a summary score and at the end of the intervention program, ratings can be obtained again for those specific problem areas. Change in each domain, or in the overall summary score, indicates that the caregivers' particular concerns were addressed.

Case Examples

In the two descriptions that follow, we describe a combination of behavioral (behavioral activation, behavioral management, relaxation training) and cognitive (challenging negative thinking, re-appraising the situation, and cognitive restructuring) interventions.

Case Study #1: Brendan

Here we describe assessment and skills training (12 sessions, taking about 20 weeks) with a male spousal caregiver. Brendan, age 77, was taking care of his wife, Alice, 76, who was diagnosed with Alzheimer's disease. He characterized her general health as poor, including impaired eyesight and unsteady walking. Brendan had the academic equivalent of an associate degree. As a retiree, he reported financial difficulty paying for the basics, including groceries, medicines, and medical care. Their married daughter Anne worked mornings and was reported to be quite helpful with her mother's care. Brendan identified bathing as the one functional domain in which Alice required significant assistance. He indicated that it was "somewhat difficult" for Alice to understand simple instructions and to find her way around the house; it was "quite a bit difficult" for her to remember recent events and speak sentences, and "very difficult" for her to know what day of the week it is and to remember her home address. On the Revised Memory and Behavior Problem Checklist, Brendan's average upset score was 2.17, meaning that he was "moderately" or more upset with the ten disruptive behaviors Alice had shown in the past week. On the Target Complaint interview, Brendan's complaint of "her slowness" had the highest average score across the four emotions (upset, sadness, frustration, and irritation); he endorsed the word "moderately" for each of the emotions.

The first three sessions focused on behavioral activation as a preventive and intervention skill for managing depressed mood in himself and in Alice. Efforts were made to link daily mood ratings to a daily pleasant event log; this was done with between-session assignments of increasing favorite activities and monitoring daily mood for himself and Alice. Secondary skills included simplifying communication with his wife (including "going along with her view of what's happening" and "avoid trying to argue or convince"), and keeping track of specific caregiving stressors with a daily monitoring form. Brendan made the connection that some specific activities (going to church, having lunch with church friends, and a visit from their daughter) were especially enjoyable and linked to higher mood ratings on those days. Time in session was also devoted to identifying and modifying thoughts that held him back from increasing pleasant events for both himself and Alice. In Sessions 4–7, skills for managing difficult behaviors were taught and reinforced, including identifying patterns of antecedents and consequences. For example, Alice's repeated questioning was discovered to be most common in the late morning and late afternoon, times when he realized that she was probably bored and he was attempting to take care of household responsibilities. To reduce the frustration he felt during these incidents, he began to plan specific times to begin a new joint activity, including sub-tasks related to household chores. He realized that it would also be important for him to plan out more activities (like listening to gospel music) for Alice to do by herself at times when he was occupied with his own chores and hobbies.

Sessions 8–10 focused on relaxation training, including development of the skills to recognize physical signs of frustration ("danger signals") and use of controlled breathing techniques to prevent or reduce strong feelings of anger and frustration. These were especially relevant for Brendan because his treatment goals included "dealing with anger." He realized that his danger signals were feelings of impatience, his thought "why can't she remember?," his mind racing, and sighing. He liked the suggested idea of visualizing a huge stop sign, saying "STOP, calm down and do controlled breathing." Between Sessions 8 and 9, Brendan completed six practice sessions using the controlled breathing exercise, and showed improved tension scores after four of the six practices. A final didactic session included home safety precautions and financial and legal issues, in addition to program summary and development of a maintenance plan. A key theme identified by Brendan was his need to give himself credit for "trial and error" approaches to problem solving, rather than expecting a single solution to always work. During his two later maintenance sessions, each scheduled 1 month apart, he identified the focus on pleasant events, and the controlled breathing relaxation exercises, as most helpful. He said that he had improved self-control and decreased anger, so he was pleased that he had moved closer to meeting the goals he had set at the beginning of the program.

Brendan's postintervention and 10-month outcome scores were compared with his baseline scores. Despite the continued progression of his wife's dementia, on the Revised Memory and Behavior Checklist Brendan's scores on all behavior problems showed improvement from "moderate" (upset or annoyance) to "a little," which we consider to be a clinically significant improvement. Brendan demonstrated both substantial decreases in his levels of distress, and uniform maintenance of these gains. While Brendan's positive affect decreased somewhat from preintervention to 10 week postintervention, his negative affect improved substantially. This posttreatment negative affect reduction is consistent with Brendan's decreased upset and annoyance over Alice's behavior problems. Brendan demonstrated steady improvement in his self-efficacy outcome scores, with a 37-point improvement on the managing behavioral problems subscale. In the preintervention "Target Complaints Interview," Brendan had chosen "her slowness" as the most stressful part of caregiving at baseline. The other two target complaints were "making meals" and "taking her out." By the time of the posttest, he no longer considered her slowness or making meals a problem, but taking her out was still stressful; on the other hand, when asked how upset, sad, frustrated, or irritated these three stressors made him feel, he said "not at all." At the 10-month assessment, making meals was the only one of the three stressors that he still considered to be a problem. For all three of the stressors, he endorsed being "a little" (2.0) distressed (upset, sad, frustrated, or irritated) on three of four of the feelings.

Case Study #2: Esther

Esther is a 52-year-old Latino woman who graduated from high school, is the mother of three teen-aged girls, lives with her husband, works part-time outside the home in an office setting, and takes care of her 75-year-old mother, who also lives with them and has been demented for about 8 years. Her mother Rosa is a Type II diabetic and not reliable about checking blood sugar or taking medication. Esther was extremely stressed at the initial evaluation: her self-reported depression on the CES-D was 20, which is at the low end of the clinical range, and she reported multiple problems taking care of her mother and dealing with ongoing stress at work and at home. She rated her overall stress as "extreme" and was very motivated to participate in the small-group intervention program we were conducting. It was offered in a community setting near her home, in the evenings so that she could attend after work, and was conducted in Spanish. Although she spoke English well, she said she preferred to be in the Spanish-speaking group, because otherwise it would be difficult to share personal information and feelings. After signing informed consent, Esther enrolled in a 12-week "Coping with Caregiving" skills training research program. An array of skills was taught, beginning with cognitive restructuring. Members of the group were asked to note thoughts that were related to stressful situations during the week and bring that information back to share next time. Esther reported many negative thoughts having to do with her inadequacy as a caregiver; these were associated with strong feelings of guilt, that she was not doing enough for her mother. She was very "hard on herself" in that she expected to come home from work and not be irritable with her mother, even though Rosa was very dependent on her and seemed to resent time she spent with her husband and daughters (who also demanded her attention). In the group she was encouraged to (gently) challenge these thoughts:

she was asked questions such as: "how much is enough?" and "what else could you be doing, that you are not doing, that would help your mother?"

In the first month this was the focus of the Coping Class: Esther learned not only to be more realistic about what she could and could not do for her mother, but she also learned to help others challenge their unrealistic thoughts as well. For example, Josefina was very tearful and depressed over how quickly her husband's dementia seemed to be progressing. She was longing for things to be "the way they used to be" with him. Esther helped her to see that, although those days were gone, Josefina could still try to make a good life with her husband as he was now. Josefina was comforted by that point of view: all was not lost. There were still things her husband could do, and things that they could do together. This in turn helped to reinforce Esther's understanding that changing negative thinking made a real difference in her life; that skill she continued to use on a regular basis.

The group then focused on the skill of learning to manage common problem behaviors in their care-receivers: either by changing antecedents, or by changing their reactions and responses to the problem behaviors. Esther, for example, reacted very strongly when her mother became upset with her and yelled at her. Esther often cried herself to sleep, and did not know what to do to handle the situation. The group suggested that she develop a plan to do shared pleasant activities with her mother. Normally this coping class emphasizes developing pleasant events for oneself, but since Esther was so busy, and had so little time to herself, we shifted the goal to include positive time with Rosa. The hypothesis was that Rosa was becoming upset because she was feeling ignored by Esther, so if Esther would take her mother's hand, for example, and the two of them go for a short walk together, perhaps that would make a difference. It did, in fact, improve the overall quality of life in the home, and enabled Esther to realize that many other things could also bring pleasure to her mother – such as looking at old photo albums together, or listening to her mother sing church hymns in Spanish (which she loved to do).

In an effort to get the teen-aged daughters involved, the group suggested this strategy: arrange for them to spend some time with their grandmother and ask her to tell them stories of her early life in Mexico. This suggestion did not always work since the girls were busy with their own lives, but they began to do it about once a week, more or less. The importance of "*la familia*" (a strong cultural value in Latino groups) reinforced their willingness to participate.

By the eighth session, the coping class focused on communication skills: with non-Latinos, we refer to this as "assertiveness training" but focus groups conducted before the intervention study began revealed that this was not the correct terminology to use with this cultural group. Instead, we called this "learning to communicate more effectively" and that was well received. The focus was, first, on communicating with the care-recipient's primary doctor, and then on communicating with others (family, friends) to get respite help. To address the first aim, caregivers were given checklists of questions to bring up, and were encouraged to fill out specifics before the next doctor's appointment. This was done as an in-class exercise. Esther reported that this was very helpful, since when she took her mother to the doctor, she usually was harried, and could not think clearly about key issues to discuss. The class then used role playing to help participants practice how to ask for "respite" help: time off from their regular caregiving to take care of other matters. Esther had a particularly difficult time with this; she was quite timid in the role play exercises, and did not ask clearly for what she needed (as was the case with most of the women in the group). Over the next 2 weeks, she became more comfortable by practicing with other class members, but Esther was still quite anxious about the "homework" assignment of actually asking someone to take care of her mother for a couple of hours so that she could get to a much-needed dental appointment. The first time she asked (a cousin who lived nearby) she was turned down; then she talked it over with her eldest daughter, and quite unexpectedly, the daughter agreed. She said she could skip some particular after-school activity that she liked to do once in awhile so that her mother could get to the dentist. This was an important experience because it taught Esther that if she did not ask, she would not get the help and support she needed; on the other hand, if she did ask, clearly and constructively, at least there was a chance that the request would be granted.

The final weeks focused on future transitions: how to recognize when the care-recipient may need more care than the family can provide. Also included was a discussion of end-of-life issues: preparing for the inevitable final transition. Most of the caregivers in this 6-person class were grateful that these issues were brought up; they said that at home, and in their culture, these kinds of things generally were not talked about openly. To facilitate discussion with other family members, class leaders provided each member with an "Advance Directive" kit in Spanish or English, and one of the homework assignments was to begin to discuss this concrete topic with another family member. Most caregivers talked with their priest or clergyperson, rather than family; once that individual approved (so to speak) of the content, then they were more able to approach family directly. Esther found that she was "too emotional" to face this, but she did encourage others in the group to do so. She said that she would pray and ask for help.

At the conclusion of the coping class, Esther's depression score on the CES-D was below 10, indicating that she was no longer having significant symptoms. More importantly, her overall stress rating was much lower (dropped by 10 points from the "pre" evaluation) and she reported being less "bothered" by her mother's behavior since she felt more confident that she could respond appropriately (most of the time). At the 6 month follow-up, Esther had maintained her gains in terms of low psychological distress, and said that she was talking with her priest about some of the end-of-life issues that were raised in the class. She was very appreciative for the opportunity to be in the project, and in fact referred several others. Overall, it appears that the "coping class" psychoeducational skill-building approach, when delivered in a culturally competent manner, and in a small-group context, was very effective for Esther.

Directions for Future Research

As noted above, this chapter focused on the use of evidence-based interventions with dementia family caregivers. Much more research is needed on interventions helpful to caregivers of adults with other chronic illnesses, such as cancer, diabetes, stroke, heart disease, and the like. Reviews by Given, Given, Kozachik, and Rawl (2003), Biegel and Schulz (1999), and Toseland et al. (2001) can be helpful in that regard. Indeed, treatment studies with other populations of caregivers appear to have fallen behind in terms of both quantity of investigations, as well as research sophistication.

Research has demonstrated a subject selection bias in caregiver studies conducted in university and medical settings (Dura & Kiecolt-Glaser, 1990). Family caregivers willing to go to a university or medical setting for assessments were less depressed and less emotionally distressed than those who were willing to be interviewed only in their home. One could extend these findings and expect that treating only those able and willing to attend weekly sessions in the community restricts the generalizability of study results. This is particularly true of group approaches in which sessions are not easily rescheduled. At this point, most skills-based interventions depend upon caregivers' ability and interest in attending consecutive weekly sessions in the community. This is an unfortunate trend because there are a number of circumstances in which this may be difficult, and caregivers are consequently prevented from receiving needed assistance. For example, it is unlikely that such interventions can routinely be offered in rural communities because of the distance and inconvenience for caregivers.

Changes in technology have opened up new possibilities for administering caregiver interventions to those whose caregiving responsibilities might hinder their participation (Buckwalter, Davis, Wakefield, Kienzle, & Murray, 2002; Steffen, Mahoney, & Kelly, 2003). The telephone has long been used by clinicians providing referrals and crisis intervention, and has more recently been used as the primary mechanism for treatment delivery (Haas, Benedict, & Kobos, 1996; Mermelstein & Holland, 1991). Likewise, investigators have explored the use of videotapes as part of cognitive-behavioral interventions. Videotapes have been used for psychoeducational purposes in primary care (Robinson, Katon, Von Korff, & Bush, 1997), and to provide education and modeling for individuals with public speaking anxiety (Ayres, Ayres, Baker, & Colby, 1993). A small preliminary study conducted by our research group suggests that a cognitive-behavioral treatment for distressed caregivers can be effective when delivered through videotapes, workbook, and telephone sessions with intervention staff (Steffen, 2000). We are currently in the process of testing a revised version of this video intervention with a larger group of dementia family caregivers, across nine states, with the goal of reducing depression levels in caregivers (Steffen & Mangum, 2003). Chang (1999) has also demonstrated the effectiveness of a videotape and telephone intervention to train caregivers in behavioral strategies. Over the next decade of caregiver intervention research, we expect to see increased attention to how technology can be used to deliver effective behavioral and cognitive treatments.

Two other issues deserve more attention in future research: first, it is not clear that any of the evidence-based interventions we have presented have any positive, discernible impact on the physical health and well-being of the caregiver. Yet caregivers can, themselves, have chronic health conditions (such as diabetes, heart disease, and hypertension) that may be exacerbated by severe stress. As Schulz and Beach (1999) reported, spousal caregivers who said they were significantly stressed were more likely to die during the course of a 4-year follow - up than caregivers with low stress. While these findings need to be replicated, they point to the very heavy negative toll that may occur in some people who are not able to manage stress well.

The second issue concerns the paucity of research on ethnic and culturally diverse caregivers: although the REACH studies have laid a good foundation, most of the ever-increasing ethnic minority groups in this country have not participated in caregiver intervention research projects. Yet caregiving expectations, coping strategies, and perceived stress are all filtered through the lens of culture (Aranda & Knight, 1997), and culturally competent research is needed to address the needs of the varied ethnic groups in our country. This is a two-sided issue: investigators and clinicians need to develop culturally relevant interventions, and then conditions need to be fostered so that diverse caregivers will enroll in the projects, or use the services provided. There are many factors that may account for the inherent difficulties in designing and executing research and clinical projects with this kind of focus, such as linguistic challenges and low literacy levels (even in the language of one's country of origin). In addition, there may be specific cultural values and beliefs that make it difficult for caregivers to seek help. In many cultural groups, signs and symptoms of dementia are regarded as "normal aging" or as part of severe mental illness, leading to stigmatization, and reluctance to seek help outside the family (Hinton, Fox, & Levkoff, 1999). Allery et al. (2004), with sponsorship from the Alzheimer's Association, have issued a "call to action" on this topic. Clearly, given the increasing longevity of the US population, and its increasing diversity, this is an issue that we can no longer ignore in our research applications and in our clinical practice.

Acknowledgments. Preparation of this chapter was supported in part by grant # MH061956-03 from the National Institute of Mental Health and grant # IRG-03-5987 from the national office of the Alzheimer's Association to Dr. Steffen; funding from the Missouri Alzheimer's and Related Disorders Program to Dr. Gant, and grants # AG 18784 and AG 13289 from the National Institute on Aging, and grant #IIRG-04-1109 from the national office of the Alzheimer's Association to Dr. Gallagher-Thompson.

References

Allery, A. J., Aranda, M. P., Dilworth-Anderson, P., Guerrero, M., Haan, M., Hendrie, H., et al. (2004). Alzheimer's disease and communities of color. In K. E. Whitfield (Ed.), *Closing the gap: Improving the health of minority elders in the new millennium* (pp. 81–86). Washington, DC: Gerontological Society of America.

Aranda, M., & Knight, B. (1997). The influence of ethnicity and culture on the caregiving stress and coping process: A sociocultural review and analysis. *The Gerontologist, 37*, 342–354.

Arguelles, S., Klausner, E., Arguelles, T., & Coon, D. W. (2004). Family interventions to address the needs of the caregiving system. In D. W. Coon, D. Gallagher-Thompson, & L. W. Thompson (Eds.), *Innovative interventions to reduce dementia caregiver distress: A clinical guide* (pp. 99–118). New York: Springer.

Ayres, J., Ayres, F., Baker, A., & Colby, N. (1993). Two empirical tests of a videotape designed to reduce public speaking anxiety. *Journal of Applied Communication Research, 21*, 132–147.

Beck, A. T., Rush, A. J., Shaw, B., & Emery, G. (1979). *Cognitive therapy of depression.* New York: Guilford Press.

Biegel, D. E., & Schulz, R. (1999). Caregiving and caregiver interventions in aging and mental illness. *Family Relations: Interdisciplinary Journal of Applied Family Studies, 48*, 345–354.

Bourgeois, M. S., Schulz, R., & Burgio, L. (1996). Interventions for caregivers of patients with Alzheimer's disease: A review and analysis of content, process, and outcomes. *International Journal of Aging and Human Development, 43*, 35–92.

Brodaty, H., Green, A., & Koschera, A. (2003). Meta-analysis of psychosocial interventions for caregivers of people with dementia. *Journal of the American Geriatrics Society, 51*, 657–664.

Buckwalter, K. C., Davis, L. L., Wakefield, B. J., Kienzle, M. G., & Murray, M. A. (2002). Telehealth for elders and their caregivers in rural communities. *Family and Community Health, 25*, 31–40.

Burgio, L. D., Hardin, M., Sinnott, J., Janosky, J., & Hohman, M. J. (1995). Acceptability of behavioral treatments and pharmacotherapy for behaviorally disturbed older adults: Ratings of caregivers and relatives. *Journal of Clinical Geropsychology, 1*, 19–32.

Burgio, L., Stevens, A., Guy, D., Roth, D. L., & Haley, W. E. (2003). Impact of two psychosocial interventions on white and African American family caregivers of individuals with dementia. *The Gerontologist, 43*, 568–579.

Chang, B. L. (1999). Cognitive-behavioral intervention for homebound caregivers of persons with dementia. *Nursing Research, 48*, 173–182.

Coon, D. W., Ory, M., & Schulz, R. (2004). Family caregivers: Enduring and emergent themes. In D. W. Coon, D. Gallagher-Thompson, & L. W. Thompson (Eds.), *Innovative interventions to reduce dementia caregiver distress: A clinical guide* (pp. 3–27). New York: Springer.

Coon, D. W., Rubert, M., Solano, N., Mausbach, B., Kraemer, H., Arguelles, T., et al. (2004). Well-being, appraisal, and coping in Latina and Caucasian female dementia family caregivers: Findings from the REACH study. *Aging and Mental Health, 8*, 330–345.

Coon, D. W., Thompson, L., Steffen, A., Sorocco, K., & Gallagher-Thompson, D. (2003). Anger and depression management: Psychoeducational skill training interventions for women caregivers of a relative with dementia. *The Gerontologist, 43*, 678–689.

Davidson, A. (2006). *A curious kind of widow: Loving a man with advanced Alzheimer's.* McKinleyville, CA: Fithian Press.

Dura, J. R., & Kiecolt-Glaser, J. K. (1990). Sample bias in caregiving research. *Journal of Gerontology: Psychological Sciences, 45*, P200–P204.

D'Zurilla, T. (1986). *Problem solving therapy: A social competence approach to clinical intervention*. New York: Springer.

Eisdorfer, C., Czaja, S., Loewenstein, D. A., Rubert, M. P., Arguelles, S., Mitrani, V., et al. (2003). The effect of a family therapy and technology-based intervention on caregiver depression. *The Gerontologist, 43*, 521–531.

Fisher, J. E., Drossel, C., Ferguson, K., Cherup, S., & Sylvester, M. (2007). Functional analysis for treating behavioral problems in individuals with dementia. In D. Gallagher-Thompson, A. M. Steffen, & L. W. Thompson (Eds.) *Handbook of behavioral and cognitive therapies with older adults*. New York: Springer.

Gallagher, D., Rose, J., Rivera, P., Lovett, S., & Thompson, L. W. (1989). Prevalence of depression in family caregivers. *The Gerontologist, 29*, 449–456.

Gallagher-Thompson, D., & Coon, D. W. (2007). Evidence-based psychological treatments for distress in family caregivers of older adults. *Psychology and Aging 22*(1), 37–51.

Gallagher-Thompson, D., Coon, D. W., Solano, N., Ambler, C., Rabinowitz, Y., & Thompson, L. W. (2003). Change in indices of distress among Latina and Anglo female caregivers of elderly relatives with dementia: Site specific results from the REACH National Collaborative Study. *The Gerontologist, 43*, 580–591.

Gallagher-Thompson, D., Haley, W., Guy, D., Rubert, M., Arguelles, T., Zeiss, L., et al. (2003). Tailoring psychological interventions for ethnically diverse dementia caregivers. *Clinical Psychology: Science and Practice, 10*, 423–438.

Gallagher-Thompson, D., & Steffen, A. (1994). Comparative effects of cognitive/behavioral and brief psychodynamic psychotherapies for depressed family caregivers. *Journal of Consulting and Clinical Psychology, 62*, 543–549.

Gitlin, L. N., & Gwyther, L. P. (2004). In-home interventions: Helping caregivers where they live. In D. W. Coon, D. Gallagher-Thompson, & L. W. Thompson (Eds.), *Innovative interventions to reduce dementia caregiver distress: A clinical guide* (pp. 139–160). New York: Springer.

Gitlin, L. N., Winter, L., Corcoran, M., Dennis, M. P., Schinfeld, S., & Hauck, W. (2003). Effects of the home environmental skill-building program on the caregiver-care recipient dyad: 6 month outcomes from the Philadelphia REACH initiative. *The Gerontologist, 43*, 532–546.

Given, B., Given, C. W., Kozachik, S., & Rawl, S. (2003). Family caregiving interventions in cancer care. In C. Given, B. Given, V. Champion, S. Kozachik, & D. DeVoss (Eds.), *Evidence-based cancer care and prevention: Behavioral interventions* (pp. 332–370). New York: Springer.

Gottlieb, B. H., Thompson, L. W., & Bourgeois, M. (2004). Monitoring and evaluating interventions. In D. W. Coon, D. Gallagher-Thompson, & L. W. Thompson (Eds.), *Innovative interventions to reduce dementia caregiver distress: A clinical guide* (pp. 28–49). New York: Springer.

Gottlieb, B. H., & Wolfe, J. (2002). Coping with family caregiving to persons with dementia: A critical review. *Aging and Mental Health, 6*, 325–342.

Haas, L. J., Benedict, J. G., & Kobos, J. C. (1996). Psychotherapy by telephone: Risks and benefits for psychologists and consumers. *Professional Psychology: Research and Practice, 27*, 154–160.

Hinton, W. L., Fox, K., & Levkoff, S. (1999). Introduction: Exploring the relationships among aging, ethnicity, and family dementia caregiving. *Culture, Medicine and Psychiatry, 23*, 403–413.

Hyer, L., & Intrieri, R. (2006). *Geropsychological interventions in long-term care*. New York: Springer.

Katz, S., Ford, A., Moskowitz, R., Jackson, B., & Jaffe, M. (1963). Studies of illness in the aged: The index of ADL, a standardized measure of biological and psychosocial function. *Journal of the American Medical Association, 185*, 914–919.

Lawton, M. P., & Brody, E. (1969). Assessment of older people: Self-maintaining and instrumental activities of daily living. *The Gerontologist, 9*, 179–186.

Lewinsohn, P. (1974). A behavioral approach to depression. In R. Friedman & M. Katz (Eds.), *The psychology of depression: Contemporary theory and research* (pp. 157–176). New York: Wiley.

Lewinsohn, P., Munoz, R., Youngren, M., & Zeiss, A. M. (1986). *Control your depression* (2nd ed.). Englewood Cliffs, NJ: Prentice-Hall.

Mermelstein, H. T., & Holland, J. C. (1991). Psychotherapy by telephone: A therapeutic tool for cancer patients. *Psychosomatics, 32*, 407–412.

Mittleman, M., Roth, D., Coon, D. W., & Haley, W. E. (2004). Sustained benefit of supportive interventions for depressive symptoms in caregivers of patients with Alzheimer's disease. *American Journal of Psychiatry, 161*, 850–856.

National Alliance for Caregiving/American Association of Retired Persons. (1997). *Family Caregiving in the U.S: Findings from a national survey*. Bethesda, MD: National Alliance for Caregiving.

Ory, M. G., Yee, J. L., Tennstedt, S. L., & Schulz, R. (2000). The extent and impact of dementia care: Unique challenges experienced by family caregivers. In R. Schulz (Ed.), *Handbook on dementia caregiving: Evidence-based interventions for family caregivers* (pp. 1–32). New York, NY: Springer.

Pargament, K., Koenig, H. G., & Perez, L. M. (2000). The many methods of religious coping: Development and initial validation of the RCOPE. *Journal of Clinical Psychology, 56*, 519–543.

Pinquart, M., & Sorensen, S. (2003). Differences between caregivers and non-caregivers in psychological health and physical health: A meta-analysis. *Psychology and Aging, 18*, 250–267.

Rabinowitz, Y., Mausbach, B., Coon, D. W., Depp, C., Thompson, L.W., & Gallagher-Thompson, D. (2006). The moderating effect of self-efficacy on intervention response in women family caregivers of older adults with dementia. *American Journal of Geriatric Psychiatry, 14*, 642–649.

Radloff, L. (1977). The CES-D scale: A self-report depression scale for research in the general population. *Applied Psychological Measurement, 1*, 385–401.

REACH II investigators (alphabetically: Belle, S., Burgio, L., Burns, R., Coon, D., Czaja, S., Gallagher-Thompson, D., Gitlin, L., Klinger, J., Koepke, K. M., Lee, C. C., Martindale-Adams, J., Nichols, L., Schulz, R., Stahl, S., Stevens, A., Winter, L. & Zhang, S.). (2006). Enhancing the quality of life of dementia caregivers from different ethnic or racial groups: A randomized, controlled trial. *Annals of Internal Medicine, 145*, 727–738.

Robinson, P., Katon, W., Von Korff, M., & Bush, T. (1997). The education of depressed primary care patients: What do patients think of interactive booklets and a video? *Journal of Family Practice, 44*, 562–571.

Schulz, R., & Beach, S. R. (1999). Caregiving as a risk factor for mortality: The Caregiver Health Effects Study. *Journal of the American Medical Association, 282*, 2215–2219.

Schulz, R., Burgio, L., Burns, R., Eisdorfer, C., Gallagher-Thompson, D., Gitlin, L. N., et al. (2003). Resources for Enhancing Alzheimer's Caregiver Health (REACH): Overview, site-specific outcomes, and future directions. *The Gerontologist, 43*, 514–520.

Schulz, R., & Martire, L. (2004). Family caregiving of persons with dementia: Prevalence, health effects, and support strategies. *American Journal of Geriatric Psychiatry, 12*, 240–249.

Schulz, R., Martire, L., & Klinger, J. (2005). Evidence-based caregiver interventions in geriatric psychiatry. *Psychiatric Clinics of North America, 28*, 1007–1038.

Schulz, R., O'Brien, A. T., Bookwala, J., & Fleissner, K. (1995). Psychiatric and physical morbidity effects of dementia caregiving: Prevalence, correlates, and causes. *The Gerontologist, 35*, 771–791.

Sorensen, S., Pinquart, M., & Duberstein, P. (2002). How effective are interventions with caregivers? An updated meta-analysis. *The Gerontologist, 42*, 356–372.

Steffen, A. M. (2000). Anger management for dementia caregivers: A preliminary study using video and telephone interventions. *Behavior Therapy, 31*(2), 281–299.

Steffen A. M. (2006). Introduction to special issue: New interventions for dementia-related concerns. *Clinical Gerontologist, 30*, 1–3.

Steffen, A. M., Mahoney, D., & Kelly, K. (2003). Capitalizing on technological advances to help distressed caregivers. In D. W. Coon, D. Gallagher-Thompson, & L. Thompson (Eds.), *Innovative interventions to reduce dementia caregivers' distress: A sourcebook and clinical guide*. New York: Springer.

Steffen, A. M., & Mangum, K. R. (2003). Distance-based skills training for depressed dementia family carers. *The Clinical Psychologist, 7*, 1–10.

Steffen, A. M., McKibbin, C., Zeiss, A., Gallagher-Thompson, D., & Bandura, A. (1997). The revised scale for caregiver self-efficacy: Two reliability and validity studies. *Journal of Gerontology: Psychological Sciences, 57*, 74–86.

Tarlow, B., Wisniewski, S. R., Belle, S. H., Rubert, M., Ory, M., & Gallagher-Thompson, D. (2004). Positive Aspects of Caregiving: Contributions of the REACH project to the development of new measures for Alzheimer's caregiving. *Research on Aging, 26*, 429–453.

Teri, L., Gibbons, L. E., McCurry, S. M., Logsdon, R. G., Buchner, D. M., Barlow, W. E., et al. (2003). Exercise plus behavioral management in patients with Alzheimer Disease: A randomized controlled trial. *Journal of the American Medical Association, 290*, 2015–2022.

Teri, L., Logsdon, R. G., Uomoto, J., & McCurry, S. M. (1997). Behavioral treatment of depression in dementia patients: A controlled clinical trial. *Journals of Gerontology: Series B Psychological Sciences and Social Sciences, 52B*, P159–P166.

Teri, L., Truaz, P., Logsdon, R., Uomoto, J., Zarit, S., & Vitaliano, P. (1992). Assessment of behavioral problems in dementia: The revised memory and behavior problem checklist. *Psychology and Aging, 7*, 622–631.

Toseland, R. W., McCallion, P., Smith, T., Huck, S., Bourgeois, P., & Garstka, T. A. (2001). Health education groups for caregivers in an HMO. *Journal of Clinical Psychology, 57*, 551–570.

Vitaliano, P. P., Scanlan, J. M., Zhang, J., Savage, M. V., Hirsch, I. B., & Siegler, I. C. (2002). A path model of chronic stress, the metabolic syndrome, and coronary heart disease. *Psychosomatic Medicine, 64*, 418–435.

Vitaliano, P. P., Young, H. M., & Zhang, J. (2004). Is caregiving a risk factor for illness? *Current Directions in Psychological Science, 13*, 13–16.

Yeo, G. & Gallagher-Thompson, D. (Eds.) (2006). *Ethnicity and the dementias* (2nd ed.). New York: Routledge/Taylor & Francis Group.

Zarit, S. H., & Zarit, J. M. (2007). *Mental disorders in older adults. Family Caregiving* (Chap. 13). New York: Guilford Press.

8
Integrated Psychosocial Rehabilitation and Health Care for Older People with Serious Mental Illness

Meghan McCarthy, Kim T. Mueser, and Sarah I. Pratt

Older people with serious mental illness (SMI) comprise a group with unique impairments and treatment needs. Despite effective pharmacological treatment, many individuals with SMI experience challenges across a variety of tasks and skills necessary for living independently in the community. These difficulties include poor basic self-care skills, community living skills, and social skills. These impairments are often compounded by lack of social support, which is strongly associated with admission to nursing home and long-term care placements (Bartels, Mueser, & Miles, 1997; Meeks et al., 1990). Functional impairments due to comorbid physical illnesses also increase the risk of placement in long-term care facilities (Burns & Taube, 1990; Meeks et al., 1990). Older adults with SMI often have multiple comorbid medical problems, which when coupled with poor health care result in poor health outcomes and earlier mortality (Bartels et al., 2004).

Psychosocial treatment is needed to improve functioning in these areas. Rehabilitative approaches such as skills training may be useful for older people with SMI because they are based on the premise that systematic teaching of life skills can reduce impairments in social and role functioning (Bellack, Mueser, Gingerich, & Agresta, 2004; Dilk & Bond, 1996; Hayes, Halford, & Varghese, 1995). In our judgment, an integrated program of health care management and psychosocial rehabilitation addresses the major deficits in older people with SMI and is essential to improving functioning in the community and avoiding restrictive residential placement.

Evidence Base

In recent years, programs have been developed specifically for older adults with SMI that focus on enhancing independent living skills and leisure skills, and expanding social networks (Pratt, Van Citters, Bartels, & Mueser, in press). However, many of these programs have not emphasized integration of both social rehabilitation and management of health care for older persons with SMI.

The Functional Adaptation Skills Training (FAST) program, developed by Patterson and colleagues (Patterson et al., 2003), is a 24-session skills training group for middle-aged and older adults with chronic psychotic disorders. The skills addressed in this intervention are based on the needs and deficits of older people, including handling finances, medication management, using public transportation, communication, and planning. In a study of FAST, four board and care homes were randomly assigned to receive either FAST or usual care. Among the 32 residents who participated in the study, the individuals who received FAST demonstrated improved community living skills compared with the individuals who received usual care.

Granholm et al. (2005) developed and tested an integrated treatment program for older adults with schizophrenia that combines cognitive-behavioral therapy and social skills training. The largest study of this intervention involved 76 outpatients who were randomly assigned to treatment as usual or treatment as usual plus the 24-session integrated treatment program. Improvements were shown at posttreatment (6-months) in cognitive insight, positive symptoms, and leisure and transportation skills.

In 1997, we developed a pilot program to provide social rehabilitation (i.e., social skills training), and health care management in order to improve the community functioning of older people with SMI. The social skills training and health management (ST + HM) intervention was delivered over 1 year with the goal of improving social functioning, communication with doctors, and health management in order to reduce long-term or permanent institutionalization (Bartels et al., 2004). The ST component consisted of weekly skills training classes in conversation skills and medication self-management based on curriculum developed by Liberman et al. (1993) and Bellack et al. (2004). The HM component involved a nurse case manager who provided management of acute and chronic medical problems and helped participants to access preventive health care services.

Participants in the pilot study of ST + HM were recruited from a community mental health center and met New Hampshire state mental health disability criteria for SMI consisting of (1) a diagnosis of schizophrenia, schizoaffective disorder, bipolar disorder, other psychotic disorder, or treatment refractory depression and (2) persistent functional impairment requiring ongoing supportive services. The final study group consisted of 24 participants, half of whom were interested in participating in the skills training component and received ST + HM, half of whom received HM only.

Participants who received ST + HM demonstrated improved independent living skills, such as personal hygiene, food preparation, and health management skills, compared with the HM only group. Clients who received ST + HM also demonstrated improved social functioning and decreased inappropriate behaviors, whereas no such changes occurred in the HM only group. All individuals in the study received HM so only a pre–post comparison of improvements in healthcare was possible. Receipt of preventive health care increased from 71 to 100%. At the start of the study, 4 of the 24 (16.7%) people did not have a primary care physician whereas 100% of the sample had obtained a primary care physician by the end of the study (Bartels et al., 2004). Results of the pilot study of the ST + HM program were encouraging and provided the basis for refining and evaluating an integrated model including skills training and health care management for older persons with SMI in a randomized controlled study of the HOPES program.

The HOPES Program

We expanded and manualized the skills training curriculum and nursing intervention to create the Helping Older People Experience Success (HOPES) program. In late 2002, we received a grant from the National Institute of Mental Health (MH R01 62324) to evaluate the program in a randomized, controlled trial in several inner-city Boston clinics and a community mental health center in New Hampshire. Social and independent living skills, health behaviors, and social, community, cognitive, and health functioning were evaluated at baseline, and at 12-, 24-, and 36-month follow-up. HOPES includes an intensive year during which participants attend weekly skills training classes led by rehabilitation specialists and monthly meetings with a nurse, and a maintenance year during which participants attend monthly skills training classes and monthly meetings with a nurse. Community trips (biweekly during year one and monthly during year two) and monthly meetings with family members or other key support individuals facilitate generalization of skills to natural settings.

One hundred eighty-three adults with SMI age 50 and older enrolled in the HOPES study. The active intervention phase of the study has concluded, with 36-month follow-up assessments continuing until the fall of 2007. Baseline characteristics of the study sample are described in another manuscript (Pratt, Bartels, Mueser, & Forester (in press)).

The primary objectives of the HOPES program were to improve social and community functioning and to decrease the use of acute health care services and long-term institution-based care. In order to achieve these goals, the original ST + HM pilot program was modified to target skill areas associated with these outcomes, including: independent living skills and functioning, social skills and social role functioning, skills for health management and medication adherence behaviors, and the use of preventive health care. The HOPES skills training program, which is detailed in a comprehensive set of instructor's manuals and participant workbooks, includes age-specific adaptations of established techniques that have been used in other standardized skills training packages and teaches skills that have been identified elsewhere as critical building blocks for successful interpersonal functioning (Bellack et al., 2004; Liberman, DeRisi, & Mueser, 1989).

Assessment

There is significant variation among older individuals with SMI in the number and severity of skill areas that are impaired. Therefore, assessment of functioning is best obtained through interviews and evaluations of clients and significant others who have first-hand knowledge about performance of social skills and behaviors (Bellack et al., 2004; Mueser & Bellack, 1998). Mental health clients aged 50 years or older with a chart diagnosis of schizophrenia, schizoaffective disorder, bipolar disorder, or treatment refractory depression and moderate to severe functional impairment were invited to participate in the study of the HOPES program. Psychiatric diagnoses were confirmed with the Structured Clinical Interview for DSM-IV (SCID; First, Spitzer, Gibbon, & Williams, 1995). In addition, the Mini-Mental State Exam (MMSE; Folstein, Folstein, & McHugh, 1975) was administered to exclude individuals with significant cognitive impairment, given the focus of the program on learning new skills. Functional impairment was defined according to New Hampshire state mental health disability criteria, including at least moderate impairment in several of the following areas: personal care skills, community living skills, social skills, education or work activities, supports and safety, and treatment self-management. In addition, a Global Assessment of Functioning Score (GAF; American Psychiatric Association, 1987) below 60 was required.

Description of the HOPES Program Components

Orientation to the HOPES Program

All participants in HOPES begin with an individual orientation with one of the skills class leaders. This meeting is essential because it sets the tone for the overall experience and gives the leader an opportunity to actively engage the participant in the program (Pratt & Mueser, 2002). During this orientation, the leader and participant discuss the goals of the HOPES program and the logistics of participation in the skills classes and nurse health management meetings. The leader asks the participant about his or her goals, and how the HOPES program can help in achieving them. When the participants have established clear goals, the leader helps them begin to break these goals down into small, manageable steps or objectives. For persons who have trouble establishing goals, the leader encourages continued consideration of meaningful goals and objectives.

During the orientation, participants also identify a supportive person in the community with whom they have regular contact, such as a friend or family member (or alternatively, their therapist, group home manager, or case manager). The class leader encourages the community support persons to take an active role in helping to support the participants in reaching their goals over the course of the HOPES program. An example of a typical goal for a HOPES participant is described in Table 1. The skills class leader and/or HOPES nurse contacts the community support persons monthly to update them on the participants' progress and to orient them to the curriculum covered in the classes.

TABLE 1. Example of a HOPES goal set by a participant.

Overall goal/objective	Steps
Module: Communicating effectively *Goal*: Improve conversational skills *Objective*: Have a conversations with a neighbor daily.	1. Start conversations with neighbors using simple greetings and small talk 2. Maintain conversations by asking questions (avoid inappropriate topics) 3. Maintain conversations by giving information (avoid medium or high levels of self-disclosure) 4. Express desire for future conversations

Skills Training Classes

HOPES classes range between 6 and 10 participants, are led by two coleaders, and are held once a week for the first year of the program. Although social skills training programs for younger adults often hold multiple classes per week (Bellack et al., 2004; Pratt & Mueser, 2002), many older people with SMI have difficulties with transportation and are reluctant to make frequent trips to their community mental health center, so the HOPES intervention is designed to deliver two classes in 1 day. Each class is 60–90 min and includes 10 min of stretching/light exercise in between the classes. During the HOPES study, a 30-minute lunch break was also held between a late morning and afternoon class and provided an opportunity for informal socialization among the participants and the coleaders.

Curriculum

The curriculum is organized into seven modules or topic areas, including: Making the Most of Leisure Time, Living Independently in the Community, Communicating Effectively, Making and Keeping Friends, Healthy Living, Making the Most of a Health Care Visit, and Using Medications Effectively. The organization of the program into modules that continually repeat permits the HOPES class to run as an open class so that clients can join at any time and receive the modules in any order.

Community Practice Trips

Every other week, HOPES participants attend a community practice trip, which provides the opportunity to practice the skills learned in the classes in community settings. For example, for the skill "Using Transportation Maps," the community trip involved taking the bus to and from an activity to practice using the public transit maps. Although the participants plan the trips, they are reminded and encouraged to select outings that will give them the opportunity to practice skills. A secondary purpose of the community practice trips is to reinforce attendance at the skills training classes. The skills training classes and the community trips both decrease to once per month during the maintenance year.

Health Management Meetings

The participants meet with the HOPES nurse once per month during the intensive year to discuss their progress on health-related goals and problems managing medical conditions, and to review the scheduling and receipt of preventive health care. In the maintenance year, the nurse continues to meet individually with participants every other month and holds group meetings on the opposite months to discuss and reinforce clients' progress toward health-related goals.

Content of the Skills Training Curriculum

The HOPES skills training program includes a unique curriculum and a teaching style especially tailored for older adults. The ST curriculum covers topic areas associated with deficits in functioning in many older adults with SMI, including community and independent living skills, communication skills, leisure skills, and social skills. The ST curriculum also addresses health behaviors, effective communication with health care providers, and medication adherence strategies.

In the intensive year, 50 classes or component skills are covered, with 6–8 skills comprising each module. In the maintenance year, all skills are reviewed and participants have the opportunity to discuss their use of these skills and to problem-solve any obstacles to using them. Table 2 outlines the seven HOPES modules and the component skills introduced in the HOPES program.

Skills Training Methods

The skills training classes are led by two coleaders and incorporate the basic components of social skills training (Bellack et al., 2004; Liberman et al., 1989). Modeling is used by the leaders both informally, when showing appropriate behavior and social skills in interactions with others, and formally, when demonstrating specific skills in role-plays. Positive reinforcement is essential to effective skills training and is provided in the form of verbal praise by both the leaders and participants for appropriate demonstration of skills, effort during role-plays, correct responses to questions, attendance, progress toward personal goals, or any other evidence of success with tasks.

Repeated practice of skills, to the point of overlearning, is the *sine qua non* of social skills training, and is critical to overcoming the cognitive impairments commonly present in people with SMI. Overlearning may be even more important for older persons with SMI, who experience further impairment due to age-related decline in cognitive functioning. The information in the skills training classes is presented in a variety of formats and is reviewed periodically through

TABLE 2. HOPES skills training curriculum.

Module	Component skills
Making the most of leisure time	Savoring the moment
	Reminiscing
	Anticipation
	Three stages of fun
	Inviting someone to share a leisure activity
	Planning for an outing
Living independently in the community	Traveling independently
	Reading maps
	Making positive requests
	Communicating on the telephone
	Making a monthly budget
	Acquiring important items from the store
	Solving community living problems
Communicating effectively	Starting a conversation
	Maintaining a conversation by asking questions
	Maintaining a conversation by giving information
	Sticking with a topic of conversation
	Appropriate self-disclosure
	Appropriate disclosure in conversations
	Ending conversations smoothly
	Putting it all together
Making and keeping friends	Building a foundation for friendship
	Meeting new people
	Learning about another person
	Giving and receiving compliments
	Making plans with a friend
	Negotiating a compromise
	Reconnecting with old friends
	Responding to requests
Healthy living	Setting and achieving goals
	Healthy sleep habits
	Major medical problems
	Healthy diet and exercise
	Preventive health care
	Anxiety management
Making the most of health care visits	Making an appointment and preparing for a health care visit
	Sharing health information with your doctor
	Reporting physical symptoms
	Asking for treatment options
	Making treatment decisions
	Making a visit to the dentist
	Making an advanced care plan
	Naming a health care agent
Using medications effectively	Medication basics
	Pros and cons of taking medications
	Strategies for remembering medication
	Reading medication labels and using pill organizers
	Using electronic devices to remember medications
	Reporting medication side effects
	Evaluating the effects of medication
	Negotiating medication issues

role-plays. Each skill is also discussed on a *community practice trip* and with the community support person. *Home practice assignments* to rehearse skills over the next week are developed at the end of each session to provide other opportunities for participants to use and practice their new skill. An example of a home practice assignment for "Making and Preparing for a Health Care Visit" might be to develop a list of questions for an upcoming doctor's appointment.

At the start of each session, following a review of personal goals set by participants during the 1:1 orientation meeting, members share their successes and problems in completing the home practice assignment. The leaders and participants provide praise to clients for success or efforts to complete the assignment, while problem solving and practicing other approaches with clients who did not try the assignment or were unsuccessful in implementing the skill. During the maintenance year of HOPES, personal goals and home practice often merge into one as the focus of the program shifts from learning new skills to reinforcing the use of learned skills, and planning to help clients continue to use skills while working toward their personal goals after the program has ended.

Steps of Social Skills Training

The steps of social skills training are consistently employed from week to week and quickly become familiar to class members. Most sessions include the following steps: establishing the rationale for the skill, introducing the steps of the skill, role-play practice, positive and corrective feedback, and discussing opportunities to practice the skill at home. These steps are fully described below.

Establish the Rationale for the Skill

The coleaders introduce a new skill by establishing the rationale for learning the skill. The leaders first elicit the rationale from the participants themselves by asking questions about how the skill might be useful or important (Pratt & Mueser, 2002). They then provide additional reasons, and illustrate how the skill might help participants by using examples and asking them how the skill relates to their personal goals.

Introducing the New Skill

The leaders invite participants to take turns reading from the HOPES workbook to introduce the new skill. Leaders periodically stop to ask participants questions to check on their understanding of the material. Leaders ask open-ended questions rather than yes/no questions given the tendency of participants to answer in the affirmative even when they might not fully understand the class material. For example, the leader may invite a participant to summarize the material and then invite others to comment on it and provide additional input. Hearing information summarized by a class member helps to reinforce concepts and facilitates retention.

Many of the skills are broken down into a sequential set of steps. The leader displays the steps on a board or easel so the participants can refer to them in the role-play. In displaying the steps, the leader facilitates learning them by asking the participants to recall them. For example, "What is the first step of the skill for, Starting a Conversation?" Often the steps are reviewed from the easel before each participant performs his or her role-play.

Practicing the Skill

The in-class practice part of the skills training class provides participants with the opportunity to practice the skill. Role-plays, which involve brief interactions with one of the leaders or a class member, are a particularly effective way to practice interpersonal skills. Skills that are not interpersonal in nature are also practiced in class. For example, in the Healthy Living module, one of the in-class practice exercises for the Healthy Diet skill is to complete a food diary for the previous day and to assign the appropriate amount of calories, fat, sodium, and cholesterol. Participants use workbook materials to find these values and compare their totals with recommended values.

Role-plays are frequently used to practice and reinforce skills. Although participants are often initially anxious about doing role-plays, they quickly become comfortable with the technique through weekly practice and receiving positive reinforcement for their efforts. The format for role-plays is the same each time and begins with a demonstration by the coleaders. The leaders then invite the participants one-by-one to participate in the role-play. As the participant prepares for the role-play, one of the coleaders reviews the steps of the skill displayed on the easel, and asks the participant to state the goal of the role-play.

Generalization is the ultimate goal of skills training and involves using newly acquired behaviors and skills in natural settings outside the class environment. In order to promote generalization, during in-class practice the coleaders help participants to tailor the role-play to a situation from their own life. This helps to ensure that they are rehearsing skills that have personal relevance to them, which will increase the likelihood of improving their social behavior and independent living skills in the community.

Immediately after the role-play has concluded, the leader initiates a round of applause for the performer. Participants genuinely enjoy and appreciate this type of reinforcement. Next, the leader elicits positive feedback from participants by asking what they liked about the individual's role-play. If the feedback is general or vague, for example, "Sue did a good job," the leader prompts for more detail by asking the person what in particular he or she liked about the person's role-play. Negative feedback is immediately redirected and corrective feedback is only provided after sufficient positive feedback is obtained. Corrective feedback involves providing brief, behaviorally specific suggestions delivered in an upbeat, noncritical manner. For example, the leader might say, "One thing that would make your performance in the role-play even better is if you looked the person directly in the eye when asking for assistance at the store." The leader then asks the participant to repeat the role-play, incorporating the suggestion for improved performance. After the repeat role-play, feedback is again elicited from class members.

Below is an example of using role-plays to demonstrate and help clients practice the skill of "Making an Appointment with a Doctor."

Leader: We have talked about the steps involved in "Making an Appointment with a Doctor," and now we are going to demonstrate it in a role-play. After Janet (coleader) and I model the skill we can talk about it and then you folks can try practicing the skill in a similar role-play. I am going to pretend to be a patient calling my doctor's office to make an appointment for a physical. Janet will pretend to be a receptionist at the doctor's office and her job is to schedule appointments for patients. My first goal is to gather the items needed to make an appointment. Can anyone remember what those items are?

Bob: The phone number and name of the doctor, a piece of scrap paper, and a pencil.

Leader: Great Bob! Those are all really important things to have with you. Mary, is there anything else that I might need to make an appointment?

Mary: A calendar so you can write the date and time. That way if you forget, you can check the calendar.

Leader: Excellent Mary! That is a great reason to write down your medical appointments. After I gather all the items, I will dial the number, and state my name and the reason for the call. After Janet gives me a date for the appointment I will check the calendar to see if I am available. I will also repeat the information for the appointment back to Janet to be sure that I have it written down correctly. Joe, can you remember the last step of this skill?

Joe: Thank the person?

Leader: Good Joe, I'll thank Janet for helping me to make the appointment. Are there any questions?

Mary: Do we all have to do it?

Leader: It would be great if everyone could try. Remember, practicing these skills makes it easier to use them in real life situations.

(The role-play is conducted, followed by questions to assess whether participants can identify performance of all steps of the skill).

Now I would like to give you each a chance to practice the skill in the role-play. Who would like to go first with Janet?

Bob: I will.

Leader: Great! Thanks Bob! Now, Bob, what is your role in the role-play?

Bob: Make a phone call to my doctor's office.

Leader: That is right and what is Janet's role?

Bob: Set up the appointment?

Leader: Right! Who will be starting the role-play?

Bob: I guess me because I will dial the phone but Janet will answer it and start talking.

Leader: Sounds good! Let's begin! The rest of us will watch to see which steps of the skill are followed.

Bob: (Bob gets items together and dials the telephone.)

Janet: Good morning, Dr. Doherty's office, how can I help you?

Bob: (Very softly, barely audible) Yes, my name is Bob Henson and I need to make an appointment for my annual physical.

Janet: Okay, could you repeat your name for me?

Bob: Bob Henson.

Janet: Hello Bob! Let's see. Dr. Doherty does not have any appointments until May. There is a 1:00 p.m. appointment on both May 15th and May 17th. Would either of these times work for you?

Bob: Let me check my calendar. Yes, the 15th at 1:00 p.m. would work.

Janet: Okay, you are all set Bob. Thank you.

Bob: See you then. Goodbye.

Leader: (The class begins clapping.) Great job with your role-play, Bob! Let's ask the class...What did people like about Bob's role-play?

Joe: He didn't say thank you before he hung up the phone.

Leader: (Ignoring negative comment...) Let's give Bob some positive feedback about his role-play. Mary, what did you like about the way Bob performed his role-play?

Mary: He was good.

Leader: Bob was good, wasn't he? Can you tell Bob, specifically, what you thought was good about his performance in the role-play?

Mary: He checked his calendar and repeated the date back to Janet.

Leader: That's right. I thought that was good too. What about the first step of the skill? Linda, did he complete the first step?

Linda: He got all of his items that he needed for the call. Yes.

Leader: Good! What about the second step?

Joe: He said his name and said he needed to make an appointment.

Leader: Excellent Joe! Mary also noticed that he completed the next steps of checking his calendar and repeating the date. What about the last step?

Bob: I forgot to thank her for making the appointment.

Leader: That's true Bob. But, you did a great job showing how to make a health care appointment. You remembered to gather all the items you needed,

> stated your name and the reason for your call, and you checked your availability and repeated the date back to the receptionist. Excellent! Now it's true that you forgot to thank receptionist for the appointment. That would have made it even better, so why don't you try the role-play one more time and this time try to add that piece?

(The same role-play is enacted again, following the same procedure of eliciting positive feedback and providing constructive criticism and additional role-playing if needed. Then, the next participant is invited to role-play.)

Home Practice

Following the role-plays, home practice is discussed and personalized assignments are designed. If individuals have trouble identifying situations in which to practice a skill, the class members and leaders offer suggestions. Participants are encouraged to share their home practice assignments with their community support person, who can often help them to complete it.

Planning for the Community Trip

At the end of every other class a community trip is planned for the following week. Leaders remind participants that the trip should involve an opportunity to practice the skills learned in class and then display potential ideas on the board or easel. The coleaders also prompt class members to consider the skills they learned for Planning an Outing including: consideration of transportation, weather, cost, time and location, and accessibility to all class members. The class members vote for the activity and plan the trip.

Age-Related Adaptations to Skills Training

The HOPES program includes several adaptations to the standard skills training approach to accommodate the specific needs of older people. First, two classes are provided on the same day in order to accommodate limitations in mobility and problems with transportation. Second, sensory impairments (e.g., hearing, vision) are accommodated with tools such as magnification devices and sound amplifiers. Third, much of the content of the class curriculum is geared toward older people. For example, there is an entire skills module dedicated to enhancing skills for accessing health care given the substantial medical comorbidity among older people with SMI. There is also a module dedicated to enhancing leisure skills given that many older people are not as interested as younger adults in finding competitive employment. Specific skills in the other modules are also tailored for older people. For example, in the Making and Keeping Friends module, participants learn the skill of "Reconnecting with Old Friends." Finally, the pace of the HOPES classes is slower than what would be expected for a typical skills training class and fewer skills are covered in each session.

Adaptations to Skills Training to Accommodate Cognitive Impairment

Cognitive impairment, including deficits in executive functioning, memory, and attention, are common in people with SMI. Some decline in cognitive functioning may also be associated with the normal process of aging. Several strategies are routinely incorporated in the HOPES classes to accommodate these impairments. Most importantly, class material is reviewed frequently and understanding of the curriculum is checked frequently with open-ended questions. For example, at the beginning of each class, after checking on participants' progress toward personal goals, class leaders ask, "Which HOPES module were we working on?" followed by, "Which specific skill did we cover the last time?" Material is presented in small segments to avoid information overload. All skills are divided into several steps, which are displayed on a white board or easel so that participants can refer to them if they need to during role-plays. Role-plays and home practice assignments are individualized to enhance understanding and personal significance. Coleaders help participants to make a specific plan to complete home assignments, which are documented on a take-home sheet that includes the steps for the skill to be practiced. Supportive individuals in the community or staff at the mental health center are also engaged in the program to help prompt and reinforce the use of skills. The steps of each skill are also provided on laminated cards that participants can carry with them to facilitate recall of the steps while practicing the skills at home. Finally, substantial positive reinforcement is used in a shaping framework during HOPES classes to encourage gradual acquisition of skills.

Health Management Procedures

The Health Management component of the HOPES program focuses on ensuring that clients receive preventive health care, improving health behaviors, and enhancing the quality of communication with health care providers. This component is delivered by a nurse who ideally has a background in general medical care and at least some training or familiarity with psychiatric nursing. The nurse begins with a comprehensive evaluation of health (complete review of systems) and receipt of preventive health care. This evaluation helps the nurse and participant to collaboratively develop a medical problem list and one or more health care/health behavior goals.

The nurse meets at least monthly with participants and tracks medical appointments and receipt of preventive health care. The nurse may also role-play with participants to help them to prepare for appointments with health care providers. The nurse attends at least one health care appointment per year with the participants, if they agree, to introduce the primary care provider to the program and to help facilitate communication with primary care providers.

In the maintenance year, the monthly contact alternates between individual meetings and group meetings in which the participants can review health care/behavior skills and progress toward their health goals. One of the class leaders is present at these meetings to facilitate the integration of the health management and skills training components.

The nurse uses many of the same social learning techniques of skills training to help participants improve their health behaviors and increase access to appropriate health care. These techniques include education, positive reinforcement, modeling, and practice in the individual and group sessions to encourage self-management of medical problems and improve health behaviors. These techniques are also used to ensure that participants are receiving preventive health care and that they are accessing appropriate health care services.

Integration of Components

The skills training coleaders and nurse meet on a weekly basis to discuss each client's participation in the program and progress toward goals. Because of the interaction between psychosocial functioning and health, it is important for the HOPES providers to discuss participants' progress in both components of the program. In addition, in the maintenance year, one of the skills leaders is present at the health management groups to help the nurse review the health-related skills introduced in the ST classes and to reinforce progress toward personal goals.

Case Example

Mary was a 63-year-old divorced woman receiving services from the community mental health center for over 30 years. She was a mother of three who, after the birth of her second child, was hospitalized for persecutory delusions and suicidal thinking. She was diagnosed at age 25 with schizophrenia and her husband died unexpectedly 5 years later in a car accident. She raised her three young children with the help of her parents. Her GAF (Jones, Thornicroft, Coffey, & Dunn, 1995) was 45 and she had moderately severe functional impairment. She had over 15 psychiatric hospitalizations since the onset of her illness. Mary had significant deficits in multiple areas of functioning including self-care skills, independent living, and social skills. She required assistance from others to pay her bills, clean her apartment, grocery shop, and communicate with her health care providers. She had limited social contact with anyone besides her children, whom she saw only three or four times per year. Mary had significant medical problems and had been diagnosed with diabetes, chronic obstructive pulmonary disease, hypertension, osteoporosis, and inflammatory bowel disease.

Mary began the HOPES program with the goal of developing new friendships and making connections with people in her apartment complex and in the HOPES class. She said that her most difficult challenges were meeting new people and opening up to existing friends and family members. She was very cautious in her interactions with neighbors and stated, "none of my neighbors likes me and the woman next door slams her door purposely so that I will move out!" Mary said that it was particularly difficult for her to start and maintain conversations. She explained, "I never know what to say after 'Good morning!' or 'Hello!' " Mary said that she only felt comfortable talking with her son or her daughters. She wanted to have closer relationships and to feel more comfortable conversing with people, explaining that she even avoided confiding in a friend whom she had known for over 10 years. She hoped that the HOPES program would give her the opportunity to learn and practice skills for communication and conversations.

Mary attended the skills training classes and the health management meetings consistently throughout the 2-year program. In the second year, she needed prompting to remember the class schedule because the frequency reduced from weekly meetings to biweekly meetings. From the start of Mary's participation, she was enthusiastic about the curriculum related to communication, leisure time, and friendships. She was able to share with class members the importance of learning and practicing these skills and how using the skills would help her reach her goals. Mary often updated the class on ways she used the skills in and outside of the class and this motivated other class members to try to use the skills in their everyday lives. Despite her anxiety and concerns about participating in role-plays, she valued the practice and was able to execute many components of the skills. In role-plays, Mary was particularly skilled at initiating conversations. However, she had difficulty maintaining conversations particularly when it came to asking people about themselves. She also had difficulty deciding on appropriate conversational topics and spent a great deal of time practicing appropriate self-disclosure at her part-time job and at a chess club that she joined while participating in HOPES.

As Mary's participation in HOPES progressed, she reported becoming more comfortable sharing personal information with her friend of 10 years. She indicated that their friendship had deepened and that she did not feel as anxious about asking her friend to share a leisure activity or for support and help if she needed it. She also made two new friends at the chess club that she joined and was able to invite them to her apartment for coffee. When Mary finished the HOPES program, she was meeting her friend three times a week for a morning walk. Mary also initiated activities with some of the HOPES class members, organizing a weekly visit to the community senior center for lunch and a bingo game. She was the leader in this venture, planning the outing and helping other class members to problem-solve methods of transportation to and from the center.

Diversity Issues

The HOPES intervention and curriculum was developed with attention to the diverse population of older adults with SMI in Boston and New Hampshire. However, limits on staff resources and time did not allow for translation of the class materials or for delivery of the intervention in languages other than English, making the program unavailable to significant populations of Spanish-speaking and Thai clients at the Boston study sites. Recruitment for HOPES was open only to clients with an adequate mastery of English. Patterson et al. (2005) noted the importance of adapting psychosocial rehabilitation programs to meet the needs of Spanish-speaking Latinos, the most rapidly growing ethnic/language minority population in the United States. During recruitment for the study, 189 individuals did not meet eligibility criteria because they did not speak English. Because the majority of these individuals spoke Spanish, a logical and worthwhile adaptation to the HOPES curriculum would be the translation of the materials into Spanish and perhaps other languages depending on the ethnic and cultural diversity of the community. Bicultural and bilingual skills trainers would of course be needed to deliver the translated curriculum. Another adaptation to consider for non-English speaking participants would be teaching skills aimed at accessing community resources and services in primarily English-speaking communities.

Summary

The HOPES program was developed to address the needs and deficits of older adults with SMI. Even with effective psychopharmacological treatment, older adults with SMI struggle with social skills, independent living skills, and managing their health and health care needs. The HOPES program addresses these critical needs by integrating a comprehensive skills training program targeting social, health care, and independent living skills with assistance from a nurse to improve access to health care. The primary goals of this model are to decrease use of acute health care services, reliance on emergency care, and the need for long-term institutional care, with the ultimate aim of helping older persons with SMI live longer, healthier, more fulfilling lives in the community.

References

American Psychiatric Association. (1987). *Diagnostic and statistical manual of mental disorders* (3rd ed., revised). Washington, DC: Author.

Bartels, S. J., Forester, B., Mueser, K. T., Miles, K. M., Dums, A. R., Pratt, S. I., et al. (2004). Enhanced skills training and health care management for older persons with severe mental illness. *Community Mental Health Journal, 40*, 75–90.

Bartels, S. J., Mueser, K. T., & Miles, K. M. (1997). A comparative study of elderly patients with schizophrenia and bipolar disorder in nursing homes and the community. *Schizophrenia Research, 27*(2–3), 181–190.

Bellack, A. S., Mueser, K. T., Gingerich, S., & Agresta, J. (2004). *Social skills training for Schizophrenia*. New York: Guilford Press.

Burns, B. J., & Taube, C. A. (1990). Mental health services in general medical care and nursing homes. In B. Fogel, A. Furino, & G. Gottlieb (Eds.), *Mental health policy for older Americans: Protecting minds at risk* (pp. 63–84). Washington, DC: American Psychiatric Press.

Dilk, M. N., & Bond, G. R. (1996). Meta-analytic evaluation of skills training research for individuals with severe mental illness. *Journal of Consulting and Clinical Psychology, 64*(6), 1337–1346.

First, M. B., Spitzer, R. L., Gibbon, M., & Williams, J. B.W. (1995). *Structured clinical interview for DSM-IV axis I disorders – Patient edition (SCID – I/P, version 2.0)*. New York: Biometric Research Department, New York State Psychiatric Institute.

Folstein, M. F., Folstein, S. E., & McHugh, P. R. (1975). Mini-mental state: A practical method for grading the cognitive state of patients for the clinician. *Journal of Psychiatric Research, 12*, 189–198.

Granholm, E., McQuaid, J. R., McClure, F. S., Auslander, L. A., Perivoliotis, D., Pedrelli, P., et al. (2005). A randomized, controlled trial of cognitive behavioral social skills training for middle-aged and older outpatients with chronic schizophrenia. *American Journal of Psychiatry, 162*, 520–529.

Hayes, R. L., Halford, W. K., & Varghese, F. T. (1995). Social skills training with chronic schizophrenic patients: Effects on negative symptoms and community functioning. *Behavior Therapy, 26*, 433–449.

Jones, S. H., Thornicroft, G., Coffey, M., & Dunn, G. (1995). A brief mental health outcome scale-reliability and validity of the Global Assessment of Functioning (GAF). *British Journal of Psychiatry, 166*, 654–659.

Liberman, R. P., DeRisi, W. J., & Mueser, K. T. (1989). *Social skills training for psychiatric patients*. Needham Heights, MA: Allyn & Bacon.

Liberman, R., Wallace, C., Blackwell, G., Eckman, T., Vaccaro, J., & Kuehnel, T. (1993). Innovations in skills training for the seriously mentally ill: The UCLA social and independent living skills modules. *Innovations and Research, 2*(2), 43–59.

Meeks, S., Carstensen, L. L., Stafford, P. B., Brenner, L. L., Weathers, F., Welch, R., et al. (1990). Mental health needs of the chronically mentally ill elderly. *Psychology and Aging, 5*(2), 163–171.

Mueser, K. T., & Bellack, A. S. (1998). Social skills and social functioning. In K. T. Mueser & N. Tarrier (Eds.), *Handbook of social functioning in schizophrenia* (pp. 79–98). Boston: Allyn and Bacon.

Patterson, T. L., Bucardo, J., McKibbin, C. L., Mausbach, B. T., Moore, D., Barrio, C., et al. (2005). Development and pilot testing of a new psychosocial intervention for older Latinos with chronic psychosis. *Schizophrenia Bulletin, 31*, 922–930.

Patterson, T. L., McKibbin, C., Taylor, M., Goldman, S., Davila-Fraga, W., Bucardo, J., et al. (2003). Functional Adaption Skills Training (FAST): A pilot psychosocial intervention study in middle-aged and older patients with chronic psychotic disorders. *American Journal of Geriatric Psychiatry, 11*, 17–23.

Pratt, S. I., Bartels, S. J., Mueser, K. T., & Forester, B. (in press). Helping Older People Experience Success (HOPES): An integrated model of psychosocial rehabilitation and health care management for older adults with serious mental illness. *American Journal of Psychiatric Rehabilitation*.

Pratt, S., & Mueser, K. T. (2002). Social skills training for schizophrenia. In S. G. Hofmann & M. C. Thompson (Eds.), *Handbook of psychosocial treatments for severe mental disorders* (pp. 18–52). New York: Guilford Press.

Pratt, S., Van Citters, A. D., Mueser, K. M. & Bartels, S. J. (in press). Psychosocial rehabilitation in older adults with serious mental illness: A review of the research literature and recommendations for development of rehabilitative approaches. *American Journal of Psychiatric Rehabilitation*.

9
Cognitive Therapy for Suicidal Older Adults

Gregory K. Brown, Lisa M. Brown, Sunil S. Bhar, and Aaron T. Beck

Suicide among older adults is a major public health problem. In 2004, there were approximately 5,198 suicides among those 65 and older in the United States according to the National Center for Health Statistics of the Centers for Disease Control and Prevention (CDC, n.d.). The rate of suicide for adults over the age of 65 was 14.3 per 100,000 and this rate increases with age. Moreover, men over the age of 65 are especially at risk for suicide given the suicide rate of 30.0 per 100,000. Given the high rates of suicide among older adults, there is an urgent need to identify the risk and protective factors associated with suicide behavior in this population. The design of effective suicide prevention strategies for older adults hinges on the identification of specific, modifiable risk factors.

Despite the public health significance of this problem, there have been relatively few prospective epidemiological studies or risk factors with this age group. Other methodologies have been used to identify possible risk factors or correlates in older adults. One method is to conduct a psychological autopsy of the suicide victim. This procedure involves a review of the victim's psychological and medical status, behavior, and other circumstances that were present prior to the death. Knowledgeable informants, such as family members and therapists, may be interviewed and medical and other social service records may be reviewed to determine the presence of specific factors that may have preceded the suicide. For example, psychological autopsy studies have found depression without comorbid psychopathology to be a common psychiatric diagnosis among older adults who commit suicide (Conwell & Brent, 1995; Conwell et al., 1996). Other correlates for suicide among older adults that have been identified have included hopelessness, suicide ideation, medical comorbidity (especially perceived poor health and chronic pain), alcohol abuse, complicated bereavement (associated with a recent death of a spouse or family member), social isolation, and loneliness (Conwell, Dubertstein & Caine, 2002; Pearson & Brown, 2000; Szanto et al , 2002).

Older adults who communicate a desire to kill themselves also have an elevated risk for suicide. One study found that nearly 40% of older adults had communicated their wish to die or to kill themselves to a health professional in the year prior to suicide (Waern, Besko, Runeson, & Skoog, 1999), and 75% of the study sample had communicated death ideation or suicide ideation to a family member or acquaintance during this period. Further evidence that death ideation or suicide ideation may be a risk factor for suicide is available from prospective studies. For example, in our psychiatric outpatient sample of older adults, patients who scored more than 1 on the *Scale for Suicidal Ideation* (SSI; Beck, Kovacs, & Weissman, 1979) were approximately 16 times more likely to commit suicide (odds ratio = 15.6) than psychiatric patients who were not suicidal (unpublished analyses from data reported in Brown, Beck, Steer, & Grisham, 2000). Because of the demonstrated connection between ideation and completed suicide, the reduction of suicide ideation is a reasonable means of lowering suicide risk.

For older adults, negative life events especially those events that involve the loss of a spouse, professional role, or a decline in health or financial status may contribute to the onset or worsening of hopelessness and depression (Byrne & Raphael, 1999; Rubenowitz, Waern, Wilhelmson, & Allebeck, 2001). Older people who live alone or who have multiple medical conditions are particularly vulnerable, endorsing significantly higher levels of psychological loss, low self-esteem, and hopelessness. Moreover, major life events such as the death of a spouse may also reinforce a preconceived belief that aging is dominated by loss.

Beck and his colleagues formulated a hopelessness theory of suicide based upon their clinical experiences with depressed suicidal patients (Beck, Kovacs, & Weissman, 1975). The construct of hopelessness refers to a set of negative, and a lack of positive, beliefs about the future (Beck 1967, 1976; Clark & Beck, 1999). They reported that suicidal crises were consistently "related to the patients' conceptualization of their situation as untenable or hopeless." Beck proposed that hopelessness was a catalytic agent in suicidal episodes. Individuals became at risk for suicide when they viewed their situation in a negative way and lacked problem-solving skills to cope with these negative events. Individuals were most at risk for suicide when they felt hopeless and believed that there was no other solution to the problem.

Numerous prospective studies have linked hopelessness to completed suicide in adults and older adults (Beck, Brown, Berchick, Stewart, & Steer, 1990; Beck et al., 1975; Brown et al., 2000; Wen-Hung, Gallo, & Eaton, 2004) or to suicide attempts in older adults (Rifai, George, Stack, Mann, & Reynolds, 1994). Using data from the Brown et al. study, we found that older adults who were seeking psychiatric treatment and who scored 9 or higher on the *Beck Hopelessness Scale* (Beck & Steer, 1988) were approximately 14 times more likely to commit suicide (odds ratio = 13.6, 95% confidence interval = 1.7–107.9, $p < .001$) than patients who scored less than 9. A recent study of institutionalized elderly patients also showed that hopelessness was strongly related to suicidal ideation (Uncapher, Gallagher-Thompson, Osgood, & Bonger, 1998), but was dependent upon the level of depression. Hopelessness predicted suicidal ideation in subjects with a high severity of depression, but was not related to suicide ideation in those with a low severity of depression. In another study, Szanto and her colleagues found that hopelessness that persisted after the remission of depression was associated with a history of suicidal behavior for older adults (Szanto, Reynolds, Conwell, Begley, & Houck, 1998). Given that hopelessness has been associated with suicidality and depression in older adults, interventions that focus on decreasing hopelessness as well as depression may help to reduce the risk of suicide.

The perception of social isolation has been linked to both depression and suicidal ideation in older adults. In a sample of depressed adults between the ages of 60–90 years, low levels of social support were found to be associated with pessimistic thinking styles, fewer social interactions, and divorce (Lynch et al., 1999). These findings suggest that therapists need to focus on relationships, social support, and the perception of social isolation when treating depressed and suicidal older patients (Lynch et al.).

Evidence-Based Treatment for Depression with Suicidal Older Adults

As previously stated, both psychological autopsy and epidemiological studies suggest that depression is a risk factor for late-life suicide (Carney, Rich, Burke, & Fowler, 1994; Conwell, Rotenberg, & Caine, 1990; Conwell et al., 1996; Draper, 1994; Turvey et al., 2002). The implication of this research is that the identification and adequate treatment of depression in older adults may reduce the risk of suicide. Cognitive behavior therapy is the most extensively studied psychotherapy for treating late-life depression (Gallagher-Thompson et al., 2000; Thompson, Gallagher, & Breckenridge, 1987). Randomized controlled trials have indicated that cognitive behavior therapy is an effective treatment for late-life depression (e.g., Campbell, 1992; Jarvik, Mintz, Steuer, & Gerner, 1982; Steuer et al., 1984; Thompson, Coon, Gallagher-Thompson, Sommer, & Koin, 2001).

Despite the evidence that cognitive behavior therapy, as well as other psychiatric treatments, is effective for reducing depression in older adults, there is limited evidence that supports the hypothesis that the treatments that have been found to reduce the severity of depression may also be effective for reducing the severity of suicide ideation. One of the few studies to have tested this hypothesis was the PROSPECT (Prevention of Suicide in Primary care Elderly: Collaborative Trial) study (Bruce et al., 2004). Specifically, the aim of this randomized controlled trial was to test the effectiveness of a depression treatment protocol that was administered in primary care settings on reducing the severity depression and suicidal ideation. The PROSPECT intervention was provided in primary care settings given that several studies have found that the majority of older adults who commit suicide have seen their primary care physician within the preceding month of death (see Conwell, 2001; Pearson, Conwell & Lyness, 1997). The PROSPECT study found that suicidal patients who were randomized to the PROSPECT intervention had a more rapid decline in the suicide ideation than patients who were randomized to the usual care control condition. Although the PROSPECT study intervention showed a greater reduction in suicidal ideation as well as depression than usual care only, unpublished analyses have found that a significant proportion (33%) of patients continued to report suicide ideation.

Despite recent advances in the development of cognitive behavioral interventions that target suicide ideation and behavior in younger adults (e.g., Brown et al., 2005), very few treatments have been evaluated that are specifically designed to reduce suicidal ideation among older adults. Further research is warranted in examining whether cognitive behavioral interventions are effective for reducing suicide ideation as well as other risk factors in older adults. We propose that interventions that focus on decreasing the severity of multiple risk factors including suicide ideation and that foster the existence of protective factors may be more efficient and efficacious in preventing suicide than interventions that mostly focus on the treatment of depression.

Suicide Risk Assessment

The assessment of the risk for suicide can be one of the most challenging tasks for a mental health professional. Detailed information on suicide risk assessment is available from the *Practice Guideline for the Assessment and Treatment of Patients with Suicidal Behaviors* (American Psychiatric Association, 2003). Epidemiological risk factors may be evaluated when assessing for the risk of suicide. Higher risk for suicide may be associated with a moderate to severe depression total score greater than 19 on the *Beck Depression Inventory – II* (BDI-II; Beck, Steer, & Brown, 1996), high hopelessness total score greater than 8 on the *Beck Hopelessness Scale* (Beck & Steer, 1988), few (if any) reasons for living could be identified by the patient, poor impulse control (e.g., anger outbursts), poor treatment compliance, previous suicide attempts, previous hospitalization for depression or suicidality, poor social support (e.g., no known neighbors or friends identified by patient as supportive), feeling like a burden on the family, any substance abuse or dependence diagnosis, any psychiatric diagnosis involving psychosis or mania, or a recent stressful life event (e.g., physician indicates that no further treatment for medical condition is available, death of loved one, bankruptcy). Protective factors or deterrents to suicide should also be assessed (e.g., religious beliefs, fear of disapproval, fantasies about their legacy if the death is by suicide, family obligations, etc.).

If any of these risk factors are indicated, the following questions may be asked: (a) Have you been thinking about death or dying? (b) Have you had thoughts that life is not worth living? (c) Have you been thinking about killing yourself?" If the answer is yes, patients may be asked whether they have any intent or desire to commit suicide and if they have thought about a specific method to do so. The availability of firearms should always be assessed and should be removed from the household for individuals who may be at risk for suicide. Although most suicide attempts in the US are by overdose of medications, the majority of completed suicides are carried out by firearms (Garand, Mitchell, Dietrick, Hijjawi, & Pan, 2006; Shenassa, Rogers, Spalding, & Roberts, 2004).

In general, the therapist should be very cautious about underestimating the risk for suicide given that older adults may be reluctant to reveal information regarding their suicide intentions. For example, a psychological autopsy study found that older adults who committed suicide were more likely to have avoided intervention and to have taken precautions against discovery and were less likely to have communicated their intent to others than younger adults (Conwell et al., 1998). When suicide is suspected by the therapist but denied by the older adult, family members or friends should be asked for collateral information on potential suicide risk factors. For example, family members may be asked whether they have observed any warning signs or clues such if the elder recently made or revised a will and gave away meaningful property.

For a more comprehensive assessment of suicide ideation, the *Scale for Suicide Ideation* may also be administered (SSI; Beck et al., 1979). The SSI is a 21-item, interviewer-administered rating scale that measures the current intensity of patients' specific attitudes, behaviors, and plans to commit suicide on the day of the interview. Each item consists of three options graded according to suicidal intensity on a 3-point scale ranging from 0 to 2. The ratings for the first 19 items are summed to yield a total score, ranging from 0 to 38. The SSI consists of five screening items: three items assess the wish to live or the wish to die and the two remaining items assess the desire to attempt suicide. If the respondent reports any active or passive desire to commit suicide, then 14 additional items are administered. Individual items assess suicidal risk factors such as the duration and frequency of ideation, sense of control over making an attempt, number of deterrents, and amount of actual preparation for a contemplated attempt. Two additional items record incidence and frequency of previous suicide attempts. The SSI takes approximately 10 minutes to administer.

The SSI has been standardized with adult psychiatric patients in psychiatric inpatient (Beck, Steer, Kovacs & Garrison, 1985) and outpatient settings (Beck, Brown, & Steer, 1997). The SSI also has been administered to elderly clinical populations (Bruce et al., 2004; Mireault & de Man, 1996). The predictive validity of the SSI for completed suicide has been established for adults (including older adults) seeking outpatient psychiatric treatment (Beck, Brown, Steer, Dahlsgaard, & Grisham, 1999; Brown et al., 2000).

In accordance with the American Psychiatric Association's practice guideline (cf. American Psychiatric Association, 2003), professionals who become aware of an individual who is in imminent danger of harming oneself will need to take action in an effort to prevent such an occurrence. If the patient is determined to be in imminent danger, specific steps for evaluating the older adult for possible hospitalization or other higher level of care should be conducted. In addition, permission to contact a family member and other health professionals who are responsible for the patient's care should be obtained and they may be informed about the patient's suicidal thoughts. Consultation with family members and professionals may also include a discussion of possible hospitalization or other treatment and the type and frequency of contact with the patient (for a more comprehensive description of high-risk management strategies for suicidal patients, see Brown et al., 2000).

In addition to conducting an initial comprehensive assessment of suicide risk, the ongoing assessment of suicide risk during treatment is crucial to ensuring the safety of the patient. One systematic way to assess for risk is for the patient to complete the BDI-II (Beck et al., 1996) at each outpatient visit. If the patient endorses scores greater than 0 on Item 2 (Hopelessness) or Item 9 (Suicidal Thoughts), then a more comprehensive assessment of suicide risk can be conducted.

Cognitive Therapy for Suicidal Older Adults and Case Example

The therapist may follow the general principles and techniques of cognitive therapy for older adults as described elsewhere in this volume. However, there are several important cognitive behavioral strategies that may be particularly useful when treating suicidal elders. Some of these strategies as described below include (a) developing a safety plan, (b) constructing a cognitive case conceptualization, (c) targeting hopelessness and increasing problem-solving skills, (d) improving social resources, (e) improving adherence to medical regimen, (f) increasing reasons for living, and (g) termination issues. A case example is provided to illustrate the application of some of these strategies.

Developing a Safety Plan

The development of the safety plan may be one of the first interventions that therapists may use when treating a patient who has made a recent suicide attempt, who has active suicide ideation, or who is at risk for becoming suicidal. A safety plan is a hierarchically arranged written list of coping strategies that can be used during or prior to a suicidal crisis. The safety plan is developed through a collaborative process by both the therapist and the patient and contains self-help strategies that may be used during a crisis. The safety plan helps to enhance patients' sense of control over suicidal thoughts and urges and helps to promote the expectation that patients can overcome the desire to commit suicide.

As a therapeutic strategy, it is important to have patients try to cope on their own with an impending crisis of suicidal thoughts and urges. Patients may be asked: "What helpful things can you do by yourself if you start to feel suicidal again?" Typical examples of self-help coping strategies for older adults may involve activities that provide a distraction such as listening to music, reading, walking, watching television (comedy), or praying or meditating. The therapist can help patients identify a few of these strategies that they may use as well as help them to prioritize which activities may be the most effective. As patients learn new skills during treatment, additional self-help coping strategies may be included in the safety plan. Although the safety plan is developed at the beginning of treatment, it is reviewed and modified in subsequent sessions as an increasingly larger number of coping skills are available to patients. As part of the safety plan, the therapist should also help the patient to identify and prioritize individuals who may be the most likely to be available and who may be the most helpful to the patient. For these individuals, the names and phone numbers of the individuals are noted. If the friend or family member is unavailable or not helpful, then the names and numbers of the mental health professional or primary care physician may be listed. If a health professional is not available (e.g., weekends) then the name, address, and phone number of the nearest psychiatric or medical emergency department should also be available. Patients should be instructed to contact a health professional or mental health professional if the patient's friends or family members, who are listed on the safety plan, are unavailable or not helpful. The safety plan should emphasize that appropriate professional help is accessible in a crisis and, when necessary, indicate how services can be obtained.

Constructing a Cognitive Case Conceptualization

One fundamental characteristic of cognitive therapy is that it is structured and time-limited. Patients understand that they will take an active, systematic problem-solving stance in session and that they will work collaboratively with their therapist to address their life problems in a goal-directed manner. Unlike cognitive therapy with other types of patients, cognitive therapy with suicidal patients involves identifying the cognitions that lead to suicide and working on life problems specifically *as they relate to their suicide ideation, hopelessness, or suicide attempt*. That is, a focus on suicidal thoughts and behavior is central to cognitive therapy with these patients. A case illustration may help exemplify how the cognitive conceptualization can be used to inform treatment.

Case Example

"Harold" entered therapy at the age of 73, 9 months after his wife died of cancer. Following her death, Harold fell into a deep depression and frequently felt guilty for not having spent more leisure time with his wife while she was alive. Specifically, he felt remorse for not taking her on vacation and spending more time at home with the family. At the time he entered therapy, he reported being indifferent about living. For several months, he contemplated ending his life by stopping his medications. Harold was taking medication for numerous health problems such as diabetes, hypertension, and cardiovascular disease. For Harold, death meant an end to his loneliness, an escape from his numerous medical problems, and reunification with his late wife. He had three adult children but he lived alone and was estranged from most of his family. He also believed that he no longer could function independently given his declining physical health (see Core Beliefs, Fig. 1). Thoughts of suicide were strongest when Harold felt hopeless about the future, when he focused on his physical limitations and when he felt ignored by his children. For example, after not receiving a phone call from his children for some days, he experienced the automatic thought "All my children want from me is money. I bet if I sent

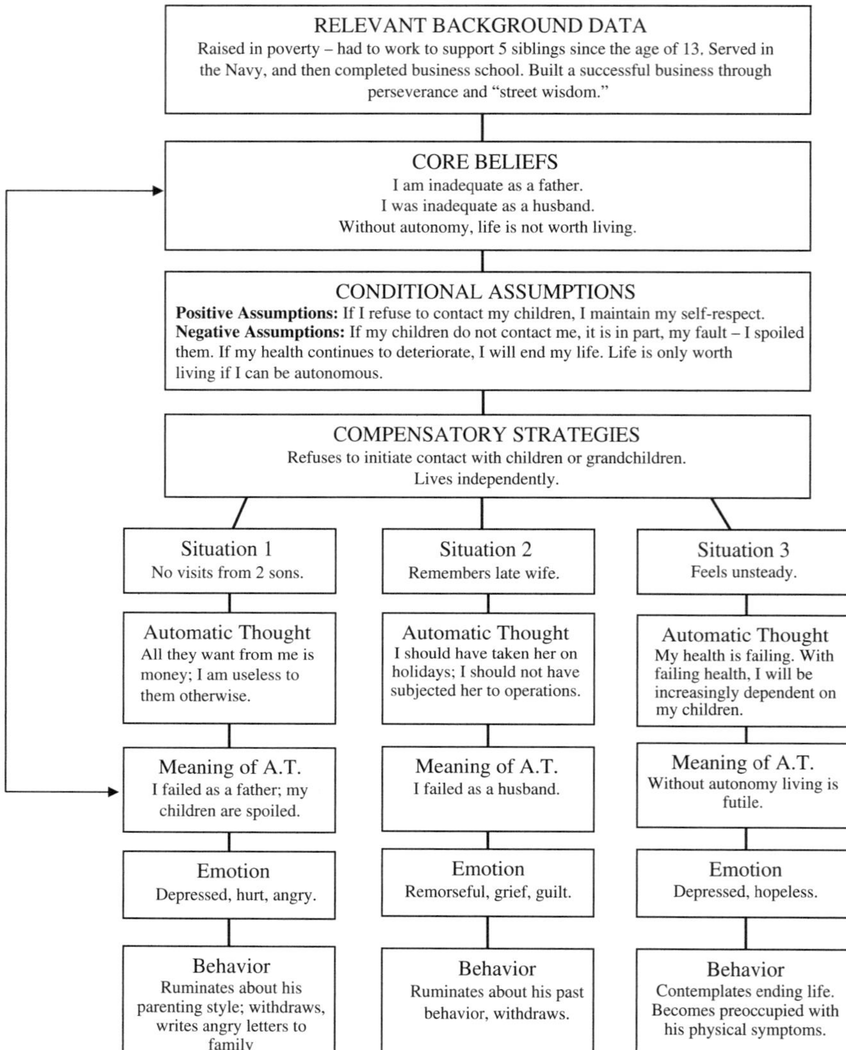

FIGURE 1. Cognitive case conceptualization diagram.

them money, I would receive a phone call." Harold would then conclude that he could not depend on his children for support and that it would be better that he were dead than to suffer the shame and humiliation of being an unwanted imposition on them. He reported thinking "I don't want to be dependent on anyone, particularly on people who do not think well of me. What's the use of living if I cannot look after myself? I will just stop taking my medication" (see Conditional Assumptions, Fig. 1).

Cognitive therapy proceeded on the basis that Harold's suicidal thoughts were activated by feelings of hopelessness, a diminished sense of autonomy, and feelings of neglect and isolation. Three goals were therefore established in therapy. First, therapy focused on reducing Harold's sense of hopelessness about the future and increasing his problem-solving skills. Second, therapy assisted Harold in developing strategies that would enable him to function more independently and to keep his medical appointments so that he would be able to retain his present health status. Third, therapy helped Harold to broaden his social and support network so that he did not feel completely reliant on his children for support and company. Based on the conceptualization that was developed for Harold, specific interventions were then identified that could address these problems.

Targeting Hopelessness and Increasing Problem-Solving Skills

As previously described, the overarching model for this therapy is based upon cognitive theory for depression, hopelessness, and suicide ideation. Given that hopelessness has been found to be one of the strongest predictors for future suicide in older men, a primary focus of therapy is to develop a sense of hope by identifying hopeless cognitions and finding evidence for refuting these thoughts. Patients may be taught to identify life stressors that precipitate or contribute to their hopelessness and generate alternative solutions to their problems. For example, one problem-solving strategy that may be used is to assist the patient in developing some potential solutions to the problem and then list the pros and cons of each solution.

These interventions were provided to help Harold feel less hopeless about his physical health and about his dependency. Through cognitive therapeutic techniques such as Socratic questioning and role-plays, Harold was able to refute negative automatic thoughts such as "There is nothing I can do about my health" and develop a more adaptive thought such as "I can exercise in order to prevent myself ending up in a wheelchair." As a way of helping Harold generate solutions for problems, the therapist used a two-pronged approach. First, the therapist encouraged Harold to talk about his previous experiences of success. Harold had a considerable history of solving problems in his business. While Harold talked about these situations, he remembered the resources and skills he did have that could be of help in his current situation. Second, the therapist asked Harold what he would advise someone else to do who was in his situation. In this context, Harold often had no difficulty generating solutions. For example, Harold identified that he felt uncomfortable relying on his son for bringing him to his therapy session because this arrangement activated for him the feeling that he was no longer independent. While talking about his achievements in the past, Harold recalled that he had been actively involved in helping patients with transportation problems. He was therefore able to then think about how he could solve his own problem, and consequently arranged for a community bus to bring him to, and pick him up from medical appointments. Harold felt a great sense of achievement at having solved that problem.

Improving Social Resources

High levels of social isolation and poor social support networks are related to suicidal ideation in older adults (Alexopoulos, Bruce, Hull, Sirey, & Kakuma, 1999; Mireault & de Man, 1996). Often, a major goal of therapy is to develop the patients' social resources. Patients may be encouraged to schedule pleasurable social activities and expand their network of social support. For some patients, interpersonal conflict may trigger suicide ideation and hopelessness. Thus, therapy may focus on resolving interpersonal conflict. For example, assertive training and anger management techniques may be used to help patients address problematic relationships.

Throughout therapy, Harold continued to be tormented by a sense of neglect by his children. He felt that his children wanted nothing to do with him and was very upset with one son in particular, because he had not spoken to him for some months. Harold was reluctant to call this son because as he put it "I have my pride. Children should come to their father, not the other way around." The therapist engaged Harold to think about the pros and cons of making the first contact and eventually, Harold decided not to pursue reunification, but rather to accept his family had "abandoned" him and that he should now focus on what he could do to improve his own situation. Therefore, respecting Harold's wishes not to initiate contact with his family, therapy focused on helping Harold feel less isolated.

Behavioral activation strategies such as activity scheduling were offered to help Harold structure his day and incorporate a mix of autonomous, pleasurable, and social activities; Harold found these strategies helpful. He prided himself to be a "man of his word" and so he often followed through with commitments that he made in therapy to do homework assignments such as visit the senior citizen's center, have a routine in the morning, read more widely, or attend to his paperwork. Gradually, his social network increased as he met people at the senior citizen's center. He was also able to arrange for someone to take him on scenic drives. In addition, he was able to use activity-scheduling methods to make and attend medical appointments. Gradually over the first few sessions, Harold felt that he was leading a more active life, and felt less isolated.

Improving Adherence to Medical Regimen

Late-life depression can often occur in conjunction with the physical illnesses associated with old age. Symptoms associated with depression, including hopelessness and suicidal ideation, may interfere with medical compliance thus worsening health outcomes (Montano, 1999). At the onset of each therapy appointment, therapists will assess the patients' adherence to their medical regimen. If adherence to their medical regimen is identified as a problem, the therapist will use cognitive strategies to examine their negative beliefs and behaviors regarding their regimen. Once these beliefs have been examined, maladaptive thoughts will be challenged and solutions will be generated to overcome the difficulties.

For Harold, the dominant beliefs that stopped him from attending medical appointments were "What's the use of going to doctors, I'm too old to change. My health problems are irreparable." Using the techniques described above (e.g., therapist role-playing the patient, while patient is playing a patient's advocate), Harold was able to challenge these thoughts, and list reasons why compliance was important. For example, Harold listed that he wanted to see his granddaughter married and another grandchild graduate from college. He also said that he enjoyed sports and wanted to be healthy enough to attend baseball games at the stadium. Further, as mentioned, Harold was also able to point out that his health problems were not as bad as some other people's problems. He could still walk, watch television, read, and hold conversations. He wanted to preserve these abilities. Toward the end of therapy, Harold was attending his medical appointments regularly.

Increasing the Reasons for Living

One strategy that may help patients identify and remind themselves of the reasons for living involves the construction of a Hope Kit. This kit consists of a container that holds mementos (e.g., photographs, letters, souvenirs, Bible vs. prayer cards, etc.) that serve as reminders of reasons to live when they are feeling hopeless. Anecdotally, older adults have found this activity to be a highly rewarding experience that often leads them to discover reasons to live that they had previously overlooked.

Since his retirement a few years back, the death of his wife some months ago and the dissolution of many of his family ties, Harold struggled to find a purpose for his life. Having taken pride in himself for being autonomous all his life, he now rarely saw himself as efficacious and able to take on life's problems, as he was adept at doing in the past. Harold summarized his sense of displacement by stating, "I no longer understand where I belong." Reflecting this sense of being adrift, he would spend most of his time watching television, reminiscing about the past, and ruminating about his current problems. The therapist asked Harold to list the reasons for living. Harold was able to identify several reasons for living such as spending time with his son, traveling, dating, attending the marriage ceremony for one of his grandchildren, and the possibility of meeting his great grandchildren. Harold then constructed a Hope Kit that included photographs of his family and pictures and brochures of places that he would like to visit.

Termination Issues

The acute phase of treatment may end when the patient has made progress toward the treatment goals and no longer experiences any suicidal ideation. When the therapist believes that the patient has made observable gains in therapy, a formal assessment of increased cognitive-behavioral skills is indicated. A relapse prevention task may be conducted that serves as an endpoint assessment of the patients' ability to apply coping skills during times of acute distress. The objective of the relapse prevention task is to prime in session as many of the thoughts, images, and feelings that have been associated with prior suicidal ideation as possible. The therapist then assesses whether patients are able to respond to problems in an adaptive way. Patients may also be encouraged to imagine future scenarios that are likely to precipitate a suicidal crisis and identify the skills that may be used to help them resolve this crisis. Once patients have successfully completed this task, the therapist may determine that patients are better equipped for dealing with situations that are likely to precipitate suicide ideation. At this point, the therapist may choose to terminate treatment or to continue treatment and address other problems in the patients' lives.

Diversity Issues

Over the last two decades, significant gender differences have been found among those who complete suicide, with men accounting for about four out of every five completed suicides among those 65 years and older (CDC, n.d.). Moreover, older men at risk for suicide may be harder to identify since they are less likely to have had a history of previous suicide attempts (CDC, n.d.). Furthermore, suicidal older men appear to be more resistant to suicide prevention interventions. A study in Gotland, Sweden, found that a suicide prevention training program for general practitioners significantly reduced female suicide rates over a 5-year period, but that male suicide rates remained impervious to the intervention (Rutz, Walinder, Von Knorring, Rihmer, & Philgren, 1997).

In Honolulu, older adults comprised the highest risk group for death by suicide, but difference existed for gender, race, and ethnicity. Male elders in the sample died by suicide more often than females by a ratio of 2.3:1 (Purcell, Thrush, & Blanchette, 1999). Strong ethnic and religious predilections of the cultures that migrated to Honolulu were reported to influence methods by which suicide occurred (Purcell et al.). Also, in this sample depression was the most common diagnosable mental disorder (Purcell et al.). However, only 13% of the group had evidence of antidepressants on their toxicology reports. This either signifies undertreatment by mental health providers or that these elderly persons were neglecting to take their prescribed medication (Purcell et al.). More attention from health-care providers should be designated for older adults with an active psychiatric illness.

A Swedish study examining the influence of demographic factors and ethnicity (defined by the study as being foreign-born) found that ethnicity was associated with an increased risk to die by suicide in both sexes and in all age groups (Johansson, Sundquist, Johansson, Qvist, & Bergman, 1997). Moreover, overcrowding was reported as a risk factor for suicide in foreign-born older adults, but financial problems were not a significant factor. Older adults, especially elderly women, were found to experience higher levels of social isolation in large cities that have a large population as compared with rural towns with a small population (Johansson et al.). Therapists should consider the effects of cultural heritage and geographic relocation when assessing for suicide risk.

Summary

Suicide among older adults is a major public health problem given the higher rates of suicide among elders compared to other age groups. More specifically, older men have higher suicide rates than other age/gender groups. However, despite the prevalence of suicide in older adults, there have been few treatments that have been found to be efficacious for reducing the risk of suicide. Given the evidence that cognitive therapy is effective for the treatment of depression in the elderly, the application of this treatment for reducing other risk factors for suicide such as suicide ideation and hopelessness appears promising. A cognitive case conceptualization may be developed for each individual patient that includes an assessment of the individual vulnerability factors that are associated with suicide ideation and acute hopelessness. A treatment plan may then be developed to target these vulnerability factors and may include specific cognitive and behavioral strategies such as developing a safety plan, teaching problem-solving skills, increasing social activities, resolving interpersonal conflict, improving adherence to medical treatment, and increasing patients' reasons for living. An important aspect of treatment involves the evaluation of whether patients can successfully use the skills learned in therapy during a suicidal crisis.

Acknowledgements. This manuscript was supported by a grant (P20 MH071905) from NIMH.

References

Alexopoulos, G. S., Bruce, M. L., Hull, J., Sirey, J. A., & Kakuma, T. (1999). Clinical determinants of suicidal ideation and behavior in geriatric depression. *Archives of General Psychiatry, 56*, 1048–1053.

American Psychiatric Association. (2003). *Practice guideline for the assessment and treatment of patients with suicidal behaviors*. Arlington, VA: American Psychiatric Association.

Beck, A. T. (1967). *Depression: Causes and treatment*. Philadelphia, PA: University of Pennsylvania Press.

Beck, A. T. (1976). *Cognitive therapy and the emotional disorders*. New York: Meridian.

Beck, A. T., Brown, G., Berchick, R. J., Stewart, B., & Steer, R. A. (1990). Relationship between hopelessness and ultimate suicide: A replication with psychiatric outpatients. *American Journal of Psychiatry, 147*, 190–195.

Beck, A. T., Brown, G. K., & Steer, R. A. (1997). Psychometric characteristics of the Scale for Suicide Ideation with psychiatric outpatients. *Behaviour Research and Therapy, 35*, 1039–1046.

Beck, A. T., Brown, G. K., Steer, R. A., Dahlsgaard, K. K., & Grisham, J. R. (1999). Suicide ideation at its worst point: A predictor of eventual suicide in psychiatric outpatients. *Suicide and Life-Threatening Behavior, 29*(1), 1–9.

Beck, A. T., Kovacs, M., & Weissman, A. (1975). Hopelessness and suicidal behavior: An overview. *Journal of the American Medical Association, 234*, 1146–1149.

Beck, A. T., Kovacs, M., & Weissman, A. (1979). Assessment of Suicidal Intention – Scale for suicide ideation. *Journal of Consulting and Clinical Psychology, 47*, 343–352.

Beck, A. T., & Steer, R. A. (1988). *Manual for the Beck Hopelessness Scale*. San Antonio, TX: Psychological Corporation.

Beck, A. T., Steer, R. A., & Brown, G. K. (1996). *Beck Depression Inventory Manual* (2nd ed.). San Antonio, TX: Psychological Corporation; Harcourt Brace and Company.

Beck A. T., Steer, R. A., Kovacs, M., & Garrison, B. (1985). Hopelessness and eventual suicide – A 10 year prospective study of patients hospitalized with suicidal ideation. *American Journal of Psychiatry, 142*, 559–563.

Brown, G. K., Beck, A. T., Steer, R. A., & Grisham, J. R. (2000). Risk factors for suicide in psychiatric outpatients: A 20-year prospective study. *Journal of Consulting and Clinical Psychology, 68*, 371–377.

Brown, G. K., Tenhave, T., Henriques, G. R., Xie, S. X., Hollander, J. E., Beck, A. T. (2005). Cognitive therapy for the prevention of suicide attempts: A randomized controlled trial. *Journal of the American Medical Association, 294*, 563–570.

Bruce, M. L., Tenhave, T., Thomas. R., Reynolds, C. F. III, Katz, I. I., Schulberg, H. C., et al. (2004). Reducing suicidal ideation and depressive symptoms in depressed older primary care patients: A randomized controlled trial. *Journal of the American Medical Association, 291*, 1081–1091.

Byrne, G. J., & Raphael, B. (1999). Depressive symptoms and depressive episodes in recently widowed older men. *International Psychogeriatrics, 11*, 67–74.

Campbell, J. M. (1992). Treating depression in well older adults: Use of diaries in cognitive therapy. *Issues of Mental Health Nursing, 13*, 19–29.

Carney, S. S., Rich, C. L., Burke, P. A., & Fowler, R. C. (1994). Suicide over 60: The San Diego study. *Journal of the American Geriatrics Society, 42*, 174–180.

Centers for Disease Control and Prevention. (n.d.). National Center for Injury Prevention and Control. Web-based Injury Statistics Query and Reporting System (WISQARS). Retrieved March 2, 2007, from http://www.cdc.gov/ncipc/wisqars/default.htm.

Clark, D. A., & Beck, A. T. (1999). *Scientific foundations of cognitive theory and therapy of depression*. New York: John Wiley & Sons.

Conwell, Y. (2001). Suicide in later life: A review and recommendations for prevention. *Suicide and Life-Threatening Behavior, 31*, 32–47.

Conwell, Y., & Brent, D. (1995). Suicide and aging I: Patterns of psychiatric diagnosis. *International Psychogeriatrics, 7*, 149–164.

Conwell, Y., Dubertstein, P. R., & Caine, E. D. (2002). Risk factors for suicide in later life. *Biological Psychiatry, 52*, 193–204.

Conwell, Y., Duberstein, P. R., Cox, C., Herrmann, J. H., Forbes, N. T., & Caine, E. D. (1996). Relationships of age and Axis I diagnoses in victims of completed suicide: A psychological autopsy study. *American Journal of Psychiatry, 153*, 1001–1008.

Conwell, Y., Duberstein, P. R., Cox, C., Herrmann, J. H., Forbes, N. T., & Caine, E. D. (1998). Age differences in behaviors leading to completed suicide. *American Journal of Geriatric Psychiatry, 6*, 122–126.

Conwell, Y., Rotenberg, M., & Caine, E. D. (1990). Completed suicide at age 50 and over. *Journal of the American Geriatrics Society, 38*, 640–644.

Draper, B. (1994). Suicidal behaviour in the elderly. *International Journal of Geriatric Psychiatry, 9*, 665–661.

Gallagher-Thompson, D., McKibbin, C., Koonce-Volwiler, D., Menendez, A., Stewart, D., Thompson, L. W., et al. (2000). Psychotherapy with older adults. In C. R. Snyder & R. E. Ingran (Eds.), *Handbook of psychological change: Psychotherapy processes and practices for the 21st century* (pp. 614–637). New York: John Wiley & Sons.

Garand, L., Mitchell, A. M., Dietrick, A., Hijjawi, S. P., & Pan, D. (2006) Suicide in older adults: Nursing assessment of suicide risk. *Issues in Mental Health Nursing, 27*, 355–370.

Jarvik, L. F., Mintz, J., Steuer, J., & Gerner, R. (1982). Treating geriatric depression: A 26-week interim analysis. *Journal of the American Geriatric Society, 30*, 713–717.

Johansson, L. M., Sundquist, J., Johansson, S. E., Qvist, J., & Bergman, B. (1997). The influence of ethnicity and social and demographic factors on Swedish suicide rates. A four year follow-up study. *Social Psychiatry and Psychiatric Epidemiology, 32*, 165–170.

Lynch, T. R., Mendelson, T., Robins, C. J., Krishnan, R., Ranga, K., George, L. K., et al. (1999). Perceived social support among depressed elderly, middle-aged, and young-adult samples: Cross-sectional and longitudinal analyses. *Journal of Affective Disorders, 55*, 159–170.

Mireault, M., & de Man, A. F. (1996). Suicidal ideation among the elderly: Personal variables, stress and social support. *Social Behavior and Personality, 24*, 385–392.

Montano, C. B. (1999). Primary care issues related to the treatment of depression in elderly patients. *Journal of Clinical Psychiatry, 60*(Suppl. 20), 45–51.

Pearson, J. L., & Brown, G. K. (2000). Suicide prevention in late life: Directions for science and practice. *Clinical Psychology Review, 20*(6), 685–705.

Pearson, J. L., Conwell, Y., & Lyness, J. M. (1997). Late-life suicide and depression in the primary care setting. In L. S. Schneider (Ed.), *Developments in geriatric psychiatry: New directions for mental heath services*, (No. 76, pp. 13–38). San Francisco, CA: Jossey-Bass.

Purcell, D., Thrush, C. R. N., & Blanchette, P. L. (1999). Suicide among the elderly in Honolulu County: A multiethnic comparative study (1987–1992). *International Psychogeriatrics, 11*, 57–66.

Rifai, A. H., George, C. J., Stack, J. A., Mann, J. J., & Reynolds, C. F., III. (1994). Hopelessness in suicide attempters after acute treatment of major depression in late life. *American Journal of Psychiatry, 151*, 1687–1690.

Rubenowitz, E., Waern, M., Wilhelmson, K., & Allebeck, P. (2001). Life events and psychosocial factors in elderly suicides: A case-control study. *Psychological Medicine, 31*, 1193–1202.

Rutz, W., Walinder, J., Von Knorring, L., Rihmer, Z., & Philgren, H. (1997). Prevention of depression and suicide by education and medication: Impact on male suicidality. An update from the Gotland study. *International Journal of Psychiatry in Clinical Practice, 1*, 39–46.

Shenassa, E. D., Rogers, M. L., Spalding, K. L., Roberts, M. B. (2004). Safer storage of firearms at home and risk of suicide: A study of protective factors in a nationally representative sample. *Journal of Epidemiology and Community Health, 58*, 841–848.

Steuer, J. L., Mintz, J., Hammen, C. L., Hill, M. A., Jarvik, L. F., McCarley, T., et al. (1984). Cognitive-behavioral and psychodynamic group psychotherapy in treatment of geriatric depression. *Journal of Consulting and Clinical Psychology, 52*, 180–189.

Szanto, K., Gildengers, A., Mulsant, B. H., Brown, G. K., Alexopoulos, G. S., & Reynolds, C. F. (2002). Identification of suicide ideation and prevention of suicidal behavior in the elderly. *Drugs and Aging, 19*, 11–24.

Szanto, K., Reynolds, C. F., Conwell, Y., Begley, A. E., & Houck, P. (1998) High levels of hopelessness persist in geriatric patients with remitted depression and a history of attempted suicide. *Journal of the American Geriatrics Society, 46*, 1401–1406.

Thompson, L. W., Coon, D. W., Gallagher-Thompson, D., Sommer, B. R., & Koin, D. (2001). Comparison of desipramine and cognitive-behavioral therapy in the treatment of elderly outpatients with mild-to-moderate depression. *American Journal of Geriatric Psychiatry, 9*, 225–240.

Thompson, L. W., Gallagher, D., & Breckenridge, J. S. (1987). Comparative effectiveness of psychotherapies for depressed elders. *Journal of Consulting and Clinical Psychology, 55*, 385–390.

Turvey, C. L., Conwell, Y., Jones, M. P., Phillips, C., Simonsick, E., Pearson, J. L., et al. (2002). Risk factors for late-life suicide: A prospective community-based study. *American Journal of Geriatric Psychiatry, 10*, 398–406.

Uncapher, H., Gallagher-Thompson, D., Osgood, N. J., & Bonger, B. (1998). Hopelessness and suicide ideation in older adults. *Gerontologist, 38*, 62–70.

Waern, M., Besko, J., Runeson, B., & Skoog, I. (1999). Suicidal feelings in the last year of life in elderly people who commit suicide. *Lancet, 334*, 917–918.

Wen-Hung, K., Gallo, J. J., & Eaton, W. W. (2004). Hopelessness, depression substance disorder, and suicidality: A community based study. *Social Psychiatry and Psychiatric Epidemiology, 39*, 497–501.

10
Cognitive Therapy for Older People with Psychosis

David Kingdon, Maged Swelam, and Eric Granholm

Psychosis is the most expensive mental disorder to treat, and the cost of care for older people with psychosis (> 65 years old) is even higher than for service users age 30–64 (Jeste et al., 1999). By the year 2050, the population over age 65 will double and the population over age 85 will increase fivefold (Jeste et al., 1999). This will lead to a dramatic increase in older people with psychosis in need of effective services.

Possible risk factors for onset of psychosis in older people can be summarized as follows: (a) *Family history* of Schizophrenia (Almeida, Howard, Forstl, & Levy, 1992; Almeida, Howard, Levy, & David, 1995); (b) *Gender*: the ratio of female to male psychosis of late onset is about 6–10:1 but this difference cannot be simply explained by the fact that women live longer (Almeida et al., 1995; Castle & Murray, 1993); (c) *Early cognitive decline*: is frequently seen in this group of people but there is no evidence of a strong relationship with developing dementia; and (d) *Sensory impairment* (Prager & Jeste, 1993).

Older people with psychosis may have enduring positive and negative symptoms which began earlier in their lives or they may present with late-onset psychosis. A number of questions remain concerning the distinction between early and late onset, but this discussion is beyond the scope of the present chapter.

Adapting Cognitive Therapy for Older People with Psychosis

Special problems common among the middle-aged and older people, including cognitive impairment, interpersonal loss, physical disability, and client beliefs about psychotherapy, can complicate treatment and create barriers to service access. Several modifications have been recommended to accommodate older people in therapy (Arean, 2003; Thompson, 1996), which are discussed in detail in other chapters. Many of these are even more applicable when treating users with psychosis, and will be considered in discussing various cognitive therapy techniques for use with the elderly schizophrenics.

Older people often lose important supports, as friends and relatives become disabled or die. Issues of loss, isolation, and improving social support, leisure activities, and interpersonal communication skills, therefore, are even more important for older service users with psychosis. Transportation and ambulation problems, which are more common among older service users with psychosis, can further exacerbate problems with isolation and withdrawal from society. It may be necessary to provide transportation to therapy sessions or conduct therapy at convenient locations in the community. Providing treatment in the community can also reduce stigma associated with going to a mental health clinic, which can especially deter older service users.

Addressing neurocognitive impairment is another important factor to consider when designing cognitive therapy interventions for older people. Due to normal aging and a lifetime of comorbid factors that can impact cognition in older clients with schizophrenia (e.g., poor education, lack of work, physical illness, poor nutrition, poverty), cognitive functioning may further decline in older clients beyond impairments due to the disorder itself. Simplifying and slowing down discussions, repeated practice of skills, and neurocognitive compensatory aids (e.g., reminder notes, treatment workbooks, writing down information in session, and acronyms to help with skill recall) then can become even more important to offset the problems of neurocognitive impairments common in psychosis and normal aging. Given common fine motor and visual difficulties in older clients, these aids and educational materials should also use large fonts and writing spaces.

Cognitive therapy targets may also change as people age. Aging in people with schizophrenia is typically associated with improvement in positive symptoms and reduced hospitalization; however, approximately 60% of older people with schizophrenia still reside in assisted care settings (e.g., board-and-care homes).

Although positive symptoms in these older service users are often fairly well controlled through pharmacologic treatment, these clients still have severe psychosocial functioning deficits (Jeste et al., 2003). An important target of cognitive therapy, therefore, in older people with psychosis is functioning. Cognitive therapy studies have only recently begun to examine functioning outcomes and a few have found improvement in this domain (Bradshaw, 2000; Granholm et al., 2005; Gumley et al., 2003; Wiersma, Jenner, van de Willige, Spakman, & Nienhuis, 2001).

Evidence for the Efficacy of Cognitive Therapy for Older People with Psychosis

The evidence for the efficacy and effectiveness of cognitive therapy in schizophrenia is now strong with over 20 randomized controlled studies attesting to its beneficial nature (Turkington, Kingdon, & Weiden, 2006). These studies have however concentrated on adult population between 16 and 65, so their generalizability to older groups needs to be considered cautiously.

There has been little research on the efficacy of any form of psychosocial intervention for older adults with schizophrenia (Van Citters, Pratt, Bartels, & Jeste, 2005). Two trials of behavioral skills training interventions for middle-aged and older people with psychotic disorders have been reported. One found improvement in independent living skills, social functioning, and identification and prevention of previously undetected medical conditions in a combined skills training and health management intervention (Bartels et al., 2004). Another trial found improvement in independent living skills and negative symptoms in a functional adaptation skills training (FAST) intervention (Patterson et al., 2003). One case study reported that cognitive therapy combined with clozapine led to complete remission of positive psychotic symptoms and significant improvement in quality of life for an 80-year-old nondemented woman with a diagnosis of paranoid schizophrenia and treatment-resistant auditory hallucinations and a persecutory delusion of being poisoned by blood tests (Pinninti & Datto, 2006). We are aware of only one published randomized controlled trial of an intervention that incorporated both cognitive therapy and social skills training, called Cognitive Behavioral Social Skills Training (CBSST) for older people with psychotic disorders (Granholm et al., 2005). The treatment manual in this study included a client workbook with homework forms. Aids to compensate for cognitive impairment common in both schizophrenia and normal aging were also added. Other age-relevant modifications included identifying and challenging ageist beliefs (e.g., "I'm too old to learn"), age-relevant role-play situations (e.g., talking to a doctor about eyeglasses), and age-specific problem solving (e.g., finding transportation, coping with hearing and vision problems). Clients in CBSST showed significantly greater skill acquisition and functioning and greater cognitive insight after treatment than clients in treatment as usual; and gains in functioning and skill acquisition were maintained one year after treatment ended (Granholm et al., 2007).

Use of Medication

Discussing the use of antipsychotics in this group of patients is beyond the scope of this chapter but it deserves a brief comment. The evidence base for use of antipsychotics in older people to target psychotic symptoms largely depends on extrapolating results from studies targeting younger populations, but adjusting for pharmacodynamics and pharmacokinetics changes in older people, e.g., using significantly smaller doses of antipsychotics. In our experience the use of cognitive therapy for psychotic symptoms forms one part of a package of care including the use of antipsychotic medications where relevant and appropriate. This generally means that cognitive therapy is used alongside medication and can facilitate compliance with medication regimes; in turn, medication effects on acute symptoms can make clients more amenable to cognitive intervention.

Use of Cognitive Therapy in Practice

In principle, the use of cognitive therapy in older people with psychosis can be expected to differ little from that in younger ones. Our experience in working with people in later middle age (50–65) and older has suggested that some additional considerations arise but not fundamental differences (Kingdon & Turkington, 1991). Thus, manuals developed for adults have applicability in this population (Chadwick & Lowe, 1994; Kingdon, Turkington, & John, 1994; Turkington et al., 2006) but relevant considerations are detailed below.

Assessment

Assessment is an important step in treating people with psychotic symptoms. In older patients this process often spans a number of sessions and may take more time than would be expected for other mental health problems. In fact, continued assessment of psychotic symptoms often occurs until therapy itself draws to a conclusion, as frequently reformulation of beliefs goes on in light of changing circumstances and developing trust.

A proper risk assessment should be carried out even before the very first interview through gathering information from the referrer. It is very important to proceed cautiously being aware of changes in client's mental state during the assessment. Any signs of agitation or irritability may necessitate a change of pace, redirection or possibly terminating the session.

Older people with psychotic symptoms, who are graduates from adult services, can still benefit from assessment of their first episode of psychosis, even when occurring many years previously, as their current symptoms are often understandable in the context of those initial events. While memories may have faded, prompting from caregivers or existing medical records can help them reconnect and allow them to reconsider their beliefs even when very strongly held. We have described four subgroups of schizophrenia (Turkington et al., 2006): (a) *sensitivity psychosis* where individuals have had a life-long sensitivity to stress; (b) *traumatic psychosis* where severe traumatic experiences possibly during childhood or wartime, still seriously impact and influence their current mental state; (c) *psychosis due to use of hallucinogenic drugs* from which some patients have failed to recover. (Some of these clients, young in the 1960s when these experiences occurred, are now reaching older age); and (d) "*anxiety psychosis*" which is especially common in patients presenting late in life, where symptoms have arisen in the context of stressful circumstances, e.g., relationship difficulties. Concerns about, but often failure to identify symptoms of anxiety, e.g., abdominal pain or persistent worry, can crystallize into delusional beliefs which dominate the individual's life and often those around them. This can be readily exacerbated by isolation – where discussion and debate of these concerns as they develop have not been possible in order to inject rationality at an early stage. Early dementia may also add to difficulties, e.g., memory lapse leading to confusion and concerns about conspiracies.

People with psychotic symptoms especially older people may be fairly isolated for the majority of time and the assessment process gives them the opportunity to talk about their problems in a nonthreatening environment, express their wishes and concerns, and in itself this catharsis can be therapeutic. As for cognitive therapy generally but especially here, the assessment is conducted in a collaborative way as many clients with psychotic symptoms feel stigmatized and misunderstood, have not been given time to talk about their beliefs and experiences, indeed are continually dissuaded from it, and experience harassment and loss of control in their lives. Therefore, giving clients the lead in expressing their needs fundamentally establishes the foundation for engaging such clients, as it is one of the rare occasions where they feel a degree of control in a human-to-human interaction and they are being listened to. The therapist assessment goes beyond understanding the clients' presenting problems and background; it helps make sense of the symptoms, beliefs, vulnerabilities, ongoing stresses, and strengths. This in turn will inform a preliminary formulation based on the *stress-vulnerability model*, and will aid the collaborative selection of interventions most appropriate at this stage of the illness.

Clients, especially those with negative symptoms, may find it difficult to talk about their problems. Intellectual impairment can also complicate this process. Therefore, it is important to get a collateral history from previous records or mental health workers, which in itself can be illuminating to the therapist as to how best she can engage the client and what interests or topics are more likely to break the ice and help the client speak about difficulties.

Clients with psychotic symptoms often don't accept that they are ill and frequently have suspicions about mental health workers, in particular. Therefore, it may be wiser to select appropriate terminology which the client agrees to, e.g., avoid using the word "illness" and ask a more general question about "problems" which wouldn't impair the engagement process. An important underlying principle in cognitive therapy for psychosis is that nearly all delusional beliefs are understandable in the context from which they arose. In older people, for example, what can be perceived as persecutory delusions may have a basis in reality and can raise issues of abuse of older people.

Therapist themselves may need to understand their beliefs about the assessment process, especially those who like to follow a tidy sequential process, as it can be far from that in treating psychotic symptoms. There may need to be variation in the pace of assessment. In older patients frequent shifting of focus may be required more often, especially when clients become unduly distressed. Timing correctly is a crucial part of the assessment process.

Therapists need to look at assessment as an exercise of gradually building trust with clients, some of whom will require a better understanding of their symptoms or become better equipped with coping mechanisms before they can open up and discuss their problems in more detail. Sensitivity on the part of the therapist will aid this process and enhance engagement of the patient into a constructive therapeutic relationship.

The therapeutic alliance is particularly important in treating psychotic symptoms in older people. Therefore, being sensitive and responding to changes in mental state and demeanor during the session will enhance this process. Using open-ended questions will give a focus on the person rather than the therapist. Direct questions in turn can be very effective if used gently and sensitively in an exploratory rather than an inquiring way.

There is an emphasis on a full assessment in cognitive therapy to understand the development of beliefs and the personal history of clients. Especially in older people when there is a rich experience and many life-long events, therapists need to be flexible but thorough in dealing with psychotic symptoms. It is always advisable to gauge the client's response to exploring certain areas of their history and certainly respect their wishes if they prefer not to discuss certain aspects. It is always possible to return to sensitive areas later; clients appreciate such sensitivity, and are more likely to disclose relevant information which can be embarrassing, often when insight is emerging as they reflect on the past.

Assessing and understanding the circumstances leading up to the onset of psychotic symptoms is crucial to later therapeutic work, as frequently the content of the beliefs or perceptual experiences has a direct link to the client's circumstances and stresses at that time. This can be done simply by collaboratively exploring the emotions, thoughts, and behaviors at the time of onset and making links to explain the emergence of the final version of symptoms. Sometimes the client's report may not be enough and collateral history may be needed to clarify the onset of symptoms. The therapist should take into consideration that the client's symptoms are also colored by several other factors in addition to the stresses at the time of onset of symptoms, e.g., adverse hospital experiences, side effects of medications, and certainly stigmatization. In assessing psychotic symptoms the therapist should be aware that there may be a number of other symptoms, which may or may not be related to the psychotic symptoms, and the task is to elicit the ones mostly relevant to the client's current condition. For example, coexisting depression, social phobia or anxiety may be as important to the client as their concern about their beliefs or hallucinations. Older people may not readily admit to feeling "depressed," and therefore careful assessment of other depressive symptoms is important to inform the risk assessment. Misuse of alcohol is a common problem among older people, though it may go undetected for a number of years; therefore it is important to ask about it in a nonjudgmental way and establish the links with psychotic symptoms if they exist.

Social circumstances need consideration before rushing into cognitive therapy, as doing so without addressing the client's social milieu may lead to difficulties. As people get older their incomes tend to decrease which may have an impact on their ability to socialize or maintain their lifestyle. Their children may be living far away, which leaves them with isolation that may feed into or can be a source of their delusional beliefs. They may need assistance with activities in daily living, e.g., mobility, personal hygiene, and domestic activities, which can have a huge impact on the therapeutic alliance. Freeing the client from worries may in turn help them to understand the psychotic symptoms, and to achieve relief from the distress associated with these.

Cognitive behavior therapy emphasizes the role of measurement to establish a baseline against which to measure change. Rating scales can give the therapist and client an idea about the global functioning of the client; e.g., Health of the Nation Outcome Scale (HoNOS) or Global Assessment of Functioning (GAF) with an additional scale which assesses specific symptoms, e.g., Psychotic symptoms rating scale (PSYRATS) in older people (Haddock, McCarron, Tarrier, & Faragher, 1999). Other scales measuring the activities of daily living and cognitive function can be valuable to inform psychosocial intervention.

Formulation and Goal Setting

Case formulation for psychotic symptoms need to be individualized so that the client can make sense of and have an overview of how personal experiences, vulnerabilities, stresses have come together to lead to the presenting problem. Therapists work with clients to clarify the emotions, thoughts and subsequent behaviors, and possible physical symptoms relevant to their experience. Clients' conviction in their thoughts or delusions may not change much during therapy, but the therapist should avoid colluding with the belief; rather the emphasis can be on developing coping mechanisms in which one or two distinct issues may be identified to work on.

In doing so clients may need support from the therapist in order to discuss the different elements of formulation in a collaborative way. The therapist's role here is to give one's own view frankly and openly on the version of events given by the client during the assessment process, but the aim is that the client makes links and points out possible areas for therapeutic intervention.

Therapists may vary the degree of complexity of their formulation depending on the individual client needs; e.g., intellectual impairments in older people. A simple diagram may be sufficient in someone with ensuing dementia, while a more complex formulation incorporating all elements, i.e., predisposing, precipitating, perpetuating and protective factors, rules for living, core beliefs, etc., may be relevant for another client. Therapists need to be aware that formulation is not about having a full overview to explain the client's problems, but rather the emphasis is on finding relevant connections that can assist in reattribution of symptoms. This in itself can be considered as therapeutic work which will pave the way to more specific work later on.

Goal setting with older patients who have been ill for 30 years or more can present unique challenges encountered less frequently with more recent onset patients. Older patients have often experienced a lifetime of failures, which can lead to low expectations for success. Information processing biases also often lead to a focus on failures while ignoring past achievements. By acknowledging a goal, chronic patients must accept the risk of yet another failure. Risk assessment is often exaggerated into symptom exacerbation and rehospitalization. By this stage of illness, older psychotic patients have also been informed that they have a severe, chronic brain disease with little hope of recovery. Typically they have had less exposure to modern recovery and psychosocial rehabilitation models that are transforming healthcare systems in order to focus more on strengths and hope for functional achievement. Hopelessness in providers can be passed on to hopelessness in older chronic patients.

These factors can create challenges for eliciting a goal in therapy. This may require patience and many sessions. Cognitive therapy tools can be used to address dysfunctional performance beliefs, negative expectancies, ignoring the positive and other attributional styles that can lead to hopelessness and reluctance to set a meaningful goal. Older patients may also initially offer a goal that they know providers want to hear (e.g., "I want to take my medications more regularly"), rather than a more personally meaningful goal. Goals may need to be revised as more meaningful goals emerge. It may be useful to ask patients to recall a time when they were functioning better in the past or imagine a time years in the future when they will be functioning much better and ask about daily behaviors. Ask, "What would you be doing if you weren't ill?" This can stimulate a helpful discussion about desired, meaningful functional behaviors (relationships, work, and school activities). Finally, for older patients who are particularly reluctant to talk about goals, we have found it useful to frame the discussion in terms of "roles," rather than "goals." This discussion focuses on eliciting all the roles played by the patient, including a person with an illness, son/daughter, roommate/neighbor, friend, partner, parent, student, employee, etc. A pie chart diagram can be useful here to illustrate the exclusion of many of these roles in thinking about the self, with the majority of the pie filled in with the attribution of the self as, "schizophrenic." The key is to expand the possible roles, and therefore goals, to arrive at a personally meaningful goal.

We have found it most helpful to focus on specific, functional behavioral goals (increasing socialization, relationships, leisure activities, living skills to promote independent housing, school, work/volunteering) chosen by patients. Emotional or symptom goals (e.g., "I want to stop being sad" and "I don't want to hear voices") are less helpful, unless framed in terms of their impact on functional behaviors. For example, one can ask, "How would you know you weren't sad anymore?" The response is often something like, "I would be getting up and doing more things." The therapist can then ask what things and frame the functional behavior as the goal (e.g., doing things with friends; taking a class). Goals should also be specific and stated as attainable behaviors. For example, the therapist can ask, "How will you know the goal has been achieved? What will you be doing differently each day if you achieve the goal?" The therapist's goal is to help older patients with psychosis set meaningful, realistic, attainable, specific goals stated in terms of functional behaviors.

It is important when setting goals with older people, as with others, to be realistic and fully collaborative so that goals are readily achievable, e.g., chronic pain or fear of being hassled and knocked over is a common problem in old people and setting a behavioral goal may need to take this into account. On the other hand goal setting in case of bereavement or permanent disability may actually facilitate acceptance and help the client to find a direction away from his distressing thoughts and psychotic symptoms.

Psychoeducation and Normalization

Psychoeducation is a key in helping clients with psychotic symptoms to make sense of their experiences, especially that these symptoms subject their sufferer to stigmatization and social isolation. Psychoeducation is a two-way process whereby the clients themselves come up with psychological explanations of their symptoms using guided discovery with inclusion of information from the therapist only where necessary. Normalization is a specific form of psychoeducation which looks at vulnerability to psychotic symptoms as lying on a continuum whereby any individual is subject to experience them if exposed to certain circumstances and stresses. Psychotic illness should be primarily related to the stress which triggered it (e.g., sensory or sleep deprivation, or unusual stresses) Bereavement in particular is relevant to older people, and can precipitate symptoms such as hallucinations or increase suggestibility leading to incorrect conclusions about events being reached. Psychoeducation needs to be individualized and sensitive to client's needs, as it has been shown in research that it is associated with suicidal thinking and increased depression especially with increased acceptance of the illness (Cunningham Owens et al., 2001).

The aim of psychoeducation should be to enhance client's acceptance that things are not right or that they are not well in some way or another and drawing from the earlier work on vulnerability and stresses can be individualized to the client's circumstances. This added to collaborative discussion of how the original stress led to secondary repercussions on his or her life will give the client a global picture and sense of his current circumstances. A negotiating approach is a key in psychoeducating clients about medications; it suggests a much more favorable long-term outcome than an authoritarian approach of health care delivery. Normalization can target automatic thoughts as well as the relationship between thoughts and actions, so that the responsibility for actions is retained. Normalization can't be complete without tackling the effects of stigmatization the individual client is experiencing, and that may involve caregivers and/or therapists to reduce the secondary damage caused by the original symptoms. This in turn improves client's self-esteem and helps both the therapist and the client to reattribute the experiences from external to internal causes without minimizing or underestimating the impact of this experience on her or his life.

Working with Hallucinations

Clients who suffer from hallucinations can benefit from a number of cognitive interventions directed at either a reattribution of the experience to internal rather than external causes; e.g., understanding hallucinations as negative automatic thoughts rather than an external phenomenon. This generally involves establishing clearly what the client perceives – are they hearing something? "Like me talking to you now" or seeing something – what does it look like? Or feeling something: "touching you?" Sometimes the touch may feel sexual or the voice be saying unpleasant things about them, that the client may be understandably reticent about discussing before a good therapeutic relationship has developed. Once the experience has been shared, asking for an explanation of it by the client is important to establish their beliefs. Often they don't know how it comes about and it is possible to "normalize" the experience by describing how such phenomena can commonly occur when under stressful circumstances, e.g., sleep deprivation, extended hostage situations, or bereavement. They often have similarity to "dreaming awake" which clients can appreciate. This helps with reattribution of the experiences to their own minds and bodies and then they may feel less pressure to act on them and more agreeable to work with them.

Dysfunctional cognitions or secondary delusions can be debated, using evidence for and against the proposed explanation, e.g., "it is the aliens communicating with me." In a lot of cases safety behaviors like avoidance can be an important factor in maintaining hallucinations, so developing and exploring alternative coping strategies may be fruitful for the client. The content of hallucinations may point to core beliefs clients hold about themselves, (e.g., "I'm evil"), and examining these beliefs may help the client to make sense of this experience, consequently reducing the distress associated with the critical content of hallucinations.

The assessment of auditory hallucinations involves careful exploration of the antecedents, i.e., the triggers, the nature of the experience and the consequences which could be affective, behavioral, or even cognitive responses; summarizing and giving feedback should follow by the therapist before proceeding to ensure that the problem has been understood correctly. A collaborative agenda then follows whereby the first step would be educational in nature so that the client wouldn't feel isolated and alien in his experience; this may happen through providing educational leaflets, e.g., "understanding voices" or through "voice hearer's groups" arranged locally by the community mental health teams.

It is important that the therapist investigates the client's understanding of their symptoms first and explores it systematically avoiding a confrontation with the client. This aids in engaging and building the therapeutic relationship, gradually through guided discovery the client's model can be elaborated using homework exercises, e.g., if they believe radio waves are interfering with their thinking, then find out current scientific knowledge about radio waves from relevant textbooks or the internet. In this way alternatives can be generated which gently impact on the degree of conviction the client has in his explanation for the hallucinatory experience, e.g., reaction to stress and evidence of hallucinations in normal subjects put under stress. As trust builds up between client and therapist, an audiotape of the voices may be attempted, that would be useful in the assessment process to cover any defects in the understanding of the origin or nature of voices, and help the client to think about it and possible alternative explanations to this phenomenon. Discussing client's symptoms can be distressing to them and it can provoke depressive cognitions, that is why instillation of hope in a realistic way is an important task for the therapist at this early stage.

The next step may involve using diaries of voices to gradually bring the antecedents and consequences to light collaboratively with the client, and in turn a diagram can be drawn linking the client's experience of hearing voices to the associated thoughts, effects and resulting behaviors, and dysfunctional coping mechanisms. A discussion about possible alternative coping mechanisms may follow and then a behavioral experiment can be done to examine the new coping strategy and compare it with the previous diaries. Clients who had a long history of voice hearing may adopt a passive approach to voices possibly resulting from learned helplessness over the years of trying several coping mechanism with no success. The therapist may try to reengage the client with the voices and explore previous coping strategies used, and future ones not yet used and design experiments to test these out, e.g., behavioral control, socialization, or cognitive control over the voices by using distraction techniques, focusing, or rational responding.

The main determinant of the distress associated with auditory hallucinations is the belief clients hold about the experience, e.g., omnipotence or omniscience of the relationship with the voices which can be explored using a number of techniques, e.g., listing the evidence, guided imagery, role-play, positive logging, or acting against the schema.

Case Formulation and Intervening with Delusions

The key to effective management of delusional beliefs lies in individualized case formulation. The first step in developing case formulation is a comprehensive assessment which includes eliciting and fully understanding the delusional belief according to the client's model because proceeding to therapeutic intervention before completing the assessment can be risky, i.e., the therapist can make false assumptions or jump to conclusions not supported by facts. Assessment provides detailed understanding of the antecedents to the onset of the delusional belief, i.e., events or circumstances that were key to clients prior to developing the belief. This enables the therapist, and client, to construct a clear, logical, and chronological account of these events. The therapist also tries to establish the onset of the first positive and/or negative symptoms.

The second step in the therapeutic process is to find connections between the antecedents, client's beliefs about them and their consequences, very similar to the ABC model suggested by Ellis and adapted by Chadwick and Birchwood (cf. Chadwick & Lowe, 1994). It is important at this stage not to focus on client's feelings about events but rather their thoughts about them. As the clients talk about their experience other cognitive errors emerge, e.g., selective abstraction or arbitrary inference, which are central in understanding the delusional belief. The therapist follows a certain sequence with the client at this stage of therapy whereby she or he collaborates with the client who leads the discussion. Once the client's model is understood it can be gently explored with the client using a normalizing approach and using the vulnerability-stress model to form alternative models where needed, to explain the client's illness. In discussing delusions the therapist starts with exploring the type of evidence the client has for the delusion and looks carefully for factors used by the client to maintain that interpretation or in other words evidence the client finds to support the belief, e.g., people's attitudes, media reports, and side effects of medications. The next step would involve exploring alternative evidence disconfirming the client's belief and in doing so, Socratic questioning and sensitivity to client's responses is paramount to avoid colluding with the client. The client leads in coming up with alternatives, but if that proves difficult gentle prompting can be appropriate. Therapists may proceed by asking clients to fill thought diaries, if they are able to do this. This can give a good idea about the context of the beliefs occurring and thereby an experiment can be designed to test that belief.

So in summary, debating with the client about their delusions involves first exploring their content and establishing the nature of evidence. This is, followed by discussions of subsequent confirmatory evidence involving significant others' opinion. Eliciting alternatives and investigating any relevant theoretical propositions can then occur, which leads one finally to wait and see if the work done brings about new developments.

If therapists are finding it difficult to make a breakthrough in discussing client's delusions, it may be advisable to revise the case formulation and look for events, which seemed to be unlinked to the delusion, but appear to be a recurring theme in discussions with the client. And there will be a point where it is better to stop reasoning with the delusion and look at the client's short- and long-term goals, despite the nature of the beliefs. The therapist may need to focus more on the function the belief serves for the client.

Inference chaining can be a very effective and nonthreatening way in dealing with resistant delusions, commencing with, e.g., "If everybody did believe exactly what you are saying, how would that affect you? Why would it be important?" This can pave the way to identifying an emotional or social need by the client, e.g., "I'd be respected" or "I would be able to go back down to the pub to see my friends" which can be identified as a goal to work toward, e.g., "OK well we've had problems getting people to believe you, but why should that stop you going and seeing your friends?" Frequently while beliefs may be very slow to change, behavior apparently consequent to them does shift and allows for an improved quality and quantity of social interaction.

Working with grandiose delusions however necessitates that therapists avoid undermining client's beliefs, as they may function in protecting self-esteem. Criticism can be counterproductive and increase their attempts to act on their grandiose beliefs. A major goal of therapy working with paranoid delusions, for example, is to reduce overgeneralization and to look closely at the process of development of the delusion, which often includes field experiments. Another way of tackling paranoid delusions is to examine the reasons behind them and then sometimes using protective measures to alleviate the client's beliefs, e.g., ensure that they are secure in their home by ensuring that they have appropriate locks or spy holes on their doors. Delusions with spiritual themes, also common in older people, can be quite resistant to direct reasoning, and the overlap with normal spiritual belief can make it difficult to establish exactly where "delusional" belief begins, although its effects on their life may be easier to determine. In such instances members of the individual's spiritual community often can be very helpful in mediation, by building trust and providing social support.

In bizarre delusions, tracing their origins may lead to misunderstood symptoms of anxiety, e.g., tingling from hyperventilation as "electricity" or sometimes to a film or a song or TV program which has been taken out of context. It is also common in older people that such delusional beliefs derived from physical symptoms of anxiety can be interpreted as "I have cancer" or "I am dementing," therefore understanding anxiety is crucial in these cases with general anxiety management techniques. Hypochondriacal delusions are fairly common in old age and can be related to relationship issues, sometimes even inadvertently reinforced by concerned caregivers. Therefore, interventions need to address these issues in a broad and comprehensive care package that addresses social needs (Kingdon, Rathod, & Turkington, 2001). People with monosymptomatic hypochondriacal psychosis are very difficult to engage in a psychological approach to their symptoms; therefore a Socratic approach with very gentle slow paced exploration is the first step to engage such clients (Jeste et al., 2003). The focus would be to reduce the distress associated with the delusion rather than removing the symptom itself, again inference chaining may prove to be an effective technique in identifying and working with that distress.

Case Study

Mr. GD is a 70-year-old retired widower who was referred by his general practitioner for the following symptoms: (a) worries that he is being followed by other people when he goes outside his house; (b) worries that his neighbors think that he is a pedophile; (c) worries that neighbors are against him and that he hears their remarks about him across the garden fence; and (d) belief that a male neighbor wants to kick him out of the area and that he follows him in his car.

The initial assessment revealed that he suffered from social anxiety and avoidant personality traits for most of his life. He had also suffered an episode of depression mixed with anxiety 4 years earlier associated with auditory hallucinations. This episode was treated with antidepressants successfully. There was a positive family psychiatric history of a persistent mental illness in the maternal grandmother.

He was born and brought up locally, and was second of nine children. He had a deprived childhood and described himself as shy and isolated from other children. He left school at the age of 15 without qualifications. His occupation was mainly as a farm worker and he had retired 8 years earlier from his job. His job had meant that he worked almost entirely on his own. He married at the age of 34, which he described as unhappy, and his wife died at a young age leaving him with two daughters. Social services provided assistance to Mr. GD in raising his children. He was socially isolated, as both his daughters were married and lived away from him, and he had minimal contact with his extended family that lived locally.

Two baseline scales were used to assess the client's symptoms and current functioning: The PSYRATS (Haddock et al., 1999) indicated that he was seriously preoccupied with his delusions. Such thoughts occurred at least once an hour, and these would persist, particularly if outdoors. His conviction that these were realistic was 100%; his level of distress over these ranged from moderate to severe, and his beliefs caused a moderate disruption to his lifestyle. This assessment was confirmed by the HoNOS, which also emphasized the presence of negative mood changes and a marked decrease in opportunities for daytime activities, thus making his problems worse.

After the initial assessment, the following agenda was agreed:

1. *Therapist's contribution to the agenda*
   (a) Encourage *Engagement* of the client to facilitate exploration of the above symptoms especially in view of their persecutory nature.
   (b) Collection of more historical data and mental state examination to facilitate the process of initial conceptualization of the client's problems and to make sense of his distressing symptoms.
2. *Client's agenda*
   (a) Identify the persecutory ideations that are the most important distressing phenomenon to him.
   (b) Determine which of these he might want to discuss during the session.

The client found it helpful to talk about his problems in depth for the first time. On questioning, he is unclear why anybody should be endangering him but nevertheless convinced of it – an incident when he had an argument with a neighbor is the only possible precipitant identified. A recent example of a persecutory experience was used as a focus to try and establish the context and possibly identify any antecedents and resulting consequences. The therapist used guided discovery as follows:

P: I have been invited to my daughter's place last weekend but I am not sure why I felt anxious and worried about going out.

T: When people experience anxiety it is usually because they are worried something unpleasant is going to happen to them.

P: I am not sure what it is though.

T: Well, is it that you find the journey to your daughter's house quite boring?

(The therapist suggests the opposite of a hypothesized fear which was non-threatening in order to encourage client disclosure.)

P: (shaking his head). Certainly not, I'll be lucky if I arrive safely there.

T: Why do you think that you may not arrive safely to your daughter's place?

P: Because of this blue car that keeps following me everywhere.

T: So is that what you're anxious about if you go out?

P: Yes, that's it.

After some discussion about being followed by the blue car it emerged that he believes he is in danger when he goes out and we started exploring whether it is his own views or whether other people as well believe he is in danger.

T: Do you think others think you are in danger?

P: I'm not sure. I suppose my daughter doesn't.

T: Ok, there's something I'm not sure about: if you think you are in danger, does that really make you in danger or can you make up your own mind about this issue?

P: Hmm, I suppose if you feel in danger, you are just in danger and you have to protect yourself.

T: And if you didn't agree with your feeling too readily, but thought carefully about it yourself, what then?
P: Well, if I really think about it, I feel safe until I approach my neighbor's house who owns the blue car.
T: So it seems that you feel safe inside your house and probably for some distance until you approach your neighbor's house and you can make up your mind whether you feel safe or not.
(The therapist summarizes the new information that has been revealed.)
P: (nods) That's right.
T: So how does this information you've just supplied fit in with your idea that you are generally in danger.
(The therapist asks a synthesizing question which applies the new information to the client's original belief of being in danger.)
P: I'm not sure. It doesn't fit in. does it? Maybe I'm not always in danger as I think I am.
(However the client is still endorsing to some extent his delusional belief.)
T: What else could this information point to if you are not as in danger as you think you are?
P: That I could begin to believe that I might actually be in danger only in certain places or at certain times.
T: Ok, then let's explore why you may be in danger in certain places or at certain times.

The discussion continues to explore in more details patient's belief and the objectivity of his assessment of risk in certain places or at certain times.

By the end of the session homework was agreed as follows:

T: We have been discussing in this session your thoughts that you are in danger and that your neighbor may be thinking of harming you, and all the problems that result from this, e.g., staying in house and not going out. On the other hand, you're becoming increasingly frustrated with yourself because you used to enjoy going out (Client nods). So what would be the first step in breaking this deadlock?
P: I'm not sure about that also.
T: OK. Do you want to stay at home for several more weeks until you start feeling safe again?
P: No. I need to do something now but I don't know what.
T: Do you want to go to your daughter's house next weekend?
P: No. that's much too soon. I need to get going on something though, but I am stuck for answers. What would you suggest?
T: I would suggest, as a first step, listing the advantages and disadvantages of not going out to, say, a nearby shop and then doing the same thing for going out.
P: That sounds reasonable and not too frightening. I can't see the harm in that.
T: In what way do you think this task will help you?
P: I think it will begin to free me from the paralysis I'm in whatever happens.
T: So are we agreed on the homework task for next week?

P: We are.

T: Okay, let me make a note of the task and give you a copy of it.
(Giving clients a copy of the homework task greatly reduces the potential disagreements at the next session over what the task actually was.)

Other items included in the homework were information leaflets about possible volunteer work, since this had been explored in therapy as a possible strategy for increasing positive activities, and a leaflet about psychotic symptoms entitled, "Understanding How Others Think," given to him to read and bring to the next session.

By the end of the session it was felt that the client's psychotic episode could be conceptualized as a sensitivity psychosis, but more sessions were needed to make sense of his symptoms. It was also believed that the client had engaged quite well in therapy and information relevant to a formulation of his problems and background issues was emerging, thus helping him begin to make sense of his symptoms and develop a negotiated plan to overcome them.

Second Session

– *Review of homework.* Client had successfully made contact with a volunteer group "Green Gym" which seems to match his original interests in agriculture and farming and he seemed motivated to follow that through by attending an assessment for suitability for the group. Client had read the leaflet.
– *Session agenda.* The client wanted to use the session for discussing an incident where he was walking along the beach with his daughters and a tall man was approaching them in fast steps. He believed that this man intended to harm him, because he thought of him as a pedophile.

Guided discovery was again used to explore the circumstances surrounding this experience. Prior to his walk his daughter informed him that she was moving abroad, and he will thus see her less often. This was upsetting to him, and had influenced his interpretation of the events on the beach. Questioning revealed that he felt sad and he thought "everyone is moving away from me," which then led to the emergence of the thought "my neighbor believes that I am a pedophile." We then discussed how he thought that the tall approaching stranger had read his earlier thought "my neighbor believes I am a pedophile" and as a result intended to harm him. The focus of the interaction then shifted to the consequences following this misattribution and misinterpretation of thoughts, and what actually happened is that the man turned round and moved away, and the client's interpretation that the presence of his two daughters prevented him from carrying his original plan. The therapist then focused on exploring possible alternative interpretations of the stranger's behavior (i.e., approaching in fast steps), and collaboratively a list was drawn whereby two more possibilities were discussed. One example was, the man was walking fast as part of a jogging or running exercise and his turning round meant he was going back to his original destination. The second example was that it was starting to rain and the stranger thought of going back before he gets wet.

Both of these possibilities were discussed with reference to information provided in the homework leaflet. The result was that client's belief in his original thought, "the man wants to attack me" was shaken slightly and possible alternative explanations had to be considered, leaving him slightly conflicted and ambivalent about his convictions. Discussing how the stranger read his mind was left aside for discussion in a future session. The therapist felt it would be overwhelming to the client who was already struggling to describe the situation. It was also felt that more engagement was needed to be able to socialize the client fully in cognitive therapy.

– *Homework.* Another leaflet was given to the client – "cognitive therapy for psychosis" – to read after the brief attempt in the second session to socialize the client to the cognitive model through discussion of emotions and thoughts.

Third Session

– *Review of homework.* Client gave positive feedback after reading the leaflet, and he related that to a self-referent experience he had when he was shopping recently. It was agreed to put that into the agenda as well as trying to make sense of his experience in a simple way collaboratively.
– *Agenda.* Discussion of the self-referent ideas he had on his visit to the supermarket. Follow this with discussion of the "making sense" formulation using the earlier material for illustration where appropriate.

Mr. GD related the experience where he was doing his own shopping, and he felt that his name was said aloud. He saw a couple shopping and looking at him, and he immediately interpreted that they were talking about him. Of particular importance was the fact that the couple had a baby and his thought was "they are concerned about their baby because they think I am a pedophile." Similar to the second session guided discovery was used to explore the client's day from the beginning. Earlier in the day his neighbor was doing a lot of work in his garden with his wife, and the client felt that they were talking about him and possibly planning to get rid of him from the neighborhood because of their original thought about him as a pedophile. We explored that the fact that he had been upset all morning as a result and as he went shopping he carried that feeling with him.

We then discussed his experience in the supermarket, noting how his feelings could have possibly overwhelmed him, thus leading him to become suspicious that people read his mind and like his neighbor think of him as a pedophile. A simple formulation was developed collaboratively that encourages the client to understand the association between his feeling upset and the thoughts that people can read his mind. This in turn leads to behavior as being wary and noticing subtle movements of people in the supermarket, which makes him more anxious and upset, thus increasing his tendency to misattribute any benign movements or eye contact from strangers. Client felt slightly relieved with the links. He still held to his belief about possibly mind reading, but on a scale of 0–100, his belief came down from 100 to 70 following the discussion.

The second agenda item was discussed namely making sense, through formulation and collaboratively the client identified that a possible genetic predisposition to his illness given the chronically mentally ill grandmother. He identified life-long social isolation as a perpetuating factor to his current problem but it was difficult to pinpoint a precipitating factor at this stage which was left to a further discussion. We were able to highlight how his thought comes about when he is upset and worried, and how this leads to becoming more alert to the environment which leads to behaviors to find more evidence to support his original thought while ignoring any other counter evidence. We used both the examples in sessions 2 and 3 to explain that in simple terms.

– *Homework.* A thought record diary was given to the client to record his thoughts about his mind being read by others and to attempt to fill the diary over the next week for discussion in the next session. Engagement seemed more evident following this session, and it appeared that the client would be cooperative during the remainder of therapy. Therapy was terminated, because the therapist moved from the area. Follow up telephone conversations established that the social intervention, i.e., attending "green gym" produced increased positive social interactions, which in turn resulted in less preoccupation and distress with his thoughts. Assessment was repeated and the scores showed that this brief CBT intervention, combined with the social intervention had improved his scores.

Conclusion

Cognitive behavior therapy is becoming much more accepted as an effective intervention for psychosis (Keith, 2006) and becoming more widely available. This has been the case in the UK for a number of years (Kingdon & Kirschen, 2006; Turkington et al., 2006) and it is now having an increasing impact on individual's life and recovery from psychosis. Older people have been included in research studies but most of the focus has been on the under 65s. Nevertheless there is good clinical reason to believe that such an intervention can be successful in reducing distress and disability in the elderly if adapted appropriately.

References

Almeida, O., Howard, R., Forstl, H., & Levy, R. (1992). Should the diagnosis of late paraphrenia be abandoned? *Psychological Medicine, 22*(1), 11–14.

Almeida, O., Howard, R., Levy, R., & David, A. (1995). Psychotic states arising in late life (late paraphrenia) the role of risk factors. *British Journal of Psychiatry, 166*(2), 215–228.

Arean, P. (2003). Advances in psychotherapy for mental illness in late life. *American Journal of Geriatric Psychiatry, 11*(1), 4–6.

Bartels, S., Forester, B., Mueser, K., Miles, K., Dums, A., Pratt, S., et al. (2004). Enhanced skills training and health care management for older persons with severe mental illness. *Community Mental Health Journal, 40*(1), 75–90.

Bradshaw, W. (2000). Integrating cognitive-behavioral psychotherapy for persons with schizophrenia into a psychiatric rehabilitation program: Results of a three year trial. *Community Mental Health Journal, 36*(5), 491–500.

Castle, D., & Murray, R. (1993). The epidemiology of late-onset schizophrenia. *Schizophrenia Bulletin, 19*(4), 691–700.

Chadwick, P., & Lowe, C. (1994). A cognitive approach to measuring and modifying delusions. *Behaviour Research and Therapy, 32*(3), 355–367.

Cunningham Owens, D., Carroll, A., Fattah, S., Clyde, Z., Coffey, I., & Johnstone, E. (2001). A randomized, controlled trial of a brief interventional package for schizophrenic out-patients. *Acta psychiatrica Scandinavica, 103*(5), 362–369.

Granholm, E., McQuaid, J., McClure, F., Auslander, L., Perivoliotis, D., Pedrelli, P., et al. (2005). A randomized, controlled trial of cognitive behavioral social skills training for middle-aged and older outpatients with chronic schizophrenia. *American Journal of Psychiatry, 162*(3), 520–529.

Granholm, E., McQuaid, J. R., McClure, F. S., Link, P. C., Perivoliotis, D., Gottleib, J. D., Patterson, T. L., & Jeste, D. V. (2007). Randomized controlled clinical trial of cognitive behavioral social skills training for older people with schizophrenia: 12-month follow-up. *Journal of Clinical Psychiatry, 68*, 730–737.

Gumley, A., O'Grady, M., McNay, L., Reilly, J., Power, K., & Norrie, J. (2003). Early intervention for relapse in schizophrenia: Results of a 12-month randomized controlled trial of cognitive behavioural therapy. *Psychological Medicine, 33*(3), 419–431.

Haddock, G., McCarron, J., Tarrier, N., & Faragher, E. (1999). Scales to measure dimensions of hallucinations and delusions: The psychotic symptom rating scales (psyrats). *Psychological Medicine, 29*(4), 879–889.

Jeste, D., Alexopoulos, G., Bartels, S., Cummings, J., Gallo, J., Gottlieb, G., et al. (1999). Consensus statement on the upcoming crisis in geriatric mental health: Research agenda for the next 2 decades. *Archives of General Psychiatry, 56*(9), 848–853.

Jeste, D., Twamley, E., Eyler Zorrilla, L., Golshan, S., Patterson, T., & Palmer, B. (2003). Aging and outcome in schizophrenia. *Acta Psychiatrica Scandinavica, 107*(5), 336–343.

Keith, S. (2006). Are we still talking to our patients with schizophrenia? *American Journal of Psychiatry, 163*(3), 362–364.

Kingdon, D., & Kirschen, H. (2006). Who does not get cognitive-behavioral therapy for schizophrenia when therapy is readily available? *Psychiatric Services: A Journal of the American Psychiatric Association, 57*(12), 1792–1794.

Kingdon, D., Rathod, S., & Turkington, D. (2001). Psychotic disorders with hypochondriacal feature: Delusions of the soma. In G. Amundsen, S. Taylor, & B. Cox (Eds.), *Health anxiety: Clinical and research perspectives on hypochondriasis and related conditions*. Chichester: Wiley.

Kingdon, D., & Turkington, D. (1991). A role for cognitive-behavioural strategies in schizophrenia? *Social Psychiatry and Psychiatric Epidemiology, 26*(3), 101–103.

Kingdon, D., Turkington, D., & John, C. (1994). Cognitive behaviour therapy of schizophrenia. The amenability of delusions and hallucinations to reasoning. *British Journal of Psychiatry, 164*(5), 581–587.

Patterson, T., McKibbin, C., Taylor, M., Goldman, S., Davila-Fraga, W., Bucardo, J., et al. (2003). Functional adaptation skills training (fast): A pilot psychosocial intervention study in middle-aged and older patients with chronic psychotic disorders. *American Journal of Geriatric Psychiatry, 11*(1), 17–23.

Pinninti, N., & Datto, C. (2006). Cognitive behavior therapy and clozapine synergy in an older adult with schizophrenia? *American Journal of Geriatric Psychiatry, 14*(8), 717–718.

Prager, S., & Jeste, D. (1993). Sensory impairment in late-life schizophrenia. *Schizophrenia Bulletin, 19*(4), 755–772.

Thompson, L. (1996). Cognitive-behavioral therapy and treatment for late-life depression. *Journal of Clinical Psychiatry, 57*(Suppl. 5), 29–37.

Turkington, D., Kingdon, D., & Weiden, P. (2006). Cognitive behavior therapy for schizophrenia. *American Journal of Psychiatry, 163*(3), 365–373.

Van Citters, A., Pratt, S., Bartels, S., & Jeste, D. (2005). Evidence-based review of pharmacologic and nonpharmacologic treatments for older adults with schizophrenia. *The Psychiatric Clinics of North America, 28*(4), 913–939.

Wiersma, D., Jenner, J., van de Willige, G., Spakman, M., & Nienhuis, F. (2001). Cognitive behaviour therapy with coping training for persistent auditory hallucinations in schizophrenia: A naturalistic follow-up study of the durability of effects. *Acta Psychiatrica Scandinavica, 103*(5), 393–399.

11
Behavioral Interventions to Improve Management of Overweight, Obesity, and Diabetes in Patients with Schizophrenia

Christine L. McKibbin, David Folsom, Jonathan Meyer,
A'verria Sirkin, Catherine Loh, and Laurie Lindamer

The relationship between schizophrenia and diabetes mellitus has been recognized in case reports since the end of the nineteenth century (Holt, Bushe, & Citrome, 2005). More recent data suggest that diabetes, obesity, and the metabolic syndrome are 1.5–2 times more common in patients with schizophrenia than in the general population (American Diabetes Association, American Psychiatric Association, American Association of Clinical Endocrinologists, & North American Association for the Study of Obesity, 2004; Henderson, 2005). Although the mechanisms of this relationship are not clearly understood, a combination of genetic and environmental factors (e.g., sedentary lifestyle, poor diet, and side effects of antipsychotic medications) may play a role in the high prevalence of metabolic dysregulation in this population (Citrome, 2004; Cohn, Prud'homme, Streiner, Kameh, & Remington, 2004; Henderson, Cagliero et al., 2006; De Hert et al., 2006; Lamberti et al., 2004; McEvoy et al., 2005). While some increased risk may be attributed to standard risk patterns (i.e., 17–50% of people with schizophrenia have a family history of diabetes; Mukherjee, Schnur, & Reddy, 1989), the high rates of obesity in patients with schizophrenia (Marder et al., 2004) may also contribute to increased risk, due, in part, to the relative weight gain liabilities of second-generation antipsychotics (Newcomer, 2005).

Poverty and poor access to healthy nutrition may exacerbate obesity and further increase diabetes risk or worsen existing diabetes. In fact, patients with schizophrenia show consistently poorer nutritional habits than their nonpsychiatric counterparts (Brown, Birtwistle, Roe, & Thompson, 1999; Davidson et al., 2001; McCreadie et al., 1998; McKibbin, Lindamer et al., 2006; Strassnig, Brar, & Ganguli, 2003). Recent literature also documents that patients with schizophrenia are physically less active than the general population (Daumit et al., 2005; Dickerson et al., 2006; Lindamer, McKibbin, Jin, Ueki, & Jeste, 2006).

Evidence Base

Exercise and diet are recommended to reduce body fat, improve blood glucose control, and lessen the complications of diabetes. Even though many cognitive-behavioral interventions to address this problem area have been conducted in patients with schizophrenia, there are only a few randomized clinical trials. Additionally, many were conducted using inpatients, and thus have utilized behavioral models not applicable for outpatient settings. Size of the samples was small in the early studies, generally including less than 25 participants (Harmatz & Lapuc, 1968; Rotatori, Fox, & Wicks, 1980). With the exception of McCreadie et al. (2005), larger studies have generally experienced moderate to high rates of attrition (cf. Faulkner, Soundy, & Lloyd, 2003; Loh, Meyer, & Leckband, 2006, for reviews of both experimental and nonexperimental trials).

Among the randomly controlled trials published to date, results are mixed. A number of studies have shown that lifestyle interventions result in greater weight loss relative to controls (Harmatz & Lapuc, 1968; Rotatori et al., 1980; Weber & Wyne, 2006). Other studies have shown that lifestyle interventions may ameliorate weight gain experienced by those on atypical neuroleptics. For example, Littrell and colleagues (Littrell, Hilligoss, Kirshner, Petty, & Johnson, 2003) found that, relative to a 16-session nutrition, lifestyle, and exercise group, their treatment as usual control group gained significantly more weight while taking olanzapine. Similarly, Evans, Newton, and Higgins (2005) found that those who participated in 6 one-on-one nutritional education sessions and who received a nutritional information booklet gained less weight than those who received the information booklet alone.

On the other hand, Brar et al. (2005) reported no significant differences between patients enrolled in their 16-week cognitive-behavioral intervention (based on the Diabetes Prevention Program) and their counterparts in the control condition. The weight loss for the experimental group was 5.4 pounds, and 1.2 pounds for the control group. The authors noted that a small sample size likely accounted for the lack of findings. Weber and Wyne (2006) found that those enrolled in their behavior therapy intervention had no greater change in weight than those enrolled in usual care where they were encouraged to lose weight, but were provided no specific strategies for doing so. Finally, no differences on weight change were found between groups offered free YMCA passes and those offered treatment as usual (Archie, Wilson, Osbourne, Hobbs, & McNiven, 2003).

In summary, there is growing interest in the development of interventions to improve health behavior and reduce or even prevent obesity in patients with schizophrenia. Studies conducted thus far show promise, but several are compromised by small sample sizes and high attrition rates. Furthermore, there is little if any work examining the efficacy of lifestyle interventions for older patients with serious mental illness who, because of age-related increases in morbidity, may experience greater obstacles in implementing health behavior changes. Diabetes is one such age-related medical morbidity that is more prevalent in patients with schizophrenia than the general population. The optimal management of this disorder requires not only ongoing medical monitoring but also active self-management on the part of the patient – a task that may be challenging for individuals with schizophrenia to achieve. In the following section, we describe a brief assessment and health promotion program, along with a promising behavioral diabetes management intervention program with specific modifications to accommodate cognitive and motivational deficits frequently experienced by patients in this population.

Diabetes Management and Rehabilitation Training

Theoretical Foundation

Our development of diabetes management and rehabilitation training (DART) was guided by the principles of social cognitive theory (Bandura, 1986, 1989). This theory has been used widely in diabetes education (Miller, Edwards, Kissling, & Sanville, 2002) and in previous intervention programs for patients with schizophrenia (Bartels et al., 2004; Patterson et al., 2003). Furthermore, the methods of behavioral change espoused by this model (e.g., identification of problems, behavioral modeling, role-plays, and reinforcement) can be applied effectively in group settings.

Basic Structure

The DART intervention was conducted in locations where patients lived, spent a majority of their day-time hours, or received some rehabilitative care. Each group was led by two group leaders, one at the doctoral level and one at the bachelor's level. A member limit of 6–8 per group was also established in order to provide some individual based feedback for each of the group members within the 90-min group duration. Because each session was designed to build upon the skills learned in previous sessions, the groups were closed and new members were not permitted after the start of the intervention. Each group began with a brief review of the previous session, homework, a semiprivate weigh-in, and problem-solving. This introduction was followed by new instruction and a subsequent break whereby healthy snacks were provided and used to demonstrate healthy eating principles (i.e., feasibility, cost, portion size, food group pairing). The break was followed by additional instruction, leader modeling, in-class exercises, feedback and problem-solving, and description of homework, termed home-practice to maximize adherence.

Sessions were categorized into three modules: basic education, nutrition, and exercise. Each module was comprised of four sessions. Basic education about diabetes and its management was taught first in order to provide basic knowledge to be used in nutrition (e.g., knowledge about hyperglycemia is required before teaching about food groups and individual food's effect on blood glucose levels). Both basic education and nutrition were taught prior to exercise for safety reasons (e.g., prevention of hypoglycemic conditions during exercise). Once the first 12 sessions were completed, the information was repeated for an additional 12 sessions. Repetition of the program enhances encoding of intervention material, provides the opportunity for patients to integrate basic education, nutrition, and exercise material into their daily routine, and also gives those patients, who missed an earlier session, the opportunity for exposure to material covered in that session.

Assessment

The treatment of diabetes involves not only control of blood glucose but also management of blood pressure, lipids, and body weight. Furthermore, diabetes is not typically managed by psychiatrists and schizophrenia is not typically managed by primary care physicians. Therefore, this type of intervention requires collaboration with both of these medical specialties. The medical assessments and actions for this intervention are based, in part, on recommendations of Marder et al. (2004) and includes the following actions (1) each patient should have a primary care physician, who is providing ongoing medical management of diabetes; (2) weight and height should be monitored and body mass index (BMI) calculated for each patient; (3) control of diabetes should be periodically assessed measuring glycosylated hemoglobin (i.e., a measure of blood glucose control over a 6- to 10-week period); (4) blood pressure should be monitored; and (5) lipid levels should be measured at baseline and at specific intervals thereafter. Other assessments that may prove useful include measures of diabetes history and existing self-care practices (e.g., Diabetes History Questionnaire (https://borc.med.umich.edu/MDRTC_surveys), Diabetes Care Profile (Fitzgerald et al., 1996)), Diabetes Knowledge Test (Fitzgerald et al.), and the participants' confidence in their ability to manage diabetes (e.g., Diabetes Empowerment Scale; Anderson, Funnell, Fitgerald, & Marrero, 2000). These and other similar measures have been previously used in patients with schizophrenia, but reliability and validity for these or other diabetes-related measures have not yet been established for this population.

Pilot Test of the DART Program

We tested the feasibility and efficacy of the DART intervention, described earlier, for middle-aged and older patients with schizophrenia, who have type 2 diabetes mellitus, using a randomized pretest, posttest control group design (McKibbin, Patterson et al., 2006). Individuals with a diagnosis of schizophrenia or schizoaffective disorder over the age of 40 were randomly assigned to either the DART (n = 32) group or Usual Care plus Information (UCI; n = 32). Control participants received three brochures regarding diabetes management from the National Diabetes Education Program. All participants in the study were recruited from board-and-care facilities and day treatment programs. Board and Care typically consists of housing, three meals per day, and some assistance with medication administration and financial management.

Fifty-seven participants completed baseline and 6-month (postintervention) assessments consisting of an interview, measures of BMI, blood pressure, and fasting blood chemistry. A mixed-model analysis of variance revealed a significant group – time interaction for body weight, with patients in the DART group losing a mean of 5 pounds and those in the UCI gaining a mean 6 pounds. Significant group – time interactions were also found for triglycerides, diabetes knowledge, diabetes self-efficacy, and self-reported physical activity, but not for fasting plasma glucose or glycosylated hemoglobin. Thus, participants in our DART program showed significant reductions in triglycerides and improvements in diabetes knowledge (i.e., from a mean of fewer than 50% items correct to over 70% items correct), self-efficacy, and self-reported physical activity level. These results indicate that lifestyle interventions, embedded within the context of a group-based cognitive-behavioral model, can produce positive health changes in middle-aged and older patients with schizophrenia and diabetes mellitus. Specific cognitive-behavioral components of the DART program are described below.

Cognitive-Behavioral Elements of the DART Intervention

Goal Setting

In the first session, some time was provided for group leaders to evaluate the patients' health-related goals and reinforcement value for behavior change. For example, group leaders identified the degree to which patients valued weight loss (i.e., our primary intervention outcome) as well as their current confidence in their ability to lose weight. This information was used to guide patients to establish preliminary short- and long-term goals.

Short-Term Goals

Because individual body weight fluctuates in the short term due to a number of factors (e.g., fluid intake, sodium intake), short term goals were consistently behavioral in nature. Group leaders assisted patients in identifying goals that were specific, observable, and realistic and that included a time-frame of approximately 1 week. Leaders also provided concrete examples of practical and impractical goals through modeling prior to setting patients' goals. Because it was difficult for some participants to conceptualize and articulate a particular goal, group leaders asked generally what they wanted their health to be like in a month or more from now (i.e., long-term goal) and used that information as a foundation for setting a realistic goal.

Behavioral Monitoring

Each DART session's home-practice involved self-monitoring, often considered one of the most essential features of behavior therapy. With self-monitoring, patients were asked to keep a record of specific behaviors. In the case of health-oriented interventions, self-monitoring often consisted of monitoring food intake and physical activity. Although patients, even in the general population, are not always accurate in reporting their diet and exercise behaviors, the primary purpose of self-monitoring is to increase the patient's and group leader's awareness of the behaviors and factors that positively and negatively influence the patient's weight management efforts. In the case of our DART participants, the purpose of self-monitoring was to increase awareness about lifestyle behaviors and provide a forum for support around healthy lifestyle behaviors. Self-monitoring was used throughout the intervention program in concert with exercises related to short- and long-term goals. Participants were asked to set a diet or physical activity goal and monitor progress toward that goal on a handout in between sessions. In addition, the DART intervention monitored weight and physical activity for each participant each week. Weight was monitored in a semiprivate setting prior to each session to improve participant insight regarding outcomes of their respective dietary and exercise behavior, and to serve as a platform for discussion and problem-solving. Physical activity was also monitored through pedometers, provided to each participant at the beginning of the exercise module (session 9).

Stimulus Control

Stimulus control involves modifying one's environment to enhance those behaviors that will support a particular goal and to reduce the likelihood of engaging in behaviors known to thwart achievement of the particular goal. In this intervention, we used stimulus control techniques primarily for nutrition. Stimulus control techniques employed included encouraging patients to keep high-fiber, low-fat snacks such as fruit in their rooms to reduce the number of trips to the vending machine, or local fast food restaurant. Additional techniques used included keeping high fat and high calorie foods out of the house or room and eating a healthful food before eating an unhealthful food. These techniques were presented along with other simple rules for healthy nutrition and were highlighted as potential solutions when unhealthful eating behavior was identified.

Problem-Solving

Problem-solving techniques are often taught in behavioral approaches to obesity and diabetes management to circumvent potential barriers to behavioral change. Common issues or problems that were observed in the DART study included a lack of financial resources to purchase healthful food, perceived lack of available healthful foods in community-based residential care facilities, concern about requesting healthy menu alternatives from residential care facility staff, binge eating or overeating in the presence of environmental cues, sedation, medical morbidity (e.g., arthritis), and environmental barriers to lifestyle exercise (e.g., neighborhood safety, temperature). Problem-solving strategies were reviewed as part of each intervention session when specific barriers to completion of home practice or goal attainment were identified. Participants were assisted in identifying the problem or barrier, which was then written on a large note pad mounted on an easel. All group members were invited to generate potential solutions. All options were written in list format on the board and group facilitators contributed to the potential solution list only if the group demonstrated difficulty generating viable options. The individual requiring problem-solving was then guided to evaluate whether each potential solution would be feasible and likely to result in an outcome that would be favorable (i.e., pros and cons). The individual was encouraged to select one option to try as a goal before the next session.

Behavioral Shaping Through Use of Incentives

One of the hallmarks of behavioral shaping through incentives is that this technique serves to accentuate the "value" associated with healthful behavior change. Behavioral techniques such as behavioral reinforcement provide an external mapping of action outcomes that may compensate for a particular patient's own limited ability to use internal representations to motivate and sustain behavior change (Velligan, Kern, & Gold, 2006). A potential concern with behavioral shaping is that gains may be lost once the reinforcement is discontinued. There are data from cognitive remediation literature, however, to suggest that gains from behavioral shaping are durable and the reinforcer may shift from the external concrete cue to other more potent reinforcers such as social praise or increased self-efficacy (Velligan et al.). Promoting self-efficacy through behavioral success is one goal of social cognitive theory.

Because people with schizophrenia may have impaired insight and motivation, the DART study incorporated a small raffle that was held at the close of the intervention each week. Participants had the opportunity to earn one raffle ticket for attending the intervention, one raffle ticket for completing behavioral monitoring home practice, and one raffle ticket for accomplishing the previous week's behavioral goal. Health-related prizes, of $5–10 value, were awarded to one raffle winner each week (e.g., water bottles, inexpensive CD players, towels, cotton socks) at the close of the intervention session. Participants in the intervention were unanimously eager to earn raffle tickets. Some collected small prizes to give as gifts to their loved ones, some shared their prize with another group member who had not won a prize before, and others retained the prize for themselves.

Graded-Task Assignments

Clients may be overwhelmed by the enormous task of losing weight. By breaking the process down into small, manageable tasks, and using a successive approximation approach, the client experiences success and enhanced self-efficacy. This is an invaluable behavioral technique in the treatment of overweight and obesity in general (Radomile, 2000), and may be just as useful for patients with serious mental illnesses. Patients learn that achieving weight loss is made up of many smaller steps. For example, an older person with schizophrenia may find it impossible to walk 30 min, 5 days a week, but many find that they can walk for 5 min without difficulty. The exercise may be gradually expanded until they reach 30 min of sustained activity 5 days per week.

Modifications for Older Adult Patients with Serious Mental Illness

There are several intervention modifications that should be considered when working with older adults with serious mental illness. First, as with all older adults, they should be assessed for sensory deficits. For example, many of those in need of glasses may not have them, the prescription may be dated, or their glasses may be in disrepair. In such instances, accommodation for visual deficits by providing written materials in large font is essential. The DART study presented materials in 16-point font and included large posters and graphics to display important intervention material. Next, it is important to administer the interventions in communities where participants live and during daytime hours. Older adults with serious mental illness, due to sensory deficits and physical health limitations, may be less inclined to use public transportation, particularly during evening hours. Although patients with schizophrenia do not show a greater slope of age-related cognitive decline than those without psychiatric disorders (Elyer-Zorrilla, Heaton, McAdams, Zisook, & Jeste, 2000), DART intervention leaders presented information in small segments and used multiple reviews and teach and query methods to encourage overlearning of material. Overlearning is often important for older persons with SMI, who may experience further impairment due to age-related cognitive decline. For example, a group facilitator, who just reviewed values for hyperglycemia, might check with each individual patient by saying "If you take your blood sugar before dinner, what number would tell you that your blood sugar is too high? Is it 100, 115, or 140?" In order to achieve mastery of the material, some patients might require multiple queries on the same material within a few minutes of one another.

With regard to interventions in populations of people affected by serious mental illness in general, it is important to evaluate and accommodate for low literacy. Only a few studies conducted document patients' reading proficiency and educational attainment. These studies suggest that reading proficiency may approximate the tenth grade level or below (Fuller et al., 2002; Harvey et al., 2000; Jablensky et al., 1999). Importantly, literature shows that, although patients may read well aloud, comprehension of written materials is poorer for patients with psychiatric illness than those without psychiatric illness (Hayes & O'Grady, 2003). Therefore, comprehension of material should be evaluated on a frequent basis (i.e., query comprehension accompanied by corrective feedback). For example, when teaching about safe glucose levels for exercise, DART group leaders provide the information and subsequently ask each member of the group what the appropriate glucose levels are. They may present the correct answer among other incorrect answers on the board or notepad in front of class. Each person is queried and re-exposed to the material and requeried periodically until they have mastered that material. Third, group facilitators should limit their use of metaphors or other abstract references. Fourth, group facilitators should limit open-ended type questions, for example, "why do you think I am asking you to track your blood sugar?" Instead, provide options from which patients may choose (i.e., "Do you think I am asking you to record your blood sugar so that you can become more aware of how your diet affects your blood sugar or do you think I am asking you to record your blood sugar so that we can see how your exercise affects your blood sugar or both?"). When intervening outside of the structured research context, additional modifications may be required (e.g., providing written materials, slowing the progression of the intervention, writing information on the board while speaking to slow the rate of presentation) depending upon the general range of cognitive functioning for each individual group.

Case Example: Ms. B

Overview

Ms. B was a 56-year-old obese female who resided alone in an independent living apartment. She had schizoaffective disorder for 22 years and was most recently prescribed risperdol, lithium, lexapro, and glucophage. She had been diagnosed with type 2 diabetes 3 months prior to enrolling in the DART study and was referred by her mental healthcare provider to a primary care physician. Following receipt of study consent and releases of information, her physician was contacted, on behalf of the participant, and alerted regarding the participant's intent to enroll in the program. Physician's approval for participation in moderate exercise (i.e., brisk walking) was obtained.

Assessment

Staff then completed a brief interview consisting of the Diabetes History Questionnaire from the University of Michigan Diabetes Research and Training Center (https://borc.med.umich.edu/MDRTC_surveys; accessed 8/31/06), Diabetes Care Profile (Fitzgerald et al., 1996), a drug record, and the Diabetes Knowledge Test (Fitzgerald et al., 1998) in order to obtain information relevant to past diabetes care, diabetes knowledge, preventive health care, current diabetes management behaviors, medication use, existing comorbidities that may complicate participation, and confidence in diabetes management skills.

On a separate visit, we completed a fasting assessment of plasma glucose, glycosylated hemoglobin, lipids, blood pressure, body weight, and BMI. Results of these tests revealed the Ms. B was an obese female (i.e., BMI = 33.5, body weight = 219 lbs; waist circumference 46 in.). She was previously assigned a primary care doctor who had not referred her to any prior diabetes management or dietary program. Her fasting blood glucose was high (i.e., 138 mg/dL), but her glycosylated hemoglobin was acceptable (6.0). Her total cholesterol was high (203 mg/dL), her high-density lipoprotein was low (29 mg/dL), and her blood pressure was 137/95. She was diagnosed with diabetes 3 months prior to enrollment in the study and was completing minimal diabetes preventive care (i.e., testing blood sugar once every 2 days, was not completing foot self-exams, and had no dilated eye exam in the past 3 years). Her overall diabetes knowledge was moderate (six of 14 correct on the diabetes knowledge test).

Intervention

The therapist, in the first session, provided an orientation to the group and general discussion of motivation and engagement. Ms. B., like everyone else, was asked the question, "What do you want your health to be like?" Ms. B indicated that she wanted to feel like she did when she was younger and that she wanted to have more energy. The group leader then presented a discussion of short- and long-term goals and helped Ms. B. to identify "having more energy" as her long-term goal. Because Ms. B. had some difficulty identifying a short-term goal independently, the therapist inquired about the participant's current health behaviors and provided potential short-term goals (i.e., walking one block to the park each morning, 15 min of housework) that reflected a slight increase above her current health behavior. Ms. B. selected one of the goals for which she had a high degree of confidence (i.e., walking to the park each morning). Ms. B. received instruction about diabetes complications, high and low blood sugar, blood sugar testing and monitoring, and problem-solving. She was provided a log to use with her diabetes testing kit, following clarification from her primary care physician, and was encouraged to test her blood sugar two times per day and present her blood sugar logs to the therapist in class.

Following the first four basic diabetes management sessions, Ms. B participated in four healthy nutrition sessions. Here, Ms. B. was asked to construct a 24-h dietary recall. She presented information suggesting that she ate one meal that day, consisting of tuna casserole. Ms. B. indicated that she had only approximately $40 per month after paying for rent, utilities, and cigarettes. Consequently, she typically ate only one meal a day and it was usually pasta with cream sauce. After reviewing the process for problem-solving, the entire group participated in generating potential solutions for Ms. B. These included advertising for a roommate to help with rent, asking her son to move in with her, quitting smoking, and obtain free groceries from a community food truck. Ms. B. then evaluated the pros and cons of each option and chose to ask her son to move in with her to share the rent. Her son, who also had mental illness moved into her apartment and as a result, she and her son pooled their available money and could then afford to incorporate fruits and vegetables into both of their diets. Through nutrition goal setting, behavioral monitoring, and behavioral incentives, Ms. B. began eating at least three fruits and vegetables per day. She also spread her food consumption out throughout the day, rather than limiting her intake to one meal.

Ms. B. was then instructed in benefits of exercise, types of physical activity, blood sugar monitoring for safe or healthy exercise, and proper foot care. She was also provided with a pedometer. Ms. B. completed an activity log for 2 separate days during the week at the outset of the exercise education and indicated that she slept or watched television for a majority of daytime hours mostly because she was tired and had nothing else to do. She did indicate that she walked for 5 min each morning to the local park reported feeling confident that she could actually walk to the park twice a day, once in the morning and once in the evening.

Self-monitoring was used in two ways. First, she completed a chart by placing an "x" in a box below the day of the week that she walked. Second, she indicated the number of steps taken at the end of each day as shown on her pedometer. Each of the sessions was then repeated with Ms. B. and her fellow participants. Ms. B., in a graduated fashion, changed her diabetes self-care, dietary, and lifestyle exercise habits throughout the intervention. She was routinely encouraged to focus on short-term goals that she felt confident that she could achieve. With the additional money obtained from combining her income with her son's income, she was able to purchase and eat 4–5 fruits and vegetables. Moreover, she eventually received food support from a local food truck, thereby allowing even more money to be available to increase her fruit, vegetable, and protein intake. She also gradually increased her level of exercise throughout the intervention, by adding an additional block each week. Toward the end of the intervention, Ms. B. reported walking a total of 16 blocks twice per day. Throughout the intervention, Ms. B. was regularly eligible to receive small incentive prizes throughout raffle drawing for attending classes and completing her homework. She was able to win t-shirts, shorts, a compact disc player, and a water bottle.

Outcomes

At follow-up her knowledge of diabetes management was significantly improved (i.e., from six of 14 correct to 12 of 14 on the diabetes knowledge test). Ms. B also lost nearly 10% of her baseline body weight and reduced her BMI by 4 points (i.e., postintervention weight = 198 lbs, waist circumference = 42 in., BMI = 31.2) and fasting glucose and glycosylated hemoglobin values were slightly reduced (i.e., 120; 5.3). According to the UK Prospective Diabetes Study, at the population level, every point decrease in glycosylated hemoglobin is associated with 35% reduction in risk for diabetes complications. These changes took place despite a medication regimen that often results in serious weight gain. There was also an improvement in the level of depressive symptoms and self-efficacy scales (Anderson et al., 2000). The Hamilton Rating Scale for Depression decreased from 12 to 8; Self-efficacy for Managing Psychosocial Aspects of Diabetes increased from 32 to 35; Self-efficacy for Setting and Achieving Goals (relevant to diabetes care) increased from 24 to 35; and Readiness to Change increased from 25 to 31.

Diversity Issues

The DART program was developed primarily for a non-Latino White population, but approximately 10% of our sample self-identified as Latino and nearly 20% self-identified as African-American. One of the entry criteria was that each participant be able to speak the English language. All sessions were conducted in

English and by non-Latino White group leaders. Patients from ethnic minorities, who had a good command of English, were able to participate effectively in the program, which suggests that others, who are not highly acculturated, could benefit if appropriate translations and bilingual/bicultural group leaders were available.

Latinos are the largest and fastest growing ethnic group in the US (Council on Scientific Affairs, 1991) currently comprising over 12% of the total US population (US Census Bureau, 2000). As the proportion of Latinos in the US increases, so too will the sheer number of Latinos in need of psychiatric care. Atypical antipsychotic medications are the mainstay of treatment for many patients with severe mental illness, and since several of these notably impact weight status, there is growing concern regarding elevated risk of metabolic side effects. This is a particular concern for Latinos (Henderson, Nguyen et al., 2005; Leslie & Rosenheck, 2004), who are at greater risk for the development of treatment emergent diabetes than non-Latino Whites (Henderson, Nguyen et al., 2005). Cultural adaptations of programs such as DART may offset negative metabolic effects in Latinos with serious mental illness. Although few in number, other treatment programs adapted for this patient group have produced favorable results (Kopelowicz, Zarate, Gonzalez Smith, Mintz, & Liberman, 2003; Patterson et al., 2005; Telles, Karno, & Mintz, 1995). However, few if any have addressed management of medical comorbidity in Latino patients with serious mental illness. Given the growing population of older Latinos in need of health and mental health-related services, a cultural adaptation of DART would serve as one step to improve the availability of culturally competent tools for use with Spanish-speaking patients with serious mental illness. The provision of bicultural and bilingual skills trainers (Mausbach et al., in press) and incorporation of family (Kopelowicz et al., 2003; Patterson et al.) in the treatment process would most likely maximize its effectiveness.

Summary

Type 2 diabetes has become a common problem for patients with serious mental illness taking antipsychotic medications. The psychiatric symptoms and cognitive deficits associated with serious mental illness may hinder patients' ability to manage another chronic and complex disease such as diabetes. Unfortunately, poor management of diabetes may contribute to serious and even life-threatening conditions and intense physical and emotional suffering. Cognitive and behavioral interventions designed specifically for patients with serious mental illness have shown promise for reduction of weight and plasma glucose in small- to moderate-sized samples. Clearly, additional studies in patients with serious mental illness and diabetes are needed. Few if any dismantling studies have been conducted to understand which behavioral intervention techniques are most useful to promote weight loss and glucose regulation in patients with serious mental illness. Clinicians or program developers may wish to include goal setting, behavioral monitoring, problem-solving, stimulus control, and graded task assignment among other useful behavioral techniques (e.g., reinforcement/incentives). The delivery of group education, however, may need to be modified to accommodate for individual patient limitations. Specifically, clinicians should consider limiting the amount of information provided at one time, using simple language to explain concepts, and incorporate techniques to accommodate for cognitive deficits to assist in comprehension of program goals and educational materials.

References

American Diabetes Association, American Psychiatric Association, American Association of Clinical Endocrinologists, & North American Association for the Study of Obesity. (2004). Consensus development conference on antipsychotic drugs and obesity and diabetes. *Diabetes Care*, *27*, 596–601.

Anderson, R. M., Funnell, M. M., Fitgerald, J. T., & Marrero, D. G. (2000). The diabetes empowerment scale: A measure of psychosocial self-efficacy. *Diabetes Care*, *23*, 739–743.

Archie, S., Wilson, J. H., Osbourne, S., Hobbs, H., & McNiven, J. (2003). Pilot study: Access to fitness facility and exercise levels in olanzapine-treated patients. *Canadian Journal of Psychiatry*, *48*, 628–632.

Bandura, A. (1986). *Social foundations of thought and action: A social cognitive theory*. Englewood Cliffs, NJ: Prentice-Hall.

Bandura, A. (1989). Perceived self-efficacy. In V. Mays, G. Albee, & S. Schneider (Eds.), *Primary prevention of AIDS: Psychological approaches*. Newbury Park, CA: Sage.

Bartels, S. J., Forester, B., Mueser, K. T., Miles, K. M., Dums, A. R., Pratt, S. I., et al. (2004). Enhanced skills training and health care management for older persons with severe mental illness. *Community Mental Health Journal*, *40*, 75–90.

Brar, J. S., Ganguli, R., Pandina, G., Turkoz, I., Berry, S., & Mahmoud, R. (2005). Effects of behavior therapy on weight loss in overweight and obese patients with schizophrenia or schizoaffective disorder. *Journal of Clinical Psychiatry*, *66*, 205–212.

Brown, S., Birtwistle, J., Roe, L., & Thompson, C. (1999). The unhealthy lifestyle of people with schizophrenia. *Psychological Medicine*, *29*, 697–701.

Citrome, L. (2004). The increase in risk of diabetes mellitus from exposure to second-generation antipsychotic agents. *Drugs Today (Barc)*, *40*, 445–464.

Cohn, T., Prud'homme, D., Streiner, D., Kameh, H., & Remington, G. (2004). Characterizing coronary heart disease risk in chronic schizophrenia: High prevalence of the metabolic syndrome. *Canadian Journal of Psychiatry*, *49*, 753–760.

Council on Scientific Affairs. (1991). Hispanic health in the United States. *Journal of the American Medical Association*, *265*, 248–252.

Daumit, G. L., Goldberg, R. W., Anthony, C., Dickerson, F., Brown, C. H., Kreyenbuhl, J., et al. (2005). Physical activity patterns in adults with severe mental illness. *Journal of Nervous and Mental Disease*, *193*, 641–646.

Davidson, S., Judd, F., Jolley, D., Hocking, B., Thompson, S., & Hyland, B. (2001). Cardiovascular risk factors for people with mental illness. *Australian and New Zealand Journal of Psychiatry*, *35*, 196–202.

De Hert, M. A., van Winkel, R., van Eyck, D., Hanssens, L., Wampers, M., Scheen, A., et al. (2006). Prevalence of the metabolic syndrome in patients with schizophrenia treated with antipsychotic medication. *Schizophrenia Research*, *83*, 87–93.

Dickerson, F. B., Brown, C. H., Daumit, G. L., Lijuan, F., Goldberg, R. W., Wohlheiter, K., et al. (2006). Health status of individuals with serious mental illness. *Schizophrenia Bulletin*, *32*, 584–589.

Elyer-Zorrilla, L. T., Heaton, R. K., McAdams, L. A., Zisook, S., & Jeste, D. V. (2000). Cross-sectional study of older outpatients with schizophrenia and healthy comparison subjects: No difference in age-related cognitive decline. *American Journal of Psychiatry*, *157*, 1324–1326.

Evans, S., Newton, R., & Higgins, S. (2005). Nutritional intervention to prevent weight gain in patients commenced on olanzapine: A randomized controlled trial. *Australian and New Zealand Journal of Psychiatry*, *39*, 479–486.

Faulkner, G., Soundy, A. A., & Lloyd, K. (2003). Schizophrenia and weight management: A systematic review of interventions to control weight. *Acta Psychiatrica Scandinavica*, *108*, 324–332.

Fitzgerald, J. T., Davis, W. K., Connell, C. M., Hess, G. E., Funnell, M. M., & Hiss, R. G. (1996). Development and validation of the diabetes care profile. *Evaluation and the Health Professions*, *19*, 208–230.

Fitzgerald, J. T., Anderson, R. M., Funnell, M. M., et al. (1998). The reliability and validity of a brief diabetes knowledge test. *Diabetes Care*, *21*, 706–710.

Fuller, R., Nopoulos, P., Arndt, S., O'Leary, D., Ho, B. C., & Andreasen, N. C. (2002). Longitudinal assessment of premorbid cognitive functioning in patients with schizophrenia through examination of standardized scholastic test performance. *American Journal of Psychiatry*, *159*, 1183–1189.

Harmatz, M. G., & Lapuc, P. (1968). Behavior modification of overeating in a psychiatric population. *Journal of Consulting and Clinical Psychology*, *32*, 583–587.

Harvey, P. D., Moriarty, P. J., Friedman, J. I., White, L., Parrella, M., Mohs, R. C., et al. (2000). Differential preservation of cognitive functions in geriatric patients with lifelong chronic schizophrenia: Less impairment in reading compared with other skill areas. *Society of Biological Psychiatry*, *47*, 962–968.

Hayes, R. L., & O'Grady, B. M. (2003). Do people with schizophrenia comprehend what they read? *Schizophrenia Bulletin*, *29*, 499–507.

Henderson, D. C. (2005). Schizophrenia and comorbid medical disorders. *Journal of Clinical Psychiatry*, *66*, 11–20.

Henderson, D. C., Cagliero, E., Copeland, P. M., Borba, C. P., Ervins, E., Hayden, D., et al. (2005). Glucose metabolism in patients with schizophrenia treated with atypical antipsychotic agents: A frequently sampled intravenous glucose tolerance test and minimal model analysis. *Archives of General Psychiatry*, *62*, 19–28.

Henderson, D. C., Nguyen, D. D., Copeland, P. M., Hayden, D. L., Borba, C. P., Louie, et al. (2005). Clozapine, diabetes mellitus, hyperlipidemia, and cardiovascular risks and mortality: Results of a 10-year naturalistic study. *Journal of Clinical Psychiatry*, *66*, 1116–1121.

Holt, R. I., Bushe, C., & Citrome, L. (2005). Diabetes and schizophrenia 2005: Are we any closer to understanding the link? *Journal of Psychopharmacology*, *19*, 56–65.

Jablensky, A., McGrath, J., Herrman, H., Castle, D., Gureje, O., Morgan, V., et al. (1999). *People living with psychotic illness: An Australian study 1997–98*. Commonwealth of Australia.

Kopelowicz, A., Zarate, R., Gonzalez Smith, V., Mintz, J., & Liberman, R. P. (2003). Disease management in Latinos with schizophrenia: A family-assisted, skills training approach. *Schizophrenia Bulletin*, *29*, 211–217.

Lamberti, J. S., Crilly, J. F., Maharaj, K., Olson, D., Wiener, K., Dvorin, S., et al. (2004). Prevalence of diabetes mellitus among outpatients with severe mental disorders receiving atypical antipsychotic drugs. *Journal of Clinical Psychiatry*, *65*, 702–706.

Leslie, D. L., & Rosenheck, R. A. (2004). Incidence of newly diagnosed diabetes attributable to atypical antipsychotic medications. *American Journal of Psychiatry, 161*, 1709–1711.

Lindamer, L., McKibbin, C. L., Jin, H., Ueki, K., & Jeste, D. V. (2006). *Physical activity in middle-age and older patients with schizophrenia.* Paper Presented at the meeting of the American Association for Geriatric Psychiatry, San Juan, Puerto Rico.

Littrell, K. H., Hilligoss, N. M., Kirshner, C. D., Petty, R. G., & Johnson, C. G. (2003). The effects of an educational intervention on antipsychotic-induced weight gain. *Journal of Nursing Scholarship, 35*, 237–241.

Loh, C., Meyer, J. M., & Leckband, S. G. (2006). A comprehensive review of behavioral interventions for weight management in schizophrenia. *Annals of Clinical Psychiatry, 18*, 18–31.

Marder, S. R., Essock, S. M., Miller, A. L., Buchanan, R. W., Casey, D. E., Davis, J. M., et al. (2004). Physical health monitoring of patients with schizophrenia. *American Journal of Psychiatry, 161*, 1334–1349.

Mausbach, B. T., Bucardo, J., Cardenas, V., McKibbin, C., Barrio, C., Goldman, S. R., et al. (in press). Evaluation of a culturally tailored skills intervention for Latinos with persistent psychotic disorders. *American Journal of Psychiatric Rehabilitation.*

McCreadie, R. G., Kelly, C., Connolly, M., Williams, S., Baxter, G., Lean, M., et al. (2005). Dietary improvement in people with schizophrenia: Randomised controlled trial. *British Journal of Psychiatry, 187*, 346–351.

McCreadie, R., Macdonald, E., Blacklock, C., Tilak-Singh, D., Wiles, D., Halliday, J., et al. (1998). Dietary intake of schizophrenic patients in Nithsdale, Scotland: Case–control study. *British Medical Journal, 317*, 784–785.

McEvoy, J. P., Meyer, J. M., Goff, D. C., Nasrallah, H. A., Davis, S. M., Sullivan, L., et al. (2005). Prevalence of the metabolic syndrome in patients with schizophrenia: Baseline results from the Clinical Antipsychotic Trials of Intervention Effectiveness (CATIE) schizophrenia trial and comparison with national estimates from NHANES III. *Schizophrenia Research, 80*, 19–32.

McKibbin, C. L., Lindamer, L., Sirkin, A., Jin, H., Ueki, K., Patterson, T., et al. (2006). *Dietary habits in middle-aged and elderly patients with schizophrenia.* Paper Presented at the meeting of the American Association for Geriatric Psychiatry, San Juan, Puerto Rico.

McKibbin, C. L., Patterson, T. L., Norman, G., Patrick, K., Jin, H., Roesch, S., et al. (2006). A lifestyle intervention for older schizophrenia patients with Diabetes Mellitus: A randomized controlled trial. *Schizophrenia Research, 86*, 36–44.

Miller, C. K., Edwards, L., Kissling, G., & Sanville, L. (2002). Evaluation of a theory-based nutrition intervention for older adults with diabetes mellitus. *Journal of American Dietary Association, 102*, 1069–1081.

Mukherjee, S., Schnur, D. B., & Reddy, R. (1989). Family history of type 2 diabetes in schizophrenic patients. *Lancet, 1*, 495.

Newcomer, J. W. (2005). Second-generation (atypical) antipsychotics and metabolic effects: A comprehensive literature review. *CNS Drugs, 19*, 1–93.

Patterson, T. L., Bucardo, J., McKibbin, C. L., Mausbach, B. T., Moore, D., Barrio, C., et al. (2005). Development and pilot testing of a new psychosocial intervention for older Latinos with chronic psychosis. *Schizophrenia Bulletin, 31*, 922–930.

Patterson, T. L., McKibbin, C. L., Taylor, M. J., Goldman, S., Davila-Fraga, W., Bucardo, J., et al. (2003). Functional Adaptation Skills Training (FAST): A pilot psychosocial intervention study in middle-aged and older patients with chronic psychotic disorders. *American Journal of Geriatric Psychiatry*, *11*, 17–23.

Radomile, R. R. (2000). Obesity. In J. R. White & A. S. Freeman (Eds.), *Cognitive-behavioral group therapy for specific problems and populations* (pp. 99–126). Washington, DC: American Psychological Association.

Rotatori, A. F., Fox, R., & Wicks, A. (1980). Weight loss with psychiatric residents in a behavioral self control program. *Psychological Reports*, *46*, 483–486.

Strassnig, M., Brar, J. S., & Ganguli, R. (2003). Nutritional assessment of patients with schizophrenia: A preliminary study. *Schizophrenia Bulletin*, *29*, 393–397.

Telles, C., Karno, M., & Mintz, J. (1995). Immigrant families coping with schizophrenia: Behavioral family intervention v. case management with a low-income Spanish speaking population. *British Journal of Psychiatry*, *167*, 473–479.

US Census Bureau. (2000). *www.factfinder.census.gov*.

Velligan, D. I., Kern, R. S., & Gold, J. M. (2006). Cognitive rehabilitation for schizophrenia and the putative role of motivation and expectancies. *Schizophrenia Bulletin*, *32*, 474–485.

Weber, M., & Wyne, K. (2006). A cognitive/behavioral group intervention for weight loss in patients treated with atypical antipsychotics. *Schizophrenia Research*, *83*, 95–101.

12
Dialectical Behavior Therapy for Personality Disorders in Older Adults

Jennifer S. Cheavens and Thomas R. Lynch

Personality disorders, by DSM-IV (American Psychological Association, 1994) definition, are long-standing/stable, first evidenced in adolescence or early adulthood, and have associated pervasive difficulties in both interpersonal and impulsive functioning. Based on these criteria, it is difficult to adequately assess personality pathology in older adults for whom adolescence and early adulthood may have occurred decades ago. Additionally, several authors have argued that the diagnostic criterion for personality disorders (e.g., impulsivity in sexual and legal domains, aggression, work-related perfectionism) are more relevant for younger adults as opposed to their older adult counter-parts (Agronin & Maletta, 2000; van Alphen, Engelen, Kuin, & Derksen, 2006; Segal, Coolidge, & Rosowsky, 2000).

Thus, identifying personality disorders in older adult samples is a daunting task. Both empirical evidence and clinical experience, however, indicate that personality disorders are present in late-life and the symptoms including unstable interpersonal relationships/loneliness, chronic feelings of emptiness, and affective instability remain costly to individuals and society. The prevalence rates for older adult personality disorders vary depending on the sampling setting. The lowest prevalence rates are reported in community settings (2–13%; Ames & Molinari, 1994; Weissman, 1993), followed by outpatient psychiatric settings (5–33%; Mezzich, Fabrega, Coffman, & Glavin, 1987; Molinari & Marmion, 1995), and inpatient settings (7–61%; Casey & Schrodt, 1989; Kunik et al., 1993; Molinari, Ames, & Essa, 1994). Two meta-analytic studies by Abrams and Horowitz (1996, 1999) suggested that the prevalence of personality disorders in adults over 50 was between 10 and 20%.

Research suggests that the majority of personality disorders evidenced in late life are Cluster C personality disorders, particularly obsessive–compulsive personality disorder, followed by Cluster A disorders (Abrams & Horowitz, 1996; Kenan et al., 2000; Kunik et al., 1994; Vine & Steingart, 1994). These disorders are typically characterized by excesses in emotional reactivity (Morey et al., 2002, 2003), particularly anxiety for the Cluster C disorders and hostility for the Cluster A disorders

(Hyler, Skodol, Kellman, Oldham, & Rosnick, 1990; Klein, Kupfer, & Shea, 1993; Nestadt, Eaton, & Romanoski, 1994; Schroeder & Livesley, 1991).

Given the impact of personality disorders on the treatment of other Axis I disorders, it is imperative to find effective ways to treat personality disorders in older adults. In older adults, personality pathology has generally been associated with poorer response to treatment for Axis I disorders (Fiorot, Boswell, & Murray, 1990; Thompson, Gallagher, & Czirr, 1988), although this association was not apparent in Kunik et al. (1993). In addition to reduced effectiveness in treating Axis I disorders, personality disorders in older adults have been associated with impaired functioning after affective symptoms improve (Abrams, Spielman, Alexopoulos, & Klausner, 1998), impaired social support (Vine & Steingart, 1994), decreased quality of life, suicide, and disability (Lyness, Caine, Conwell, King, & Cox, 1993). Older adult depressed patients with comorbid PD are four times more likely to experience maintenance or reemergence of depressive symptoms than those without PD diagnoses (Morse & Lynch, 2004).

Evidence Base for Treatment of Older Adults

The literature on the treatment of personality disorders in older adults is sparse. Given some of the debate regarding the diagnosis of personality disorders in older adults described previously, it is not surprising that the science related to treating these disorders is underrepresented. To date, two randomized clinical trials (RCTs) have been conducted in an effort to develop an adaptation of standard dialectical behavior therapy (DBT) for older adults. The initial trial was conducted with the group format of standard DBT and the primary aims were to gain experience with the clinical population and determine the feasibility and acceptability of the group component of DBT for older adults. The aim of the second RCT was to conduct a trial of standard DBT in order to develop a manual for $\text{DBT}^{\text{D+PD}}$ with an emphasis on treatment of personality disorders in older adults.

Study 1

In this trial, 34 participants with major depressive disorder (MDD) of age 60 or older were randomly assigned to either an antidepressant medication condition (MED) or a combination of standard DBT skills training plus antidepressant medication condition (MED + DBT; see Lynch, Morse, Mendelson, & Robins, 2003 for a detailed description of the study and results). Personality diagnosis was not an inclusion/exclusion criterion. All participants were provided with clinical medication management using physician choice of an SSRI by a board-certified psychiatrist. The active phase of the treatment lasted for 6 months with a 6-month follow-up period.

Results indicated that participants in both conditions improved in terms of depressive symptoms. However, a greater proportion of individuals in the MED + DBT, as compared to the MED group, were classified as in remission on interviewer-rated depression scores at both posttreatment (71% compared to 47%) and follow-up (75% compared to 31%). Participants in the DBT + MED group showed significant reductions in dependency (i.e., sociotropy), perfectionism, and self-criticism (i.e., autonomy) from pretreatment to posttreatment, but this improvement was not evidenced in the MED group (Lynch et al., 2003). There were no treatment drops from the MED condition and only one treatment drop from the DBT + MED condition, which suggests that older adults with psychiatric conditions can tolerate the group treatment component of DBT^{D+PD}.

Study 2

After 8 weeks of antidepressant treatment, 37 participants aged 55 or older with a diagnosis of MDD and at least one comorbid personality disorder were randomized to either standard DBT (both group and individual; Linehan, 1993a, 1993b) plus antidepressant medication (DBT + MED) or antidepressant medication with clinical management for a period of 24 weeks (MED; see Lynch et al., 2007 for a more detailed description of the study and results).

Again, participants in both the DBT + MED and MED groups demonstrated significant reductions in depressive symptoms. Findings suggest that the DBT + MED group reached the level of remission more quickly than the MED group. Additionally, individuals in the DBT + MED group demonstrated significantly lower Interpersonal Sensitivity and Interpersonal Aggression scores than individuals in the MED group at both posttreatment and follow-up. These findings suggest that the DBT condition impacted important constructs of personality pathology in older adults.

Assessment Considerations

Assessment is important in the treatment of personality disorders in older adults for at least two reasons. First, some form of assessment must enable the clinician or researcher to make accurate diagnoses to include those participants most likely to benefit from the treatment. Second, the assessment must include constructs that are likely to change over the course of treatment (i.e., mechanisms of change or mediators of change) in order to determine the usefulness of the treatment for a particular individual.

Structured Clinical Interview for DSM-IV, Axis II (SCID-II) (First, Spitzer, Williams, & Benjamin, 1997). Personality disorder diagnoses for older adults, at this point in time, are based on the DSM-IV-TR (American Psychological Association, 2000) criteria for Axis II disorders. Thus, the currently most reliable and valid way to assess whether or not an individual meets criteria for any personality disorder is to use the SCID-II interview. To ease in the administration of this interview, participants can first make yes/no responses to a 119-item questionnaire assessing specific symptoms of each personality disorder.

Coolidge Axis II Inventory (CATI) (Coolidge, 1993). The CATI is a 225-item self-report instrument that has been used frequently to assess personality disorders in older adults. This instrument is based on DMS-IV (1994) personality disorder criteria and includes items to assess disorders for further study from the DSM-IV (1994). In addition to the 13 personality disorder scales, the CATI includes two validity scales, three Axis I scales (i.e., brain dysfunction, depression, anxiety), and critical items to assess substance abuse, suicidal ideation, and problems in thinking. Although original reliability and validity estimates were based on a college student sample (Coolidge & Merwin, 1992), several other publications have used the CATI with older adult samples (e.g., Coolidge, Burns, Nathan, & Mull, 1993; Segal, Hook, & Coolidge, 2001).

Inventory of Interpersonal Problems – Personality Disorders (IIP-PD) (Pilkonis, Kim, Proietti, & Barkham, 1996). In addition to making a categorical diagnosis, it is important to have some dimensional measures of personality constructs. The IIP-PD is a 47-item version of the Inventory of Interpersonal Problems (IIP, Horowitz, Rosenberg, Baer, Ureno, & Villasenor, 1988) developed by Pilkonis and colleagues, which has been shown to distinguish between individuals with and without personality disorders. According to Pilkonis et al., factor analyses of the scale have revealed five subscales: interpersonal sensitivity, interpersonal ambivalence, aggression, need for social approval, and lack of sociability. The scale is a self-report instrument and thus, requires little training for administration.

Assessment Issues. We are not aware of any diagnostic or assessment measures of personality disorders that have been developed specifically for older adult samples. Thus, researchers and clinicians must be mindful of the limitations inherent in generalizing assessments designed for and normed on younger adult samples to older adults. It is likely that the behavioral presentations of older adults with personality disorders differ in significant ways from the behavioral presentations of younger adults with the same personality pathology (e.g., less sexual promiscuity, theft, aggressive outbursts, work-related perfectionism). This is an important consideration for clinicians and researchers when assessing personality disorders in older adults.

DBT^{D+PD} for Older Adults with Personality Disorders

Originally, DBT was developed for the treatment of borderline personality disorder (BPD) by Linehan (1993a, 1993b). In the adaptation of DBT^{D+PD}, most of the fundamental aspects of DBT (e.g., treatment format, transactional theory of onset and maintenance, focus on skills, etc.) remained the same. The adaptations to the treatment focused on two primary changes. The first was widening the scope of the DBT to include treatments for all personality disorders. The second change was to incorporate findings from geropsychology literature to create a special emphasis for the treatment of older adults diagnosed with personality disorders. For a more in-depth description of the treatment, please refer to Lynch, Cheavens, Cukrowicz, and Linehan (2006) and Lynch et al. (2007). In this section, we will briefly describe the treatment and focus our discussion of DBT^{D+PD} on the components specific to older adults.

DBT^{D+PD} consists of four modes of treatment: individual therapy, group skills training, phone consultation, and team consultation. Each of the four modes of therapy is described briefly below.

Individual Therapy

The individual therapy component in DBT^{D+PD} consists of weekly sessions for approximately 6 months, at minimum. In our experience, older adults with personality disorders are typically involved in fewer overtly life-threatening behaviors (e.g., intentional self-injury) and fewer treatment-interfering behaviors (e.g., calling too often, missing sessions) than younger, BPD patients. Thus, individual therapy sessions can focus on enhancing motivation to change active problem solving and increasing flexibility in the behavioral repertoire. In individual sessions, therapists work to maintain a balance between accepting and change-oriented stances. As in standard DBT, treatment targets are arranged in a hierarchical order as follows (1) reduce high-risk suicidal and life-threatening behaviors; (2) reduce therapy-interfering behaviors (e.g., noncompliance, noncollaboration, nonattendance); and (3) reduce quality-of-life interfering behaviors, with reducing depression and increasing openness to experience as the top priority in DBT^{D+PD} quality-of-life targets. Behaviors to be increased or decreased are tracked daily on a diary card with daily emotions, sleep patterns, and self-care habits. Older adults have successfully completed this card at a high rate during pilot studies (Lynch et al., 2003).

In the individual therapy mode, therapists are encouraged to use a variety of dialectical and cognitive behavioral strategies to target problem behavior and assist the patient in moving toward increased emotional experiencing and regulation. In terms of dialectical strategies, therapists are encouraged to balance acceptance and change, utilize metaphors and stories in session to demonstrate teaching points, and to look for synthesis in seemingly disparate options. As such, therapists are encouraged to work with patients to move from an either/or stance that is typical of cognitive rigidity and move toward a both/and stance that incorporates the useful aspects from divergent perspectives. In terms of cognitive behavioral strategies, therapists are encouraged to use cognitive restructuring to reduce judgmental thinking and other cognitive distortions, behavioral activation to increase pleasant events, and goal-setting strategies to increase problem-solving abilities. In conducting behavioral analyses with older adult patients, it may be of increased importance to understand historical antecedents that serve as vulnerability factors for maladaptive behaviors as these behaviors likely have a long reinforcement history.

Group Skills Training

The group skills training mode of DBT^{D+PD} occurs concurrently with the individual therapy mode. We recommend that patients begin the individual mode of therapy (2–4 weeks) before engaging in the group skills training mode so that the rationale for various modes of therapy can be provided and the relationship with the individual therapist can begin to be forged. Participation in the DBT^{D+PD} skills training group is recommended for a minimum of 28 weeks. The treatment manual for the group (Lynch et al., 2006) is very similar to the standard DBT manual (Linehan, 1993b) but highlights additional skills theorized to be of clinical importance for older adults with personality disorders. Thus, the DBT^{D+PD} skill manual includes skill modules for mindfulness, interpersonal effectiveness, distress tolerance, and emotion regulation. An additional module, Radical Openness has been added to target cognitive rigidity, a construct often present in older adults diagnosed with personality disorders that has negative implications for treatment (e.g., Ehrlich & Bauer, 1966; Ogrodniczuk, Piper, Joyce, McCallum, & Rosie, 2002, 2003).

One example of a change to the treatment manual that targets cognitive rigidity and behavioral inflexibility is the inclusion of an additional "states of mind" handout. Standard DBT introduces the concepts of "emotion, reasonable, and wise mind states" to help individuals with BPD synthesize the seemingly disparate forces of emotion and reason. In working with older adults, our participants told us, and we noticed, that their decision making often suffered due to an over-reliance on past problem-solving strategies, behaviors, and information. Thus, we introduced the concepts of fresh, fixed, and fluid mind to assist in synthesizing what one knows to be true based on history and past experiences with the truth of the present moment and new information.

Additionally, in the Radical Openness module that was added to the manual, we introduce skills that focus on reducing bitterness through forgiveness of self, others, and the environment. Reviewing and making meaning of past experiences has long been theorized to be an important developmental milestone (Butler, 1963; Erikson & Erikson, 1997; Staudinger & Pasupathi, 2000) and related skills are included in DBT^{D+PD}.

Telephone Consultation

In DBT^{D+PD}, telephone consultation between the therapist and the patient is included as a mode of treatment. Our experience, however, suggests that this mode of treatment is utilized infrequently with older adults who report "feeling bad" about contacting their therapist between sessions and/or deny the need for intersession contact. Following standard DBT guidelines, consultation calls can be for (1) coaching in use of skills in crisis situations and crisis intervention, (2) contact with the therapist to maintain or strengthen the therapeutic relationship, and (3) relationship repair. Thus, in working with personality-disordered patients, it is important to include phone access as a safety and relationship-enhancing mechanism, although at this point it is difficult to draw conclusions about the efficacy of this mode with older adult patients.

Team Consultation

As in standard DBT, therapist team consultation is a part of the treatment itself rather than an ancillary to the treatment. Therapists meet weekly to assist each other in the implementation of the treatment. One of the functions of team consultation is to reduce therapist burn-out and hopelessness. Although older adult, personality-disordered patients are less likely than younger BPD patients to present with severely life-threatening or impulsive behaviors on a regular basis, the chronicity of symptoms and difficult nature of therapeutic change in older adults with personality disorders calls for a supportive environment for the therapist.

Case Example

Assessment

Ms. P presented at the clinic in response to an advertisement for a treatment study for older adults with "emotional and personality problems." Ms. P was interviewed with the SCID-II and Hamilton Rating Scale for Depression (HAM-D). Additionally, Ms. P completed several self-report instruments, including the IIP-PD.

According to the initial assessment, Ms. P met criteria for BPD and her HAM-D was elevated beyond the study cut-off for significant depressive symptoms. The BPD criteria that Ms. P endorsed included a pattern of intense and unstable relationships, a markedly and persistently unstable sense of self, affective instability, inappropriate, intense anger, and chronic feelings of emptiness. Of note, Ms. P reported that she had a history of self-harm when she was much younger and stated that this behavior had not occurred in approximately three decades. Ms. P endorsed some passive suicidal ideation although she denied plan and intent.

Treatment Interventions

Ms. P completed a full course of both individual and group DBT. In addition, she met on a monthly basis with the study psychopharmacologist and reported routinely taking her prescribed antidepressant. During the course of treatment, Ms. P completed a diary card daily to track her mood, sleep, eating, suicidal ideation, and her primary treatment targets.

Increasing interpersonal relationships. In the initial stages of therapy, Ms. P reported that she was very lonely and that the most consistent difficulty in her life was a lack of social support. Although Ms. P had three sons and several grand-children, her relationships with her sons were strained and her access to her grand-children was restricted as a result. Ms. P had recently moved to the area to get a "new start" and thus did not have a strong support system in the area. Additionally, although Ms. P was active in her church, she had already experienced some "run-ins" with other women in the church and thus was hesitant about cultivating relationships in this domain.

To address this target, treatment focused on increasing the DBT interpersonal effectiveness skills related to asking appropriately for some things needed and saying no to others not needed or wanted, validating others and their perspectives, and cultivating self-respect. These skills were presented in the group format and reinforced in individual therapy. In addition, individual therapy focused on increasing exposure to interpersonal interactions in order to cultivate success experiences. Specifically, individual sessions often focused on planning potential interpersonal interactions and role-playing various iterations of such interactions. In the role-play scenarios, the therapist was attentive to the role of distorted thoughts (e.g., "This person is going to be a jerk") in heightening emotion and cognitive restructuring was a key component of the role-play exposures.

Reducing judgment. One of the main elements that contributed to Ms. P's social isolation was her vacillating judgment of others and of herself. Reducing judgment was not a treatment goal that Ms. P brought to therapy but it became evident in the initial stages of therapy that judgmental thinking was a barrier to Ms. P's goals of increasing interpersonal relationships and increasing value-congruent activity. Thus, the therapist noted the link of judgmental thinking to Ms. P's other goals and the two collaboratively agreed to target this as well.

To target judgmental thinking, Ms. P was asked to track judgments between sessions and to record any emotions, behaviors, sensations, and other thoughts that directly preceded or followed the judgmental thought. A pattern soon became clear that indicated that Ms. P was most likely to feel anger, accompanied by sensations of heat, tightness, and shallow breathing, when she had judgmental thoughts of others. These thoughts were often preceded by thoughts and feelings associated with feeling rejected or hurt. Additionally, Ms. P noticed that judgmental thoughts about herself were often associated with feelings of guilt, accompanied by sensations of heaviness around the eyes and a pit in the stomach, and the urge to escape a situation. Once these patterns became clear, individual therapy focused on cognitive restructuring to reduce judgment – including working on providing alternative interpretations of behaviors – and dialectical thinking to emphasize a both/and stance in place of an either/or stance.

Increasing value-congruent activities. When she came to treatment, one of Ms. P's primary goals was to figure out "who I am." Ms. P reported that over the years what she did in her personal life had largely been a function of what she "should" be doing. She reported that she was a "completely different" person with one group of people than she was with another group of people. In order to address this goal, Ms. P and the therapist worked on increasing value-oriented activities. The consultation team agreed that this might work to reduce feelings of emptiness and an incoherent sense of self.

The first step in addressing this target was to help Ms. P identify her personal values. The therapist provided lists of values and goals from the DBT^{D+PD} skills. Ms. P worked on identifying her values and then on generating behaviors that were either congruent or incongruent with these values. The therapist and Ms. P utilized activity scheduling to increase the value-congruent behaviors. Additionally, when Ms. P engaged in value-incongruent behavior, this was treated as a problem behavior and a behavioral analysis was conducted to determine what variables increased the likelihood of this occurring. When value-congruent behaviors were planned and not completed, a behavioral analysis was completed to determine what variables were barriers to completing these behaviors.

Treatment Outcome

At the end of the treatment, Ms. P reported improvement in all of the targets outlined in the previous section. The target that proved most resistant to change was judgmental thinking. Ms. P reported that she did not feel as though she "should have to" give up judgments. Although she had a more complete awareness of the impact of her judgmental thinking, she had a difficult time catching these thoughts in the moment and replacing them with nonjudgmental thoughts. Ms. P indicated that she felt more connected to other people and stated that she had a decrease in loneliness. Ms. P reported that she felt aware of her values for the first time that she could remember and stated that working to increase value-congruent behavior had helped in broadening her social circle as well as decreasing her sense of emptiness.

The objective measures that were completed at the initiation of treatment were repeated at treatment termination. Ms. P's depressive symptoms, as measured by the HAM-D, had decreased to a nonsignificant level of impairment. Additionally, Ms. P had reductions in her interpersonal aggression and interpersonal sensitivity scores on the IIP-PD. Personality disorder diagnoses do not lend themselves to being "resolved" in the course of 6 months of treatment due to the "pervasive pattern" inherent in the definition. Nonetheless, Ms. P noted that her sense of emptiness, inappropriate anger, and unstable relationships were less problematic since completing her treatment. She also consistently denied any suicidal ideation after the first 2 months of treatment.

DBT^{D+PD} in Combination with Psychopharmacological Treatment

To this point in time, all of the research conducted to examine modifications of DBT with older adults has been done in conjunction with psychopharmacological treatment. This is for two reasons. First, personality disorders are often experienced as congruent with one's character and thus, individuals with personality disorders often present to treatment with significant distress related to some Axis I disorder, such as MDD. Thus, providing an older adult with the first-line pharmacological treatment for the Axis I disorder will likely help with initial symptom reduction. In an effort to treat a personality disorder, it may be useful to remove as much variance as possible from any comorbid Axis I diagnosis. However, as noted previously, a comorbid personality disorder will likely decrease the impact of pharmacological treatments alone for Axis I disorders. Second, in addition to clarifying the personality disorder, reducing distress through pharmacotherapy may increase a patient's ability to participate in therapy and the associated skills training. One of the core targets in DBT^{D+PD} is to increase emotional experiencing and reduce emotional inhibition as emotional inhibition has been shown to be related to suicidal ideation (Lynch, Cheavens, Morse, & Rosenthal, 2004) and poor treatment response in older adults (Rosenthal, Cheavens, Compton, Thorp, & Lynch, 2005). Reductions of negative affect intensity through psychopharmacological treatment may increase a patient's ability to initially tolerate painful emotions.

Diversity Issues and Personality Disorders in Older Adults

To date, there have not been any trials of DBT^{D+PD} conducted solely or primarily with ethnically or culturally diverse older adults (although approximately 12% of the sample from the Lynch et al. (2007) study identified as either African-American or Asian-American). Thus, it is difficult to speculate as to how the intervention might need to be adapted to meet the needs of older adults in ethnic minority groups. Nonetheless, there are some areas in which we could imagine cultural and ethnic diversity issues arising.

We have explicitly included family/caregiver/or significant other (in the broadest sense of the word) sessions in DBT^{D+PD}. The manual for DBT^{D+PD} suggests that up to 6 sessions, within the 28 individual sessions, be utilized as family or friend sessions. This component of the treatment may be particularly important for older adults in various cultural or ethnic groups in which collectivistic ideals are espoused. If these family and friends sessions are utilized in therapy, it would be important to get a clear sense from the patient, before the session occurs, as to what he or she expects from the session as well as a description of important interpersonal dynamics that might be influencing the patient's behavior. These sessions are useful in that the therapist is offered the opportunity to understand the context of behavior at a broader level and is also provided with a chance to increase generalization of behaviors through working on increasing the reinforcement of adaptive behaviors in the natural environment.

Another area in which ethnic and cultural diversity may impact the effectiveness of treatment is in the group skills mode of therapy. There are some skills that might be interpreted as inconsistent with certain religious or cultural beliefs. For example, one Distress Tolerance skill is prayer. This may be more or less useful for any given individual given their religious and/or cultural background. Additionally, patients often initially brace against nonjudgmental stance because they feel that removing judgment is inconsistent with their cultural and/or religious value system. An overarching theme in DBT^{D+PD} is to increase effectiveness; that is to increase a focus on what works in a given situation. Thus, just as we would ask a client to embrace effectiveness, we would ask the same of therapists. There is not even one skill in the manual that is sufficient and necessary for behavior change. If a particular skill is experienced as incompatible with values, we suggest moving to another skill for that particular individual.

Conclusion

In this chapter, we have attempted to provide a brief introduction to the rationale, evidence base, and practical applications associated with DBT^{D+PD}. The treatment of personality disorders in older adult patients has long been neglected by the psychological and psychiatric communities for myriad reasons. Although the treatment of older adults with personality disorders can be a daunting process, we are hopeful that DBT^{D+PD} may provide a framework for those ready to take on the task.

References

Abrams, R. C., & Horowitz, S. V. (1996). Personality disorders after age 50: A meta-analysis. *Journal of Personality Disorders, 10*, 271–281.

Abrams, R. C., & Horowitz, S. V. (1999). Personality disorders after age 50: A meta-analytic review of the literature. In E. Roscowsky, R. C. Abrams, & R. A. Zweig (Eds.), *Personality disorders in older adults* (pp. 55–68). New Jersey: Lawrence Erlbaum Associates.

Abrams, R. C., Spielman, L. A., Alexopoulos, G. S., & Klausner, E. (1998). Personality disorder symptoms and functioning in elderly depressed patients. *American Journal of Geriatric Psychiatry, 6*, 24–30.

Agronin, M. E., & Maletta, G. (2000). Personality disorders in late life: Understanding and overcoming the gap in research. *American Journal of Geriatric Psychiatry, 8*, 4–18.

van Alphen, S. P. J., Engelen, G. J. J. A., Kuin, Y., & Derksen, J. J. L. (2006). The relevance of a geriatric sub-classification of personality disorders in DSM-IV. *International Journal of Geriatric Psychiatry, 21*, 205–209.

American Psychological Association. (1994). *Diagnostic and statistical manual of mental disorders* (4th ed.). Washington, DC: Author.

American Psychological Association. (2000). *Diagnostic and statistical manual of mental disorders* (4th ed., Test revision). Washington, DC: Author.

Ames, A., & Molinari, V. (1994). Prevalence of personality disorders in community-living elderly. *Journal of Geriatric Psychiatry and Neurology, 7*, 209–215.

Butler, R. (1963). The life review: An interpretation of reminiscence in the aged. *Psychiatry, 26*, 65–76.

Casey, D. A., & Schrodt, C. J. (1989). Axis II diagnoses in geriatric inpatients. *Journal of Geriatric Psychiatry and Neurology, 2*, 87–88.

Coolidge, F. L. (1993). *The Coolidge axis II inventory manual.* Clermont, FL: Synergistic Office Solutions.

Coolidge, F. L., Burns, E. M., Nathan, J. H., & Mull, C. E. (1993). Personality disorders in the elderly. *Clinical Gerontologist, 12*, 41–55.

Coolidge, F. L., & Merwin, M. M. (1992). Reliability and validity of the Coolidge Axis II Inventory: A new inventory for the assessment of personality disorders. *Journal of Personality Assessment, 59*, 223–238.

Ehrlich, H. J., & Bauer, M. L. (1966). The correlates of dogmatism and flexibility in psychiatric hospitalization. *Journal of Consulting Psychology, 30*, 253–259.

Erikson, E. H., & Erikson, J. M. (1997). *The life cycle completed*. New York: Norton.

Fiorot, M., Boswell, P., & Murray, E. J. (1990). Personality and response to psychotherapy in depressed elderly women. *Behavior, Health, and Aging, 1*, 51–63.

First, M. B., Gibbon, M., Spitzer, R. L., Williams, J. B. W., & Benjamin, L. S. (1997). User's guide for the structured clinical interview for DSM-IV axis II personality disorders (SCID-II). Washington, DC: American Psychiatric Press.

Horowitz, L. M., Rosenberg, S. E., Baer, B. A., Ureno, G., & Villasenor, V. S. (1988). Inventory of interpersonal problems: Psychometric properties and clinical applications. *Journal of Consulting and Clinical Psychology, 56*, 885–892.

Hyler, S. E., Skodol, A. E., Kellman, H. D., Oldham, J. M., & Rosnick, L. (1990). Validity of the personality diagnostic questionnaire – Revised: Comparison with two structured interviews. *American Journal of Psychiatry, 147*, 1043–1048.

Kenan, M. M., Kendjelic, E. M., Molinari, V. A., Williams, W., Norris, M., & Kunik, M. E. (2000). Age-related differences in the frequency of personality disorders among inpatient veterans. *International Journal of Geriatric Psychiatry, 15*, 831–837.

Klein, M. H., Kupfer, D. J., & Shea, M. (1993). *Personality and depression: A current view*. New York: Guilford Press.

Kunik, M. E., Benoit, M. H., Rifai, A. H., Sweet, R. A., Pasternak, R., & Zubenko, G. S. (1994). Diagnostic rate of co-morbid personality disorder in elderly psychiatric inpatients. *American Journal Psychiatry, 151*, 603–605.

Kunik, M. E., Mulsant, B. H., Rifai, A. H., Sweet, R. A., Pasternak, R., Rosen, J., et al. (1993). Personality disorders in elderly inpatients with major depression. *American Journal of Geriatric Psychiatry, 1*, 38–45.

Linehan, M. M. (1993a). *Cognitive-behavioral treatment of borderline personality disorder*. New York: Guilford Press.

Linehan, M. M. (1993b). *Skills training manual for treating borderline personality disorder*. New York: Guilford Press.

Lynch, T. R., Cheavens, J. S., Cukrowicz, K., & Linehan, M. M. (2006). *Dialectical behavior therapy for depression with co-morbid personality disorders: An extension of standard DBT with a special emphasis on older adults. Unpublished treatment manual.* Durham, NC: Duke University Medical Center.

Lynch, T. R., Cheavens, J. S., Cukrowicz, K. C., Thorp. S., Bronner, L., & Beyer, J. (2007). Treatment of older adults with co-morbid personality disorder and depression: A dialectical behavior therapy approach. *International Journal of Geriatric Psychiatry*.

Lynch, T. R., Cheavens, J., Morse, J. Q., & Rosenthal, M. Z. (2004). A model predicting suicidal ideation and hopelessness in depressed older adults: The impact of emotion inhibition and affect intensity. *Aging & Mental Health, 8*, 486–497.

Lynch, T. R., Morse, J. Q., Mendelson, T., & Robins, C. J. (2003). Dialectical behavior therapy for depressed older adults: A randomized pilot study. *The American Journal of Geriatric Psychiatry, 11*, 1–13.

Lyness, J. M., Caine, E. D., Conwell, Y., King, D. A., & Cox, C. (1993). Depressive symptoms, medical illness, and functional status in depressed psychiatric inpatients. *American Journal of Psychiatry, 150*, 910–915.

Mezzich, J. E., Fabrega, H., Coffman, G. A., & Glavin, Y. F. (1987). Comprehensively diagnosing geriatric patients. *Comprehensive Psychiatry, 28*, 68–76.

Molinari, V., Ames, A., & Essa, M. (1994). Prevalence of personality disorders in two geropsychiatric inpatient units. *Journal of Geriatric Psychiatry and Neurology, 7*, 209–215.

Molinari, V., & Marimon, J. (1995). Relationship between affective disorders and Axis II diagnoses in geropsychiatric patients. *Journal of Geriatric Psychiatry and Neurology, 8*, 61–64.

Morey, L. C., Gunderson, J. G., Quigley, B. D., Shea, M. T., Skodol, A. E., McGlashan, T. H., et al. (2002). The representation of borderline, avoidant, obsessive-compulsive, and schizotypal personality disorders by the five-factor model. *Journal of Personality Disorders, 16*, 215–234.

Morey, L. C., Warner, M. B., Shea, M. T., Gunderson, J. G., Sanislow, C. A., Grilo, C., et al. (2003). The representation of four personality disorders by the schedule for nonadaptive and adaptive personality dimensional model of personality. *Psychological Assessment, 15*, 326–332.

Morse, J. Q., & Lynch, T. R. (2004). A preliminary investigation of self-reported personality disorders in late life: Prevalence, predictors of depressive severity, and clinical correlates. *Aging & Mental Health, 8*, 307–315.

Nestadt, G., Eaton, W. W., & Romanoski, A. J. (1994). Assessment of DSM-III personality structure in a general-population survey. *Comprehensive Psychiatry, 35*, 54–63.

Ogrodniczuk, J. S., Piper, W. E., Joyce, A. S., McCallum, M., & Rosie, J. S. (2002). Social support as a predictor of response to group therapy for complicated grief. *Psychiatry, 65*, 346–357.

Ogrodniczuk, J. S., Piper, W. E., Joyce, A. S., McCallum, M., & Rosie, J. S. (2003). NEO-five factor personality traits as predictors of response to two forms of group psychotherapy. *International Journal of Group Psychotherapy, 53*, 417–442.

Pilkonis, P. A., Kim, Y., Proietti, J. M., & Barkham, M. (1996). Scales for personality disorders developed from the inventory of interpersonal problems. *Journal of Personality Disorders, 10*, 355–369.

Rosenthal, M. Z., Cheavens, J. S., Compton, J. S., Thorp, S. R., & Lynch, T. R. (2005). Thought suppression and treatment outcome in late-life depression. *Aging& Mental Health,* 9, 35–39.

Schroeder, M. L., & Livesley, W. J. (1991). An evaluation of *DSM-III-R* personality disorders. *Acta Psychiatrica Scandinavica, 84*, 512–519.

Segal, D. L., Coolidge, F. L., & Rosowsky, E. (2000). Personality disorders. In S. K. Whitbourne (Ed.), *Psychopathology in later adulthood* (pp. 89–115). Hoboken, NJ: John Wiley and Sons, Inc.

Segal, D. L., Hook, J. N., & Coolidge, F. L. (2001). Personality dysfunction, coping styles, and clinical symptoms in younger and older adults. *Journal of Clinical Geropsychology, 7*, 201–212.

Staudinger, U. M., & Pasupathi, M. (2000). Life-span perspectives on self, personality, and social cognitiion. In F. I. M. Craik & T. A. Salthouse (Eds.), *The handbook of aging and cognition* (pp. 633–688). Mahwah, NJ: Erlbaum.

Thompson, L. W., Gallagher, D., & Czirr, R. (1988). Personality disorder and outcome in the treatment of late-life depression. *Journal of Geriatric Psychiatry, 21*, 133–153.

Vine, R. G., & Steingart, A. B. (1994). Personality disorder in the elderly depressed. *Canadian Journal of Psychiatry, 39*, 392–398.

Weissman, M. M. (1993). The epidemiology of personality disorders: A 1990 update. *Journal of Personality Disorders, Supplement 1*, 44–62.

13
Treating Persons with Dementia in Context

Jane E. Fisher, Claudia Drossel, Kyle Ferguson, Stacey Cherup, and Merry Sylvester

The term "dementia" indicates impairment in cognitive functions due to neuropathology. Many conditions cause degenerative dementia, including Alzheimer's disease, multiple infarcts (vascular dementia), frontotemporal pathology, Huntington's disease, Lewy body disease, and Parkinson's disease. An increasing inability to meet situational demands as well as memory, verbal, and learning deficits are the hallmarks of dementia. Family and friends of the person with dementia may complain about "personality changes" as uncharacteristic emotions, cognitions, and behaviors emerge. Strained social interactions, social withdrawal, and difficulty is completing multistep tasks (e.g., managing finances) are often the first detectable signs. As neuropathology progresses, significant impairment in communication and daily living skills are typical. Persons with degenerative dementia eventually may become bedridden and dependent on others for the provision of all needs.

Age is the single greatest risk factor for dementia. According to the Federal Interagency Forum on Age-Related Statistics (2006), 5.1% of persons between the ages of 65 and 69 demonstrate moderate to severe memory impairment. This number steadily rises to 13.6% for persons 75–80 years of age and is estimated to range from 32.1% to as high as 47% (Evans et al., 1989) for persons over 85.

Dementia is a serious health concern across all ethnic groups. Cultural factors related to dementia significantly influence access to services (Milne & Chryssanthopoulou, 2005), particularly the decision to institutionalize the family member with dementia (Mausbach et al., 2004).

Cognitive Decline, Behavioral, and Psychological Symptoms

Currently, there are no treatments available for stopping or reversing the effects of degenerative forms of dementia. The absence of treatment options for reversing or halting neuropathology requires that health-care providers adjust their conceptualization of treatment goals to focus on slowing decline by maintaining current skills and preventing excess disability (i.e., greater impairment than that predicted by the disease). This chapter describes a contextual, functional analytic (FA) conceptualization of dementia and applies the FA strategies to the promotion of a high quality of life and the prevention of excess disability in persons with dementia. However, since pharmacological interventions constitute the current standard of care for cognitive symptoms and are also an important contextual feature in the care of the person with dementia, first pharmacological interventions will be reviewed briefly.

Pharmacological Treatment

Treatment guidelines emphasize the use of acetylcholinesterase inhibitors to slow the verbal or cognitive deterioration of mild to moderate dementia, and/or *N*-methyl-D-asparate (NMDA) receptor agonists for moderate to severe dementia. In contrast, the behavioral and psychological symptoms of dementia (BPSD) should be treated with psychotropic medication *only in case of an emergency* and *only after nonpharmacological interventions*, including a thorough and qualified assessment, have failed (Cummings, 2004). No drugs are currently approved by the US Food and Drug Administration (FDA) for managing BPSD; and in 2005, the FDA issued a black box warning concerning the off-label use of antipsychotics (US Food and Drug Administration, 2005). BPSD, although traditionally interpreted as a primary consequence of progressive neurodegeneration and topographically similar to psychiatric presentations, may constitute a normal response to the loss of cognitive and verbal competence and function as a novel means to communicate (Fisher, Harsin, & Hadden, 2000). Surveys have consistently documented the overmedication of elderly persons with dementia ("chemical restraint"). Such restraint-based interventions produce a unique iatrogenic effect in that behavior is eliminated within the repertoire of individuals who already are losing communicative behaviors due to the underlying degenerative dementia (Fisher, Drossel, Yury, & Cherup, 2007; Yury & Fisher, 2007). In essence, restraint-based interventions – rather than reducing disability – promote *excess* disability. In contrast, first-line contextual nonpharmacological interventions, particularly the FA model, focus on this communicative nature of BPSD and provide the individual with socially appropriate response alternatives (Fisher et al., 2007).

Behavioral Treatment

In addition to the first-line use of behavioral interventions for BPSD as a standard set by the American Academy of Neurology (2001), the American Academy of Neurology Ethics and Humanities Subcommittee (1996) recommends "care over cure" and maximizing the care recipient's quality of life by preventing excess disability:

1. Assuring proper hearing and vision to enhance communicative ability; assessing nutrition, hydration, sleep, safety, bowel, and urinary functioning; preventing restlessness (e.g., by access to physical activity).

2. Identifying and treating depression: providing access to meaningful events.
3. Effectively treating comorbidity (e.g., diabetes, hypertension, chronic lung disease).
4. Carefully limiting and closely monitoring medications to avoid adverse effects.

The functional analytic (FA) model of behavior change in dementia detailed below thus meets both the standards and the guidelines put forth by the American Academy of Neurology (1996, 2001).

The Functional Analytic Model

The FA model assumes that all behaviors are influenced by a person's physiological and psychological history and his or her current social and physical context. Increasing neuropathology may lead to a declining or inappropriate verbal mediation of behavior (e.g., stating rules of proper conduct; following instructions) (see Fisher et al. (2007), for a detailed discussion), which in turn results in a failure to meet contextual (i.e., social and environmental) demands. Thus, many of the behavioral, cognitive, and affective changes observed in persons with dementia may be a function of or exacerbated by the strenuous person–environment and interpersonal interactions, especially within the relationship with the caregivers. Caregivers exert a particularly powerful influence on the behavior and well-being of persons with dementia, and therefore effective treatment planning requires arranging emotional and instrumental support for caregivers (see Steffen et al., this volume).

Generally, the FA model conceptually integrates and systematizes the use of psychosocial and context-altering strategies that traditionally have been considered separate entities (e.g., behavioral interventions, structured activities, or environmental design, see Cohen-Mansfield, 2001). It thus renders a coherent and principled system of person-centered care in which interventions are directly prescribed by the preceding individualized assessments. These nonpharmacological interventions are the first-line treatments of BPSD and should be professionally administered before resorting to pharmacological interventions (American Geriatrics Society and American Association for Geriatric Psychiatry, 2003).

Evidence Base for the FA Model

Numerous studies have documented the effectiveness of functional analytic approaches for reducing and preventing behavior problems and promoting high quality of life for persons with cognitive disabilities (see Austin & Carr, 2000; Sturmey, 2007). Moreover, experimental investigations in the *Journal of Applied Behavior Analysis* (1–40) exemplify the approach and provide empirical support. The FA model is also well-suited to investigate drug and combination treatment effects (for a review, see Poling & Byrne, 2000).

Assessment Issues Unique to Dementia

While there are global changes that occur in dementia, the rate and topography of cognitive and behavioral changes across persons vary dramatically. Premorbid psychological history and current contextual features significantly influence the course of degenerative dementias (e.g., Aalten, de Vugt, Jaspers, Jolles, & Verhey, 2005; Cohen-Mansfield, 2000; Teri, Hughes, & Larson, 1990). Consequently, an idiographic approach to assessment will be most effective in designing interventions for individuals with dementia.

The deterioration of behavioral repertoires caused by dementia also presents unique assessment and treatment challenges. Because the repertoires of persons with degenerative dementia eventually may be thoroughly comprised, *all* behaviors (including behaviors perceived as "problems" by others) should be considered worthy of *preservation* rather than elimination. The goal of interventions should be the preservation of challenging behaviors and the concomitant increase or maintenance of alternative behaviors that are adaptive for the care recipient and, only secondly, deemed more appropriate in the social context. As the disease progresses, the interaction between the person and his or her environment continues to change. For these reasons, the FA model emphasizes maintaining functional behaviors as the primary treatment goal and continued assessment of the person–environment interaction to adjust treatment goals and interventions to the current skill level.

The contextual FA model of treatment holds that behavior is the result of the interaction between an individual and his or her environment. Consequently, analyzing behavior under consideration of the circumstances within which the behavior occurs and assessing possible functions of (or the purpose or motivation for) the behavior is imperative. Developing a clear operational definition of a behavior problem is the first step in assessment. To illustrate, the commonly used umbrella terms "agitation" and "aggression" are ambiguous. "Aggression," for example, may include a care recipient's self-protective behavior in situations subjectively perceived as dangerous. To avoid error and confusion, specific behaviors should be identified (e.g., "biting," "kicking," etc.). Table 1 provides a list of instruments designed to assess the topography and severity of behaviors perceived as challenging.

Although these instruments may be helpful for assessment of baseline levels and monitoring treatment effects, the direct observation of behaviors in the natural environment will be a more reliable and accurate means of determining the context, frequency, severity, rate, and duration of the behavior (Fisher, Ferguson, & Drossel, 2006).

After behaviors have been clearly defined, determining their function is the next step. An experimental functional analysis (EFA) is the most direct means of empirically determining the function that challenging behaviors serve for a particular care recipient. EFA involves a systematic manipulation of the context in which behaviors occur in order to determine their function, as in Iwata, Dorsey, Slifer, Bauman, and Richman (1994/1982). However,

TABLE 1. Instruments designed for the assessment of topographical features of problem behaviors of persons with dementia.

Instrument	Purpose
Cohen-Mansfield Agitation Inventory (CMAI) (Cohen-Mansfield, 1986)	Short and long forms available; caregiver rates the frequency of challenging behaviors on a Likert scale
Behavioral Pathology in Alzheimer's Disease Rating Scale (BEHAVE-AD) (Reisberg et al., 1989)	Caregivers rate their level of distress in response to 25 behaviors
Empirical Behavioral Pathology in Alzheimer's Disease Rating Scale (E-BEHAVE-AD) (Auer, Monteiro, & Reisberg, 1996)	Developed for use with the BEHAVE-AD as an observational method of measuring behavior
Pittsburgh Agitation Scale (PAS) (Rosen et al., 1994)	Assesses agitation and aggression and relies on the direct observation of 1–8 h

in addition to its labor-intensive nature, EFA may be impractical for some problem behaviors (e.g., if they are dangerous or occur at very low rates and as such are difficult to observe directly). For these behaviors, a descriptive functional assessment may be more appropriate. A *descriptive functional assessment* involves directly observing the behavior and recording the antecedents and consequences to determine a pattern of when the behavior is most likely to occur (see below). Tools that have been developed to assist in generating hypotheses regarding the function of behavior include the Motivational Assessment Scale (MAS) (Durand and Crimmins, 1988) and the Functional Analysis Screening Tool (FAST) (Iwata, 1995) (see Table 2 for an example of a descriptive functional assessment tool).

Note that topographically similar behaviors may serve different functions for different individuals; alternatively, different functions may maintain the behavior of one individual across time or situations.

TABLE 2. Descriptive assessment card for an institutional setting (adapted from Pyles & Bailey, 1990).

Resident's initials	Date/time	Activity/location
Antecedents	Behaviors	Consequences
Ignored by staff, or staff walked away	Here, a specification of the target behavior would be provided	Attention (talking, praising, chiding)
Leisure material or food		Redirected to other area or activity
Other request denied		Leisure material/food given
Given instruction/task demand		Leisure material/food taken away
Physical contact by staff member		Task demand terminated
Provoked by other resident		Physical contact terminated
New staff person entered area		Staff walked away
Nothing (alone, doing nothing)		Staff did nothing

Behavior Change Strategies

The FA model of behavior change assumes that all behavior has a function (i.e., serves a purpose) and, as such, may be altered by changing situational demands (Fisher et al., 2007; Stokes, 2000). In other words, challenging behavior may constitute an alternative form of communication or an adaptive way to meet environmental demands in the presence of cognitive or behavioral deficits. The functional relationship between the environment and behavior is called a contingency (Sulzer-Azaroff & Mayer, 1991). "Contingency" means that the occurrence of certain aspects of the social or nonsocial environment and a person's behavior form detectable patterns when observed across time (see "Assessment" above). These contingencies are described linearly in three terms, i.e., by patterns of antecedents, behaviors, and consequences (Bijou, Peterson, & Ault, 1968). Antecedent situations and consequences will be discussed in greater detail later.

From a contextual perspective, therefore, an effective intervention may modify the situation such that more acceptable behavior is promoted while unacceptable behavior – namely, the "challenging behavior" – is more or less "left alone." After observing the target behavior and obtaining baseline data, the next step in developing a behavior-change treatment plan is to specify treatment goals.

Treatment Goals

The top priorities of treatment goal setting are preserving the functional repertoire and promoting a higher quality of life for the person with dementia. Steps in specifying treatment goals include: defining the target behavior, as mentioned earlier (see Barlow & Hersen, 1984; Hawkins & Dobes, 1977); specifying how the behavioral target should change as a result of treatment (i.e., the expected outcomes with respect to behavioral excesses or deficits); the setting(s) in which the change should occur; and whose behavior is expected to change (i.e., the care recipient's and/or staff's) (see Pinkston & Linsk, 1984). Another key point of clarification is to specify *for whom* the behavior is problematic.

Three core strategies. There are three core strategies to managing BPSD (see Plaud, Moberg, & Ferraro, 1988; Teri et al., 1992): (1) modifying the features of the antecedent situation in which the challenging behavior occurs most frequently; (2) modifying consequences that maintain the behavior; and (3) reinforcing alternative or incompatible socially acceptable behaviors. Either of these or a combination of two or more strategies can be used for any given behavior.

(1) *Modify antecedents of problem behavior*. "Antecedents" are the contextual (social or environmental) features in the presence of which a behavior may produce a particular reinforcing consequence. If, over a period ranging from days to months or years, the behavior reliably is followed by that consequence at least some of the time, the probability that the behavior will occur more frequently in the "antecedent" situation increases. Thus, when a particular behavior occurs

more frequently in the presence of one situation than another, the behavior is said to be "discriminated" (see also the literature on *antecedent* or *stimulus control*). Antecedent contexts may include noise, inactivity, demands of activities of daily living, or social demands. If behavior occurs across a wide variety of situations, one speaks of "generalization" (Sulzer-Azaroff & Mayer, 1991, p. 249ff.) Antecedent interventions thus may focus on (a) discrimination training – limiting the occurrence of behavior to specific contexts or (b) arranging generalization – broadening the occurrence of the behavior to a range of situations.

(a) *Discrimination training*. Persons with cognitive impairments may engage in inappropriate sexual behavior such as public masturbation (Alagiakrishnan et al., 2005). An FA intervention entails discrimination training assuring that behavior will be bound to its appropriate context – the privacy of one's room. Masturbation should be interrupted only when it occurs publicly, and the care recipient should be gently and patiently redirected to his or her room. In the absence of the reinforcer (orgasm), the behavior eventually will become less likely in the public setting (Hussian, 1981). Behavior that is limited to a certain context may also be established by employing orientation or memory aids (Bourgeois & Mason, 1996). Examples of such aids include coloring doors differently with respect to function (e.g., blue for bedrooms); using pictures or symbols in lieu of printed text on cupboards and drawers; and positioning signposts in hallways.

(b) *Arranging generalization*. Targeting the social behavior of multiple residents of a long-term care facility, Melin and Gotestam (1981) examined the effects of an antecedent intervention on the behavior of dementia care recipients. Specifically, they arranged the furniture to make the setting more conducive for conversation, and they increased the duration and the choice of meals. As a result, the frequency of communication in the experimental group exceeded that of the control (i.e., treatment as usual) as well as baseline levels.

(c) *Other antecedent interventions*. While wandering per se is not a problematic behavior, accessing potentially dangerous areas is. Hussian (1981) observed that paths of the people who wandered became predictable when simple rather than complex antecedents were used. For example, individuals with dementia who may not be able to read a map of a long-term care facility are able to follow simple arrows or track lighting. Correspondingly, Feliciano, Vore, LeBlanc, and Baker (2004) added a cloth barrier to prevent residents from wandering into an unsafe area. This simple intervention reduced entry into a restricted area and was deemed an acceptable intervention by all parties involved.

(d) *Facilitating communication by modifying antecedents*. Antecedent interventions may also be individualized. Bourgeois (1993), for example, used memory aids (i.e., personally relevant pictures and narratives) to generate situations that then reliably evoked on-topic statements; decreased the frequency of nonproductive utterances; lengthened conversational turns; and/or increased the frequency of turns.

Behaviors in response to verbal and nonverbal messages (e.g., vocal speech, gestures, facial expressions) provide another example for antecedent interventions

(Baum, 1994). Communicative deficits are ubiquitous in persons with dementia, particularly when the disease has advanced (Beers & Berkoff, 2000). An individual's behavior may remain unaffected by complex verbal constructions (e.g., multisyllabic words, etc.). Hence, such individuals would be unable to follow complex commands or logical arguments, multistep instructions, and so forth.

Regardless of the level of impairment, persons with dementia are still affected by nonverbal communication. Care recipients respond well to persons who assume a welcoming and nonthreatening stance (Hendryx-Bedalov, 2000; Kato, Hickson, & Worrel, 1996), e.g., by approaching from the front (to not startle the person); establishing eye contact (to ensure attention); lightly touching, if appropriate and not contraindicated by a person's history; using a soothing voice; speaking slowly; and providing ample time to respond.

Persons with dementia prefer engaging in verbal communication (Williams, Kemper, & Hummert, 2004). Given that many care recipients have difficulty tracking conversations, one may minimize confusion by simplifying communication: being specific and direct with requests; using short, simple words and sentences; providing few, simple choices; and eliminating distractions from the environment (Hendryx-Bedalov, 2000).

(2) *Modify consequences that maintain the behavior* and (3) *reinforce socially acceptable behaviors*. All behaviors are followed by consequences. Reinforcers, the third term in the three-term contingency, are events that increase the probability of the behavior they follow if they do so reliably, albeit sporadically (see Catania, 1992). *Positive reinforcers* are defined by the *presentation* of events that may be social in nature and include touching, providing attention; alternatively, they may be environmental, and the person may gain access to objects, other events, or activities. When the *termination, removal,* or *postponement* of particular events makes behavior more likely, these events are termed *negative reinforcers*. Reinforcers differ across people and over time for the same individual. Note that while punishment is used to decrease the rate of a particular behavior, reinforcement *always increases* the rate of behavior.

The discontinuation of reinforcement is called "extinction." "Extinction" denotes that the behavior has stopped to produce its usual reinforcing consequences and that its rate is decreasing as a result. Because extinction may produce highly variable and possibly emotional behavior ("extinction bursts"), the withdrawal of preferred consequences might not be indicated. Instead, we recommend reinforcing an appropriate behavioral alternative that then produces the same consequence, or presenting the consequence regardless of behavior (see "noncontingent reinforcement" below).

Depression is one of the most frequently diagnosed psychiatric disorders in dementia, and the diagnosis correlates with poorer quality of life, greater disability in activities of daily living, excess disability, and higher mortality (Starkstein & Mizrahi, 2006). A major reason for depression in persons with dementia is a marked reduction in access to positive reinforcement (leading to a generalized decrease in the rate of activity) and a marked increase in extinction and punishment due to ever-mounting disabilities (followed by decreased social engagement). Reduced access to reinforcement ("extinction") may be related to increasing verbal deficits. Individuals with dementia often have difficulty in describing situations, including correlating emotions or needs. Thus, in the presence of social antecedents care recipients are not able to generate the verbalizations (behaviors) that could prompt understanding or effective actions by others (reinforcers). Because of this absence of reinforcement, verbalizations may decrease in rate; if the overall level of activity decreases, clinically depressed behavior may occur.

The behavioral solution for depression, therefore, entails increasing access to reinforcement (called *behavioral activation*, see Jacobson, Martell, & Dimidjian, 2001) and decreasing punishment. Because low motivation is the hallmark of depression, involving care recipients in activities can be difficult. Engelman, Altus, and Mathews (1999) employed prompts (i.e., "reminders") and social reinforcement to increase the engagement of five older adults with dementia in daily living activities. The researchers evaluated their intervention using a multiple-baseline design across morning and afternoon work shifts. Compared to baseline rates of behavior, all five participants showed marked increases in activity levels.

Even when the overall level of impairment is severe, the behavior of persons with dementia, like all behavior, is sensitive to contingencies of reinforcement (Pinkston & Linsk, 1984). Consequently, treatment plans should attempt to increase access to reinforcement for socially acceptable or "benign" behaviors (this may prevent depression), while decreasing access to reinforcement for challenging behavior (see Cohen-Mansfield, 2001; Turner, 2005).

After selecting appropriate reinforcers for the care recipient (see Fisher & Buchanan, 2003, for a discussion on preference assessment and reinforcer identification in severe dementia), the next step is arranging their delivery. Reinforcers may be delivered noncontingently (NCR), i.e., in fixed time intervals and without a predictable relationship to behavior, or they may be delivered contingent on some behavior. On the other hand, one may reinforce alternative or incompatible behavior while withholding reinforcement for a challenging behavior (see Sulzer-Azaroff & Mayer, 1991).

Although NCR schedules of reinforcement are widely adopted in settings serving persons with developmental disabilities (e.g., Kahng, Iwata, DeLeon, & Wallace, 2000), they are seldom used with cognitively impaired elderly populations. However, Buchanan and Fisher's (2002) study suggests that NCR schedules might reduce BPSD. After conducting a functional analysis, the authors implemented an intervention that involved providing reinforcers in fixed time intervals, regardless of whether disruptive vocalizations were occurring (i.e., non contingently). As a result of providing reinforcers on a noncontingent basis, the frequency of disruptive vocalizations was significantly reduced for both participants. NCR's may decrease a person's motivation to engage in a particular target behavior: The problem behavior is no longer "needed" because reinforcement is available independent of the behavior (Marcus & Vollmer, 1996). NCR interventions are restraint free in that they maintain a person's repertoire (i.e., the person is still able to vocalize) as well as access to reinforcement. For these reasons NCR interventions have ethical advantages for elderly person with dementia.

Wandering is also sensitive to the effects of reinforcement. Individuals with dementia may wander to increase stimulation, escape discomfort, or to satisfy basic physical needs (i.e., exercise; Algase, 1999). Accordingly, Heard and Watson (1999) delivered the reinforcers that had been found to maintain care recipients' wandering (identified via a functional assessment) only if wandering had not occurred. Thus, they differentially reinforced all behavior other than wandering. Significant reductions in wandering behavior occurred in the intervention relative to the baseline conditions.

(3) *Combining antecedent and consequent interventions.* Distraction is one of the most useful strategies in managing behavior (see Gwyther, 1994) and consists of a combination of the aforementioned strategies. The care recipient's attention is redirected by quickly changing the situation through the introduction of events, activities, materials, or communicative topics familiar to and preferred by the care recipient; subsequently, the care recipient's engagement in the situation is differentially reinforced. The key to using distraction successfully is knowledge of what captures or maintains the care recipient's attention (Stokes, 2000).

Case Example

Initial Contact

Jim, age 78, was referred to the Nevada Caregiver Support Center (NCSC) after seeking services through an anxiety treatment program. He had been diagnosed with Alzheimer's disease 6 months earlier. At the time of intake, Jim had been married for 20 years to his 76 -year-old wife, Marilyn. Jim had adult children from a previous marriage who lived out of state. Marilyn was childless. Jim retired from his work as an accountant in his early 60s and had not been employed since. Throughout service provision, Jim was in the care of a family physician and a neurologist. He was taking donepezil, memantine (dementia), and sertraline (anxiety). He wore glasses and one hearing aid and had regular visual and auditory exams (see "prevention of excess disability," above).

The intake was conducted during a home visit, at which both spouses met with NCSC coaches in separate rooms. Consent forms and the limits of confidentiality were explained to both individuals, who identified Jim's anxiety as a factor decreasing the couple's quality of life. To understand the history as well as contextual factors contributing to the presenting problem, the assessment included taking histories of (a) the presenting problem; (b) the couple's relationship; (c) general psychosocial functioning and coping; and (d) family and individual medical or psychiatric problems. Moreover, specific descriptions of the presenting problem as well as the conditions that are most frequently associated with its occurrence were obtained. Psychometric instruments were administered repeatedly during the course of service provision (see "Assessment" above).

Case Conceptualization

Jim's functioning. Initially, Jim's score on the MMSE (Mini-Mental Status Exam; Folstein, Folstein, & McHugh, 1975) was unremarkable at 29. He reported that he initiated the diagnostic process for dementia when he noticed difficulties performing electrical or mechanical repair tasks and simple arithmetic. He reported anxiety when: (1) he was not able to complete electrical and mechanical repair tasks around the home or the usual car maintenance and (2) he was required to make financial purchases or other decisions. Jim had lost his father during World War II and subsequently taken care of his mother and younger siblings. Historically, he had assumed the caregiver role in his relationships with women. Jim's increases in anxiety correlated with his inability to complete chores and subsequent arguments with his spouse.

Marilyn's functioning. Marilyn perceived herself as physically frail due to a bout of cancer two decades earlier and strongly relied on Jim's care, including his versatility with electrical and mechanical equipment. She only drove short distances and turned to Jim for all household work other than light chores. She attributed Jim's growing failure to perform requested tasks to "meanness," rather than to a decline in cognitive functioning. Similarly, Marilyn did not interpret Jim's increasing worries about finances as signs of his cognitive impairment but as "chauvinist" and "controlling" behaviors. Marilyn's relationship history included physical abuse in her family of origin and domestic violence in earlier intimate relationships.

Presenting problem. When Marilyn asked Jim to perform certain tasks he chose to not disclose his increasing difficulties. Instead, he simply failed to perform the tasks, or he presented his wife with rationales for postponing or not undertaking the requested tasks. In addition to experiencing feelings of abandonment and rejection, Marilyn engaged in verbally argumentative behavior with angry facial expressions and a raised voice. Jim, feeling overwhelmed, tended to turn away from Marilyn and not respond to her at all anymore. Consequently, she engaged in hitting his back or pushing him, to which he responded by turning and holding her away from him. While Jim identified the argument per se as anxiety provoking, Marilyn characterized Jim's anxiety (rather than his functional impairment) as causing havoc with his chore completion.

Descriptive Functional Assessment

Jim's behavior functioned to provide "dignified exits" from overwhelming task demands (Table 3).

All of his topographically dissimilar responses to Marilyn fulfilled the same function. Marilyn's behavior, on the other hand, failed to engage Jim in tasks he used to be able to do for her (Table 4).

The nonoccurrence of the historical consequence (task completion) constituted an "extinction" condition for Marilyn (see "Modifying consequences" above).

TABLE 3. Descriptive assessment card: arguments function to enable escape from demand.

Antecedents	Jim's behaviors	Consequences
Marilyn asks Jim to perform task beyond his current skill set	Jim ignores requests Jim argues with Marilyn against task performance and presents rationales	Jim escapes from task
Marilyn expresses anger and gives Jim negative feedback on his behavior	Jim turns away and ignores Marilyn	Jim escapes from Marilyn's negative feedback
Marilyn hits Jim's back	Jim turns toward Marilyn and holds her away from him	Jim escapes further hits
Marilyn cries and threatens to call the police	Jim lets go off Marilyn	Marilyn stops threatening Jim and removes herself from the situation. Jim escapes further interaction

TABLE 4. Descriptive assessment card: Marilyn's behavior is extinguished.

Antecedents	Marilyn's behaviors	Consequences
Marilyn has a domestic need	Marilyn asks Jim to perform task beyond his current skill set	Jim ignores request. Jim argues with Marilyn against task performance and presents rationales
Marilyn still has domestic need	Marilyn gives Jim negative feedback on his behavior	Jim turns away and ignores Marilyn
Marilyn hits Jim's back	Marilyn hits Jim's back	Jim turns away from Marilyn and holds her away from him. Marilyn still has domestic need

Initial Coaching Plan

1. *Crisis management.* A report to elder protective services (EPS) was made. Marilyn was instructed to walk away from Jim and to call the NCSC's 24-h Coaching Helpline at any time an argument was about to ensue. A coach would then assist the Marilyn in reassessing the situation and engaging in problem-solving behavior. To decrease reliance on Jim, she was also instructed to obtain regular third-party housekeeping services and to hire maintenance and repair staff for other tasks. The arrangement of third-party housekeeping services constituted an antecedent intervention for Jim (Marilyn would not pose task demands beyond his current skill set) and a consequent intervention for Marilyn (asking for help was reinforced; the extinction condition was removed).
2. *Anxiety intervention.* Jim received direct instruction in progressive relaxation techniques (see Bernstein, Borkovec, & Hazlett-Stevens, 2000). Relaxation

strategies were conceptualized as an alternative to worrying, arguing, presenting rationales, and engaging in discussions, and reinforced.

3. Dementia education and prevention of problem behaviors.
   a. Marilyn was provided with "The best friends approach to Alzheimer's care" (Bell & Troxel, 2003) to promote her understanding of Jim's experience. She also attended educational workshops provided by the NCSC to enhance her ability to cope with a variety problems commonly experienced by family caregivers. Individual sessions with Marilyn focused on being able to complete individualized versions of the "Descriptive Assessment Card" (see Tables 3 and 4 above) thereby conceptualizing Jim's behavior in terms of its antecedents and consequences. Empathy and understanding for Jim's behavior increased. Marilyn learned that feelings of not being able to think were accompanied by terror and helplessness, and that Jim hid his deficits and his fear of "losing his mind" by withdrawing. At the same time, emotional coping and distress tolerance strategies selected from Linehan (1993) were taught to the caregiver.
   b. Jim's concerns about Alzheimer's disease were addressed as they arose, on a one-on-one basis to reduce his anxiety and problem-solve around specific concerns.
4. *Prevention of caregiver and care recipient depression.* Preferred activities were assessed and scheduled for both the caregiver and the care recipient (see "Modifying consequences" above).

The First Two Years

After initiation of services, arguments involving physical contact ceased entirely and verbal arguments became rare events. Two coaches (one focused on the caregiver; one on the care recipient) conducted individual sessions at the couple's home for the first six months. In addition to scheduling regular third-party housekeeping services, the caregiver proved quite resourceful at enlisting help from neighbors and her local community. Rather than "taking over" for Jim, tasks were presented as collaborative endeavors during which Jim tended to watch or participate to the best of his abilities. After half a year, the coaches' home visits became unnecessary. Contact was maintained via telephone coaching and workshop attendance for the next year and a half with one exception: A home visit was scheduled after the caregiver had reported abrupt cognitive decline. This decline was a function of a medication interaction effect and reversed after the prescription was changed to a more tolerable medication by the prescribing physician.

The Third Year

By the third year, Jim's MMSE score had decreased significantly, and he started having more difficulty with conversations. Particularly, he noticed that it took effort to maintain the topic of a conversation; to follow current events; and to understand jokes or light banter. For the first time, he was not able to report events accurately and merged recent and past episodes into one narrative. He also struggled to complete sequential tasks, such as dressing. Jim sometimes lost his way while driving in his hometown. When he noticed that he was not able to park the car without effort, Jim stopped driving completely. Overall, he was less able to care for others and for himself, and his concerns were future-oriented. While Jim became more dependent upon Marilyn and his disabilities more evident, Marilyn began a process of social isolation accompanied by intense feelings of sadness, loss, and regret.

Second Coaching Plan

1. Prevention of caregiver and care recipient depression.
   a. In-home services were increased and a senior companion was arranged for Jim once a week. Additionally, Jim and Marilyn started walking 3 miles per day (see "Modifying consequences" above).
   b. Marilyn's feelings of sadness, loss, and regret were normalized, and emotional coping and distress tolerance skills were revisited.
2. *Dementia education.* Dementia education for Marilyn focused on the effects of Jim's loss of verbal sophistication and his deteriorating sequential task performance. In addition to revisiting the antecedent, behavior, consequent (ABC) conceptualization of Jim's behavior, Marilyn also learned to break down tasks into their components.
3. *Increasing support to prevent problem behaviors*
   a. Instrumental. Marilyn was coached to notice shifts in the duration needed to complete activities of daily living. With the help of a coach, she designed signs and instructions to support and maintain Jim's independence (see "Modifying antecedents" and "discrimination training" above).
   b. Emotional: Marilyn was coached to respond to signs of frustration, anger, or anxiety by generating "comfort zones" for Jim and to redirect his attention to familiar and pleasant events (see "Combining antecedent and consequent interventions" above).
4. Anxiety reduction
   a. Jim's concerns about the future were addressed and long-term care planning activities initiated, while Jim was still able to participate.
   b. Pursuant to Jim and Marilyn's request, Jim's adult children were contacted and provided with education about Alzheimer's disease to encourage social support. As the assistance was increased both for the couple and for Jim specifically, they were able to settle into a comfortable routine. For a period of 3 months, home visits occurred once a week. Telephone coaching and Marilyn's attendance at workshop were maintained throughout the yearlong period.

The Fourth Year

In the fourth year, Jim had a series of falls related to dizziness and also some weight loss. When medical exams did not produce physical evidence, Jim's dizziness was attributed to potential dehydration. Further in-home assessment showed that Marilyn tended to ask Jim to describe his needs. Unable to specify them, Jim denied them emphatically and categorically. Marilyn interpreted this denial as "rude" or "mean" and did not follow-up. Generally, Jim who still engaged in conversational turn-taking and reason-giving was less likely to present accurate verbal content that was appropriate to current context. Instead, he tended to repeat narratives. Moreover, when faced with vague requests that he could not comprehend (e.g., "Get ready to go"), he responded with vehement refusal, which served a self-protective escape function (see *negative reinforcement*).

Third Coaching Plan

1. *Nutrition and hydration.* Marilyn kept a "food and drink" journal, established with the guidance of a physician and regularly monitored by the coaches of the NCSC, to ensure Jim's proper food and water intake.
2. Dementia education. "When a family member has dementia: steps to becoming a resilient caregiver" (McCurry, 2006) was read by and discussed with Marilyn. The focus was on Jim's loss of verbal skills: because Jim was less likely to initiate behaviors or to follow complex verbal instructions, simple sentences, and modeling (e.g., eating, drinking, or personal care such as brushing teeth) were emphasized. Marilyn was also trained to observe the global function of Jim's verbal behavior (e.g., reason-giving, seeking reassurance, complaining) rather than its specific content, using specifically tailored descriptive assessment cards (see Table 2). The communicative function of other behaviors and their adaptive replacement of verbal skills were stressed, and so was Jim's reliance on nonverbal environmental cues. Accordingly, Jim's fear and restlessness related to the loss of verbal abilities, especially when in unfamiliar surroundings, were also normalized and strategies for the prevention of self-protective behavior were taught.
3. *Increasing instrumental support*
   a. *Dementia care.* Care in a dementia-specific unit and related public transport were arranged. The duration of care in the day unit was increased from 2 to 5 days per week over the course of half a year. Jim received breakfast and lunch in this unit.
   b. Prevention of stressful episodes: In-home care staff was arranged to assist Jim with personal care in the morning and help him to enter the bus to the care unit. Jim was more likely to refuse personal care and transportation by bus in Marilyn's presence, for she had a history of following his instructions and tended to keep him at home upon refusal.
4. *Prevention of depression.* Preferred events were again assessed and novel tasks suggested for individuals with memory impairments were introduced to provide Jim with the opportunity to engage in his historical role as a caregiver or "helper" at home (see "Modifying consequences" above).

As Jim's and Marilyn's new routine involving the adult day center became established, Marilyn again reported decreased depression and corresponding increases in quality of life. Direct observation by NCSC coaches showed that Jim participated in the day center's activities and positively interacted with staff as well as residents. After Jim had regained his strength, he and Marilyn resumed their 3-mile walks. When and if the transition to the assisted care or nursing care facility becomes necessary, Jim's new caregivers will be provided with Jim's history and will be able to receive similar support services focusing on

1. Thorough individualized assessment using the Descriptive Assessment Card
2. Dementia education
3. The adaptive and communicative role of behaviors of individuals with dementia, including the need to "listen" to the individual with dementia by carefully observing patterns of behavior and verbalization that convey functional themes (e.g., need of comfort or safety; escape from pain or distress, etc.)
4. Strategies to generate "comfort zones" for individuals with dementia and thereby reduce anxiety and prevent self-protective behaviors (e.g., during personal care)
5. Strategies to appropriately motivate individuals with dementia and thereby reduce depression or apathy
6. Enhancement of caregiver coping and stress management skills, including access to pleasant events and community resources.

In this fashion, care recipient behaviors are not only assessed but also treated in an individualized manner taking into account the preferences, the needs, and the skill level of both caregiver and care recipient. It is important to note that caregiver skills often do not generalize but have to be taught specifically to each respective situation, possibly because of the long history of interaction between a family caregiver and care recipient.

References

Aalten, P., de Vugt, M. E., Jaspers, N., Jolles, J., & Verhey, F. R. J. (2005). The course of neuropsychiatric symptoms in dementia. Part II: Relationships among behavioural subsyndromes and the influence of clinical variables. *International Journal of Geriatric Psychiatry, 20*, 531–536.

Alagiakrishnan, K., Lim, D., Brahim, A., Wong, A., Wood, A., Senthilselvan, A., et al. (2005). Sexually inappropriate behaviour in demented elderly people. *Postgraduate Medical Journal, 8*, 463–466.

Algase, D. L. (1999). Wandering: A dementia-comprised behaviour. *Journal of Gerontological Nursing, 25*, 10–16.

American Academy of Neurology: Ethics and Humanities Subcommittee. (1996). Ethical issues in the management of the demented patient. *Neurology, 46*(4), 1180–1183.

American Academy of Neurology: Practice Standards Subcommittee. (2001). Practice parameters: Management of dementia (an evidence-based review). *Neurology, 56*, 1154–1166.

American Geriatrics Society and American Association for Geriatric Psychiatry. (2003). Consensus statement on improving the quality of mental health care in U.S. nursing homes: Management of depression and behavioral symptoms associated with dementia. *Journal of the American Geriatric Society, 51*, 1287–1298.

Auer, S. R., Monteiro, I. M., & Reisberg, B. (1996). The empirical behavioral pathology in Alzheimer's Disease (E-BEHAVE-AD) rating scale. *International Psychogeriatrics, 8*(2), 247–266.

Austin, J., & Carr, J. E. (Eds.). (2000). *Handbook of applied behavior analysis*. Reno, NV: Context Press.

Barlow, D. H., & Hersen, M. (1984). *Single case experimental designs: Strategies for studying behavior change* (2nd ed.). New York: Pergamon Press.

Baum, W. M. (1994). *Understanding behaviorism*. New York: HarperCollins.

Beers, M. H., & Berkoff, R. (Eds.) (2000). *The Merck Manual of Geriatrics, Third Edition*, Rahway, NJ: Merck Publishing Group.

Bell, V., & Troxel, D. (2003). *The best friends approach to Alzheimer's care (rev.)*. Baltimore, MD: Health Professions Press.

Bernstein, D. A., Borkovec, T. D., & Hazlett-Stevens, H. (2000). *New directions in progressive relaxation training: A guidebook for helping professionals*. Westport, CT: Greenwood Publishing.

Bijou, S. W., Peterson, R. F., & Ault, M. H. (1968). A method to integrate descriptive and experimental field studies at the level of data and empirical concepts. *Journal of Applied Behavior Analysis, 1*, 175–191.

Bourgeois, M. S. (1993). Effects of memory aids on the dyadic conversations of individuals with dementia. *Journal of Applied Behavior Analysis, 26*, 77–87.

Bourgeois, M. S., & Mason, L. A. (1996). Memory wallet intervention in an adult day-care setting. *Behavioral Interventions, 1*, 3–18.

Buchanan, J. A., & Fisher, J. E. (2002). Functional assessment and noncontingent reinforcement in the treatment of disruptive vocalization in elderly dementia patients. *Journal of Applied Behavior Analysis, 35*, 99–103.

Catania, A. C. (1992). *Learning* (3rd ed.). Englewood Cliffs, NJ: Prentice Hall.

Cohen-Mansfield, J. (1986). Agitated behavior in the elderly II. Preliminary results in the cognitively deteriorated. *Journal of the American Geriatrics Society, 38*, 785–792.

Cohen-Mansfield, J. (2000). Heterogeneity in dementia: Challenges and opportunities. *Alzheimer Disease and Associated Disorders, 14*(2), 60–63.

Cohen-Mansfield, J. (2001). Nonpharmacologic interventions for inappropriate behaviors in dementia: A review, summary, and critique. *American Journal of Geriatric Psychiatry, 9*, 361–381.

Cummings, J. L. (2004). Alzheimer's disease. *New England Journal of Medicine, 351*, 56–67.

Durand, M. V., & Crimmins, D. B. (1988). Identifying the variables maintaining self-injurious behavior. *Journal of Autism and Developmental Disorders, 18*, 99–117.

Engelman, K. K., Altus, D. E., & Mathews, R. M. (1999). Increasing engagement in daily activities by older adults with dementia. *Journal of Applied Behavior Analysis, 32*, 107–110.

Evans, D. A., Funkenstein, H. H., Albert, M. S., Scherr, P. A., Cook, N. R., Chown, M. J., et al. (1989). Prevalence of Alzheimer's disease in community populations of older persons: Higher than previously reported. *Journal of the American Medical Association, 262*, 2551–2556.

Federal Interagency Forum on Age-Related Statistics. (2006). *Older Americans Update 2006: Key indicators of well-being*. Washington, DC: Government Printing Office.

Feliciano, L., Vore, J., LeBlanc, L. A., & Baker, J. C. (2004). Decreasing entry into a restricted area using a visual barrier. *Journal of Applied Behavior Analysis, 37*, 107–110.

Fisher, J. E., & Buchanan, J. A. (2003). Stimulus preference assessment. In W. T. O'Donohue, J. Fisher, & S. C. Hayes (Eds.), *Empirically supported techniques of cognitive behavioral therapy: A step-by-step guide for clinicians* (pp. 402–406). New York: John Wiley & Sons, Inc.

Fisher, J. E., Drossel, C., Yury, C., & Cherup, S. (2007). A contextual model of restraint-free care for persons with dementia. In P. Sturmey (Ed.), *Functional analysis and clinical treatment*. San Diego, CA: Elsevier.

Fisher, J. E., Ferguson, K. E., & Drossel, C. (2006). Evidence-based practices for dementia and schizophrenia. In S. E. Levkoff, H. Chen, J. E. Fisher, & J. S. McIntyre (Eds.), *Evidence-based behavioral health practices for older adults* (pp. 135–158). New York: Springer.

Fisher, J. E., Harsin, C. W., & Hadden, J. E. (2000). Behavioral interventions for persons with dementia. In V. Molinari (Ed.), *Professional psychology in long term care*. New York: Hetherleigh.

Folstein, M. F., Folstein, S. E., & McHugh, P. R. (1975). Mini-mental state: A practical method for grading the cognitive state of patients for the clinician. *Journal of Psychiatric Research, 12*, 189–198.

Gwyther, L. (1994). Managing challenging behaviors at home. *Alzheimer's Disease and Associated Disorders, S3*, 100–112.

Hawkins, R. P., & Dobes, R. W. (1977). Behavioral definitions in applied behavior analysis: Explicit or implicit. In B. C. Etzel, J. M. LeBlanc, & D. M. Baer (Eds.), *New directions in behavioral research: Theory, methods, and applications* (pp. 167–188). Hillsdale, NJ: Erlbaum.

Heard, K., & Watson, T. S. (1999). Reducing wandering by persons with dementia using differential reinforcement. *Journal of Applied Behavior Analysis, 32*, 381–384.

Hendryx-Bedalov, P. M. (2000). Alzheimer's dementia: Coping with communication decline. *Journal of Gerontological Nursing, 26*, 20–24.

Hussian, R. A. (1981). *Geriatric psychology: A behavioral perspective*. New York: Van Nostrand Reinhold.

Iwata, B. A. (1995). *Functional Analysis Screening Tool (FAST)*. Florida: Florida Center on Self-Injury, University of Florida.

Iwata, B. A., Dorsey, M. F., Slifer, K. J., Bauman, K. E., & Richman, G. S. (1994/1982). Toward a functional analysis of self-injury. *Journal of Applied Behavior Analysis, 27*, 197–209.

Jacobson, N. S., Martell, C. R., & Dimidjian, S. (2001). Behavioral activation treatment for depression: Returning to contextual roots. *Clinical Psychology: Science and Practice, 8*, 255–270.

Kahng, S., Iwata, B. A., DeLeon, I. G., & Wallace, M. D. (2000). A comparison of procedures for programming noncontingent reinforcement schedules. *Journal of Applied Behavior Analysis, 33*, 223–231.

Kato, J., Hickson, L., & Worrel, L. (1996). Communication difficulties of nursing home residents: How can staff help? *Journal of Gerontological Nursing, 22*, 26–31.

Linehan, M. (1993). *Skills training manual for treating borderline personality disorder*. New York: The Guilford Press.

Marcus, B. A., & Vollmer, T. R. (1996). Combining noncontingent reinforcement and differential reinforcement schedules as treatment for aberrant behavior. *Journal of Applied Behavior Analysis, 29*(1):43–51.

Mausbach, B. T., Coon, D. W., Depp, C., Rabinowitz, Y. G., Wilson-Arias, E., Kraemer, H. C., et al. (2004). Ethnicity and time to institutionalization of dementia patients:

A comparison of Latina and Caucasian female family caregivers. *Journal of the American Geriatrics Society, 52*, 1077–1084.

McCurry, S. (2006). When a family member has dementia: Step to becoming a resilient caregiver. Westport, CT: Praeger.

Melin, L., & Gotestam, K. G. (1981). The effects of rearranging ward routines on communication and eating behaviors of psychogeriatric patients. *Journal of Applied Behavior Analysis, 14*, 47–51.

Milne, A., & Chryssanthopoulou, C. (2005). Dementia care-giving in Black and Asian populations: Reviewing and refining the research agenda. *Journal of Community and Applied Social Psychology, 15*, 319–337.

Pinkston, E. M., & Linsk, N. L. (1984). *Care of the elderly: A family approach*. New York: Pergamon Press.

Plaud, J. J., Moberg, M., & Ferraro, F. R. (1988). A review of Alzheimer's disease and dementia: Applied behavioral assessment and treatment procedures. *Journal of Clinical Geropsychology, 4*, 269–300.

Poling, A., & Byrne, T. (Eds.). (2000). *Behavioral pharmacology*. Reno, NV: Context Press.

Pyles, D. A. M., & Bailey, J. S. (1990). Diagnosing severe behavior problems. In A. Repp & N. Singh (Eds.), *Perspectives on the use of nonaversive and aversive interventions for persons with developmental disabilities*. Sycamore, IL: Sycamore Publishing Co.

Reisberg, B., Borenstein, J., Salob, S. P., Ferris, S. H., Franssen, E., & Georgotas, A. (1989). Behavioral symptoms in Alzheimer's disease: Phenomenology and treatment. *Journal of Clinical Psychiatry, 48*, 9–15.

Rosen, J., Burgio, L., Kollar, M., Cain, M., Allison, M., Fogleman, M., et al. (1994). The Pittsburg Agitation Scale: A user friendly instrument for rating agitation in dementia patients. *American Journal of Geriatric Psychiatry, 2*, 52–59.

Starkstein, S. E., & Mizrahi, R. (2006). Depression in Alzheimer's disease. *Expert Review of Neurotherapeutics, 6*, 887–95.

Stokes, G. (2000). *Challenging behaviour in dementia: A person-centered approach*. Bicester, UK: Speechmark.

Sturmey, P. (Ed.). (2007). *Functional analysis and clinical treatment*. San Diego, CA: Elsevier.

Sulzer-Azaroff, B., & Mayer, G. R. (1991). *Behavior analysis for lasting change*. New York: Harcourt Brace College Publishers.

Teri, L., Hughes, J. P., & Larson, E. B. (1990). Cognitive deterioration in Alzheimer's disease: Behavioral and health factors. *Journals of Gerontology, 45*(2), P58–P63.

Teri, L., Rabins, P., Whitehouse, P., Berg, L., Reisberg, B., Sunderland, T., et al. (1992). Management of behavior disturbance in Alzheimer disease: Current knowledge and future directions. *Alzheimer's Disease and Associated Disorders, 6*, 77–88.

Turner, S. (2005). Behavioural symptoms of dementia in residential settings: A selective review of non-pharmacological interventions. *Aging and Mental Health, 9*(2), 93–104.

US Food and Drug Administration. (2005). *Health advisory: Deaths with antipsychotics in elderly patients with behavioral disturbances*. Available at http://www.fda.gov/cder/drug/advisory/antipsychotics.htm.

Williams, K., Kemper, S., & Hummert, M. L. (2004). Enhancing communication with older adults: Overcoming Elderspeak. *Journal of Gerontological Nursing, 30*, 17–25.

Yury, C. A., & Fisher, J. E. (2007). Meta-analysis of the effectiveness of atypical antipsychotics for the treatment of behavioural problems in persons with dementia. *Psychotherapy and Psychosomatics, 76*, 213–218.

14
Cognitive Behavioral Case Management for Depressed Low-income Older Adults

Patricia A. Areán, George Alexopoulos, and Joyce P. Chu

The purpose of this chapter is to introduce the reader to cognitive behavioral case management (CB case management) for low-income elderly patients. This intervention was developed at San Francisco General Hospital and has been researched by the Over-60 Team at UCSF. CB case management arose out of the recognition that older adults living in financial distress often have difficulty utilizing the important therapeutic strategies that are present in CBT (Areán & Miranda, 1996). It also arose out of the fact that income status appears to be an important factor in treatment outcomes for late-life depression; older adults living in poverty appear to have poor and unstable response to antidepressant medication and other psychotherapies, such as interpersonal therapy (Cohen et al., 2006). In addition, our clients told us that while they valued the utility of depression interventions such as CBT, the interventions themselves did not address complex and urgent social problems that the clients were at a loss to manage themselves; in essence, they needed another person's help to negotiate the complex social service system to have their needs met. Our team at UCSF piloted the combination of CBT and clinical case management in treating major depression in low-income elderly, and found that the combination of these two interventions was more powerful in alleviating depression than the two interventions on their own. These data were recently published in *Psychology and Aging*. This chapter discusses the prevalence of poverty in late life, the consequences of poverty in an aging population, and the rationale and process by which CB case management alleviates depression.

Background

According to federal definitions of poverty, approximately 10% of people over the age of 65 live at or below the poverty line, and are at high risk for depression (US Census, 2004). Poverty level older adults have higher rates of major depression (9%) and other mental illnesses than do community dwelling older adults (3.8%) (Areán & Alvidrez, 2001). Moreover, low socioeconomic level is associated not only with higher new onset and lifetime prevalence of mental illness but also with persistence of these symptoms (Gilman, Kawachi, Fitzmaurice, & Buka, 2003; Lorant et al., 2003).

The clinical context of late-life depression, coupled with financial limitations, creates a formidable challenge for care. Low-income elders are likely to receive inadequate health services both because of their limited resources and because of their high medical burden and disability after many years of neglect. Along with disappointment in the health care system and the paucity of resources, depression and the resultant feelings of resignation may further limit their access to clinical and social services. Further, cognitive impairment, which afflicts approximately 40% of nondemented depressed elders (Alexopoulos, Kiosses, Klimstra, Kalayam, & Bruce, 2002) and is a common problem in low-income elders (Jagust et al., 2002; Lin, Lai, Tai, Lin, & Liu, 2004; Lopez et al., 2003), may further limit these persons' ability to advocate for themselves and even utilize resources. Finally, psychiatric treatment may help symptoms associated with mental illness, but does not address the social problems related to poverty that contributes to mental health issues.

Although the literature is scant and mixed with regard to the cause and effect of poverty on mental illness, the existing literature does suggest a mental health model that helps in understanding how to best address mental health issues in low-income elderly (Alexopoulos, 2005; Areán & Reynolds, 2005). The model focuses on contributors to disability commonly occurring in poverty level older persons that are likely to respond to the combination of case management and cognitive behavioral approaches. In this model, poverty promotes depression by increasing the burden of medical illnesses because of poor health care and health related behaviors (Alexopoulos, Buckwalter et al., 2002; Alexopoulos, Kiosses, Murphy, & Heo, 2004; Angel, Frisco, Angel, & Chiriboga, 2003; Areán & Alvidrez, 2001; Bruce et al., 2002; Butler, 1991; Cairney & Krause, 2005; Corney, 1984; Dohan, 2002; Epping-Jordan, Bengoa, Kawar, & Sabate, 2001; Frojdh, Hakansson, Karlsson, & Molarius, 2003; Garrett, 2003; Gilman et al., 2003; von Goeler, Rosal, Ockene, Scavron, & De Torrijos, 2003; Krause, 2005; Mellor & Milyo, 2003; Nunez, Armbruster, Phillips, & Gale, 2003; Ostir, Eschbach, Markides, & Goodwin, 2003), and it exposes individuals to chronic adversity as even their basic needs are often unsatisfied (Bruce, 2001; Bruce & Hoff, 1994; Butler; Cairney & Krause). Further, research suggests that many people in poverty have limited coping skills, which amplifies the experience of adversity; poverty level older adults often experience difficulties in negotiating obstacles and utilizing the few available resources (Ashby, Isen, & Turken, 1999; Kraaij, Arensman, & Spinhoven, 2002; Moos, Brennan, Schutte, & Moos, 2006). Adversity from unsatisfied needs or from repeated failures (in part due to coping skill deficits) may contribute to poor self-efficacy, hopelessness, and eventually negative affect, thus further fueling the experience of adversity and promoting depression (Ashby et al.; Fry, 1993; Kraaij et al.; Moos et al.) (see Fig. 1).

In our experience with poverty level older adults, we have found addressing both access to social services and improving coping skills can interrupt the spiral of interacting factors contributing to depression in this population. Case

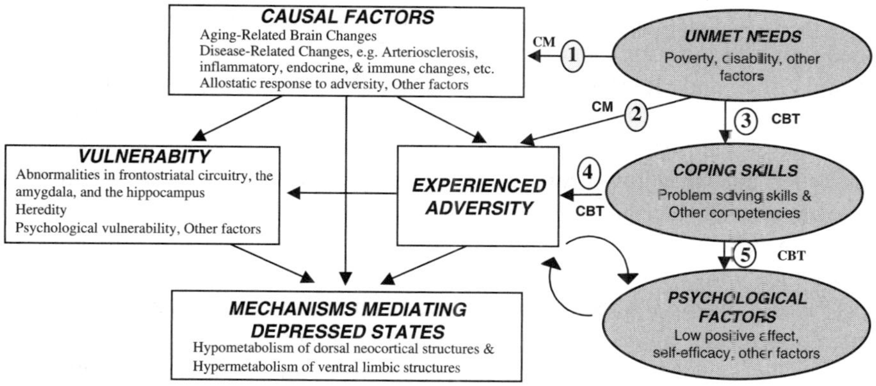

FIGURE 1. Biopsychosocial model of poverty and depression.

management (CM) may improve access to social services, reduce the impact of medical disorders promoting depression, and through the provision of other services, reduce "real" hardship. Resolving previously untenable problems can impart hope. CBT can make elders better managers of their lives, and besides directly reducing adversity, may improve self-efficacy and impart hope and positive affect thus further reducing the experience of adversity (Artistico, Cervone, & Pezzuti, 2003; Coleman & Newton, 2005; Phillips, Smith, & Gilhooly, 2002). However, the poor encounter problems that require additional knowledge of services and their regulations, and thus, CBT in and of itself may be insufficient in managing all their problems. Thus an intervention enabling patients to manage the problems they can address while also improving access to services maximizes the chances of overcoming depression (see Fig. 1).

Evidence Base

Research on the management of mental illness in low-income elderly is scarce. Despite CBT's and antidepressant medication's efficacy in nonimpoverished elders, these interventions alone may not be enough to offset depression in low-income elders. A secondary analysis of pooled data ($N = 248$) from two NIMH-supported trials documented that low socioeconomic level is associated with less favorable course of geriatric depression receiving combined pharmacotherapy and interpersonal therapy (IPT) (Cohen et al., 2006). During a 26-week period, depressed, low-income elders had a lower likelihood of responding to treatment than middle-income elders. These findings are consistent with two other treatment studies of depression in which low education was associated with poor response (Hirschfeld et al., 1998) and limited improvement in social adjustment (Spillmann et al., 1997).

A potential explanation as to why low-income elderly have a poor response to medication and therapy alone is that these individuals often experience financial strain, crime, and violence and live in an unstable housing environment (Angel et al., 2003; Krause, 1987, 1993; Ostir et al., 2003), preventing them from making good use of the psychotherapy skills they are being taught. Psychotherapy may not be feasible when low-income elders struggle for their next meal, are distressed over the loss of their home, or experience discomfort and disability because of limited access to medical care. A qualitative study of low-income elderly clients' experience with CBT indicated that while the clients found the philosophy of CBT useful, the need to complete assignments related to mood management was often superceded by crises in the home and neighborhood (Areán & Miranda, 1996). Clients in this study reported that if they had someone help them with their immediate needs, they would have had more time to use the techniques they were learning in treatment.

Based on results from this qualitative study, we created an intervention that included case management to address chronic adversity and CBT to address depressive symptoms. We then investigated the feasibility and compared the efficacy of CBT alone, case management (CM) alone, and combination of CM and CBT (CM-CBT) in 70 low-income (<15,000 per year) adults (>59 years) with major depression, who were randomly assigned to the three study arms (Areán et al., 2005). Depression severity was assessed with the 17-item HDRS at baseline, 16 weeks (treatment end) and 6 months after completion of treatment. The results indicated that CBT alone was ineffective in alleviating depressive symptoms. Interestingly, participants who received the CM condition resulted in greater improvements in depression than those who received the CBT alone condition; those who received the combination of CM and CBT had the best clinical outcomes.

Assessment Considerations

Mental health assessment in low-income elderly is similar in many ways to assessment in other older populations; issues related to fatigue, education, disability, and stigma that have been discussed in other chapters in this book also apply to low-income elderly, and thus will not be reiterated here. Assessment in this population does vary in one significant way, specifically, that in addition to assessing mental health symptoms, history, cognitive ability and disability, assessment of low-income elderly must include a social needs and environmental resource assessment. These issues are particularly important when applying CB case management. One of the best studied needs assessment tools is the Camberwell Assessment of Needs for the Elderly (CANE; Hoe, 2001). This is a psychometrically tested instrument used to assess older adults' needs, service linkage and satisfaction with services. The CANE has satisfactory test-rest reliability ($r = 0.85$) and interrater reliability (kappa = 0.94; Hoe). The instrument takes no more than 5 min to administer and covers several domains, including health, finances, family, social support, housing, legal, and safety. Identified need can then be used to generate a problem list that identifies urgent and complex issues. These needs can be ordered along a hierarchy to develop an action plan that the client and provider will work on together. Action plans are discussed in more detail in the next section. The CANE is also useful in identifying client resources. It includes questions about how well needs are being met and who is meeting those needs. This process helps to identify other collateral help the client may have who can also be enlisted to assist with social problems.

As with any other geriatric population, our assessments with low-income elderly are generally structured and can take 60–90 min depending on the degree of mental health and social issues the client presents with. We typically conduct our assessments in clients' homes, as long as they are amenable. Home assessments allow for a thorough view of client living environments and a clarification of the degree of environmental strain. A view of a client's neighborhood and housing hazards can help in defining immediate goals to work on as well as understand the limitations of CBT for a poor client. For instance, pleasant activities for a client who lives in a safe neighborhood may include health promoting activities such as small walks outside or visits with neighbors; for a client living in an unsafe neighborhood, more planning around activities of this nature will be necessary.

In our experience with low-income elderly, most clients prefer to discuss their case management needs first. In many ways, discussing those issues that the client feels are pressing opens the door for discussion of other issues. Low-income clients often feel they are better heard and often report to us that they feel that the provider is interested in their problems, rather than getting through an assessment. Once a needs assessment is completed, then assessment of mental health, cognitive and health symptoms is useful.

Treatment Model

In order to understand how the CB case management model works, it is important for the reader to be familiar with case management. Different types of case management exist but all share the theme of helping individuals cope with their illnesses through linkage to social services, advocacy, rehabilitation, and ongoing support during recovery from illnesses (Johnsen et al., 1999). Mueser, Bond, Drake, and Resnick (1998) have divided case management into three categories, i.e., rehabilitation models, intensive case management, and clinical case management. Rehabilitation and intensive models have been specifically developed for patients with severe mental illness and involve an array of team-based services. Clinical case management has largely been used for less severe mental illnesses, such as major depression, and for chronic health problems. Clinical case management is typically offered by workers acting as advocates for the service user and as "purchasers" of services (Mueser et al.); the case manager makes resources available through advocacy and linkage. Thus, problem resolution is achieved through a joint effort by clients and case managers, but the case managers do most of the problem solving on behalf of the client. In addition, standard case management offers to its clients education about their treatment needs, encouragement in utilizing resources, especially those related to their rehabilitation and in dealing with day-to-day problems, and crisis intervention. Case managers maintain long-term relationships with their clients so that they can monitor their clients' changing needs and help them by providing additional resources and encouraging their utilization (Kanter, 1989; Simpson, Miller, & Bowers, 2003).

It is not uncommon for case managers to also deliver some form of psychotherapy in addition to brokering services. As Mueser et al. 1998 have indicated, more often than not the intervention is supportive in nature and not an active mental health intervention. We have added CBT to case management because of its known mood regulating properties, its focus on current problems, and its efficacy as an intervention in older adults.

The Structure of CB Case Management

Treatment consists of 12 weekly meetings in which the provider teaches the client the CBT model and also updates him/her on case management activities during these sessions. The CBT approach used in this model is one developed by Gallagher-Thompson and Thompson, and includes modules on behavioral activation, cognitive restructuring, time management and assertiveness training. Based on feedback from our clients in research and practice, we have also added a short module on safety and health maintenance. Case management consists of a needs assessment and action plan developed by the provider in collaboration with the client. CM activities and CBT are delivered concurrently and both provider and client employ action plans between sessions. The provider typically addresses case management needs between sessions, while clients engage in CB techniques between sessions. Thus, both have homework to report on at each session. During the course of treatment, the sessions are organized in the following manner: (1) review of provider's progress on case management; (2) discussion of any new case management needs; (3) overview of CBT homework; and (4) review of new CBT strategy.

The *first session* consists of a needs assessment and education on the CB-case management process. The case manager explains the basic concepts of CB case management and divides problems into those that the case manager will address and problems that the patient will work on using the CBT model. During this introductory discussion, the case manager focuses needs assessment guided by the CANE, prioritizing problems and dividing them into problems the case manager will work on and problems the patient will work on. In most cases, case managers work on issues related to health, housing, legal aide, and other social issues; clients work on more interpersonal issues, such as affect regulation, communication skills, and health behaviors. The remaining time is spent introducing the client to behavioral activation strategies and pleasant activities. By the end of the first session, both the case manager and the patient each have an action plan to implement.

We have found in our experience that low-income elderly tend to stick with the CB part of CB case management if less complex CB skills are introduced first; it is for this reason that session 1 focuses on behavioral activation. Session 2 focuses on time management, session 3 on safety, and session 4 on health care. Starting with less complex skills empowers and encourages the client to make changes, and potentially leads to better self-efficacy. Additionally, by session 5 many of the crises that low-income clients have at the beginning of treatment are no longer urgent problems, because of the case management activities. Thus, by the time more complex skills such as thought records and assertiveness training are introduced, the client is in a better frame of mind and able to take on bigger interpersonal tasks. Sessions 5–7 then focuses on cognitive restructuring skills, using a thought record. Sessions 8–11 focuses on assertiveness training. The final session is focused on relapse prevention and discussion of any additional, outstanding case management needs.

On occasion, clients finish the CB part of CB case management before all their case management needs are met. In our research, only 5% were left needing additional advocacy after treatment ended. In those cases, we typically finish the job we set out to do, although we may not meet with the client on a weekly basis after 12 weeks.

Case Example

Mr. Z is a 75-year-old, widowed Caucasian man who suffers from macular degeneration and diabetes, which had resulted in above the knee amputation. He lives on the second floor of a walk-up building and has difficulty leaving his apartment because of his poor eyesight and because his prosthesis no longer fits his leg after a weight loss of 15 pounds in the past year. His only income is social security with which he can barely make ends meet. Since his eyesight began to fail and his disability increased, Mr. Z has become increasingly depressed. His coverage by Medicare leaves him with a co-payment for mental health services that he is unable to afford. Mr. Z was connected with case management services but he was unable to find the motivation to work with his case worker, and he continued to miss his medical appointments frequently. He failed to receive help with his medical bills, find more amenable housing or obtain a usable prosthesis for his leg. In addition, his eldest son temporarily needed a place to stay and had been asking Mr. Z persistently to move in with him, a situation which in the past has led to friction. The Home-Delivered Meals Program referred Mr. Z to us for assessment of depression and treatment. Mr. Z met DSM-IV criteria for major depression. A case worker trained in CB case management visited Mr. Z at his home and worked with him to identify problems that were above Mr. Z's ability to deal with as well as problems that Mr. Z could work on using the CB techniques approach. Mr. Z's case worker was able to link him to paratransit, which enabled him to go to his senior citizen center and to his medical appointments. The case worker also made arrangements for Mr. Z to obtain a new prosthesis and introduced Mr. Z to a legal aid, whom Mr. Z could engage if his son moved into Mr. Z's apartment against his wishes. Furthermore, Mr. Z and the case worker used the CB case management approach to identify problems that Mr. Z could address. The case worker helped Mr. Z with decisions related to his daily routine and some of his affairs, including setting limits with his son. Mr. Z identified the series of steps he needed to take prior to his visit to the senior citizen center, i.e., notify paratransit and prepare his clothes a day ahead of his visit, get up from bed well ahead of the time of his departure, and establish a routine for his personal hygiene. He used a similar approach for his medical visits and developed a plan for injecting his insulin and for taking his medication. Mr. Z also prepared a written list of contingency actions that he could undertake if challenged by his son. Twelve weeks after this intervention, Mr. Z's depressive symptoms subsided and his HDRS dropped to 7 from a baseline of 32. Mr. Z reported feeling connected with the world, supported by the services he had received, and empowered by his increased ability to prioritize his problems, think of alternatives, and decide on a course of action.

Cultural Considerations

Because so many people of color are also low-income, cultural considerations are particularly important when working with low-income, elderly clients. In fact, poverty rates for ethnic minority groups are among the highest in the United States. According to the 2005 Current Population Survey Annual Social and Economic Supplement, poverty rates were 9.8% for Asians, 21.9% for Hispanics, and 24.7% for Blacks compared to 8.6% for non-Hispanic Whites. Ethnic minority individuals also comprise the fastest growing segment of the elderly population. Though in 2000 the elderly US population included 5% Hispanic, 8% Black, and 4% of other ethnic groups, the elderly population in 2050 is projected to include 16% Hispanic, 12% Black, and 7% other ethnic groups (Himes, 2001). In total, over the next 30 years, the minority population 65 years and older is expected to increase from 16.5% (5.7 million) in 2000 to 25.6% in 2030 (US Census, 2000). Clearly, we must consider the special issues of ethnic minorities when addressing CB case management as treatment for depression in low-income older adults.

For many ethnic minority individuals, specific cultural and minority status considerations may, like poverty, promote depression via increased exposure to chronic adversity in the form of language barriers, family burdens, transportation problems, racial discrimination, acculturative stress, unequal access to health care, and paucity of culturally responsive mental health services (e.g., Gilmer et al., 2005; Mio, Nagata, Tsai, & Tewari, 2006; Sanchez-Lacay et al., 2001). Several of these factors are ethnic-specific and unique to ethnic minority groups. For example, racial disparities in hiring, salary, promotion, and race-related discrimination are well documented and serve as additional sources of social and financial hardship. Also, many ethnic minority older adults (especially Asian and Hispanic individuals) are immigrants with limited English proficiency who struggle to navigate their basic needs in a largely monolingual nation. Other culture or race factors exacerbate many of the same social issues affected by poverty status. In particular, low-income minority elderly not only experience paucity of access to health care resources because of low socioeconomic status and lack of insurance, but there exist race-related disparities in health care access. For immigrant and non-English speaking elderly, the language and/or ethnic-specific health services necessary to serve their health care needs are severely limited (e.g., Chin, 1998; Ruiz, 2002; Woloshin et al., 1995), often available only in larger cities. Indeed, studies have found that perceived discrimination, acculturative stress, and fewer formal health care providers predict depressive symptoms (e.g., Mui, 1993; Nyborg & Curry, 2003).

These ethnic-specific considerations, combined with the clinical picture of late-life depression and financial considerations of poverty, make treatment of depression in the low-income ethnic minority older adult a complex challenge. Of particular importance are ethnic-specific factors that affect willingness to seek services and dictate how treatment should be delivered. Mistrust of the health care system serves as a formidable barrier to help-seeking among Blacks (e.g., Whaley, 2001). Stigma and loss of face about mental illness contribute to reluctance among ethnic minorities to utilize mental health services (e.g., Gerber, Nguyen, & Bounkeua, 1999; Lawson, Kahn, & Heiman, 1982; Sirey et al., 2001) and may serve as moderators of treatment effectiveness, particularly in treatments involving disclosure (Zane & Yeh, 2002). This cultural and stigma and mistrust may have important effects on service utilization. Asian Americans, for example, report a time delay for seeking mental health treatment and among the lowest rates of mental health service utilization (Zane et al., 2004).

In our experience, the CB case management approach may ameliorate the influences of stigma and mistrust as hindrances to help-seeking. Studies have found ethnic minority clients to view CBT as a credible and preferred treatment modality, speculating that the structure provided in CBT, combined with a skills-acquisition focus and view of the therapist as active and "expert" are amenable to cultures that value hierarchical, authority, and respect-oriented practices (e.g., Chen & Davenport, 2005; Leong, 1986; Wong et al., 2003). Additionally, because CBT does not focus on exploring feelings and achieving insight as primary vehicles of change, it may not hold the same level of stigma associated with a more psychodynamic treatment. Indeed, a very limited amount of evidence has shown that CBT is efficacious in ethnic minority groups (e.g., Dai et al., 1999; Miranda et al., 2003). Though CBT is a preferred treatment modality among some ethnic minorities that may attenuate some concerns of mistrust and stigma, it does little to ease the initial transition of an ethnic minority client into treatment. Stigma barriers to seeking mental health services are considerations particularly in the initial stages of help-seeking – getting the ethnic minority individual to agree to receive services, promoting psychoeducation about the utility of treatment, and increasing the likelihood of commitment to a full course of treatment.

With its inclusion of case management, CB case management may address these cultural considerations of stigma early in treatment. The emphasis on CM and problem-solving of practical issues from the very first session provides an impetus for ethnic minority clients to commit to treatment and later engage in the more psychological aspects of CBT. Indeed, elderly immigrants have reported preference for tangible treatments and material intervention (e.g., Dhooper & Tran, 1998; Gerber et al., 1999). Compared to treatments that start with discussion of depressive symptoms, less cultural stigma may be associated with problem-solving a tangible social issue, and less mistrust may be associated with a provider who takes an active stance in addressing a practical concern. In addition, CB case management can address race-related discrimination issues, disparities in access to care, and language barriers through the provision of resources such as connection with legal aid or health care provider contact. The psychological aspects of acculturative or discrimination stress can be addressed through the CBT portion of CB case management. We posit that the addition of CM to CBT results in a treatment that comprehensively addresses the cultural, social, financial, and depression-related concerns of a low-income ethnic minority older adult. Research is needed to test the efficacy and develop culture-specific modifications of CB case management for low-income ethnic minority older adults.

Conclusions

The addition of case management to CBT not only increases the likelihood that low-income elderly will participate in CBT, but also enhances treatment. By simultaneously addressing instrumental needs and teaching coping strategies to manage depression and stress associated with poverty, the quality of life for those who are having trouble making ends meet can greatly improve. Moreover, the combination of these two approaches allows for an efficient method of addressing the various needs that low-income elderly have.

References

Alexopoulos, G. S., Buckwalter, K., Olin, J., Martinez, R., Wainscott, C., & Krishnan, K. R. (2002). Comorbidity of late life depression: An opportunity for research on mechanisms and treatment. *Biological Psychiatry, 52*(6), 543–558.

Alexopoulos, G. S., Kiosses, D. N., Klimstra, S., Kalayam, B., & Bruce, M. L. (2002). Clinical presentation of the "depression-executive dysfunction syndrome" of late life. *American Journal of Geriatric Psychiatry, 10*(1), 98–106.

Alexopoulos, G. S., Kiosses, D. N., Murphy, C., & Heo, M. (2004). Executive dysfunction, heart disease burden, and remission of geriatric depression. *Neuropsychopharmacology, 29*(12), 2278–2284.

Alexopoulos, G. S. (2005). Depression in the elderly. *Lancet, 365*(9475), 1961–1970.

Angel, R. J., Frisco, M., Angel, J. L., & Chiriboga, D. A. (2003). Financial strain and health among elderly Mexican-origin individuals. *Journal of Health and Social Behavior, 44*, 536–551.

Areán, P. A., & Alvidrez, J. (2001). The prevalence of psychiatric disorders and subsyndromal mental illness in low-income, medically ill elderly. *International Journal of Psychiatry in Medicine, 31*(1), 9–24.

Areán, P. A., & Miranda, J. (1996). The treatment of depression in elderly primary care patients: A naturalistic study. *Journal of Clinical Geropsychology, 2*(3), 153–169.

Areán, P. A., Gum, A., McCulloch, C. E., Bostrom, A., Gallagher-Thompson, D., & Thompson, L. (2005). Treatment of depression in low-income older adults. *Psychology and Aging, 20*(4), 601–609.

Areán, P. A., & Reynolds, C. F. III. (2005). Psychosocial risk factors in late life depression. *Biological Psychiatry, 58*, 277–282.

Artistico, D., Cervone, D., & Pezzuti, L. (2003). Perceived self-efficacy and everyday problem solving among young and older adults. *Psychology and Aging, 18*(1), 68–79.

Ashby, F. G., Isen, A. M., & Turken, A. U. (1999). A neuropsychological theory of positive affect and its influence on cognition. *Psychological Review, 106*(3), 529–550.

Bruce, M. L. (2001). Depression and disability in late life: Directions for future research. *American Journal of Geriatric Psychiatry, 9*(2), 102–112.

Bruce, M. L., & Hoff, R. (1994). Social and physical health risk factors for first on-set major depressive disorder in a community sample. *Social Psychiatry and Psychiatric Epidemiology, 29*, 165–171.

Bruce, M. L., McAvay, G. J., Raue, P. J., Brown, E. L., Meyers, B. S., Keohane, D. J., et al. (2002). Major depression in elderly home health care patients. *American Journal of Psychiatry, 159*(8), 1367–1374.

Butler, R. N. (1991). The poverty of living alone. *Geriatrics, 46*(10), 13.

Cairney, J., & Krause, N. (2005). The social distribution of psychological distress and depression in older adults. *Journal of Aging and Health, 17*(6), 807–835.

Census, U. (2004). http://www.census.gov/hhs/ww/poverty.html

Chen, S. W., & Davenport, D. S. (2005). Cognitive-behavioral therapy with Chiense American clients: Cautions and modifications. *Psychotherapy: Theory, Research, Practice, Training, 42*, 101–110.

Chin, J. L. (1998). Mental health services and treatment. In L. C. Lee & N. W. S. Zane (Eds.). *Handbook of Asian American psychology* (1st ed., pp. 485–504). Thousand Oaks, CA: Sage.

Cohen, A., Houck, P. R., Szanto, K., Dew, M. A., Gilman, S. E., & Reynolds, C. F., III. (2006). Social inequalities in response to antidepressant treatment in older adults. *Archives of General Psychiatry, 63*(1), 50–56.

Coleman, M. T., & Newton, K. S. (2005). Supporting self-management in patients with chronic illness. *American Family Physician, 72*(8), 1503–1510.

Corney, R. H. (1984). The mental and physical health of clients referred to social workers in a local authority department and a general practice attachment scheme. *Psychological Medicine, 14*(1), 137–144.

Dai, Y., Zhang, S., Yamamoto, J., Ao, M., Belin, T. R., Cheung, F., et al. (1999). Cognitive behavioral therapy of minor depressive symptoms in elderly Chinese Americans: A Pilot study. *Community Mental Health Journal, 35*, 537–542.

Dhooper, S. S., & Tran, T. V. (1998). Understanding and responding to the health and mental health needs of Asian refugees. *Social Work in Health Care, 27*(4), 65–82.

Dohan, D. (2002). Managing indigent care: A case study of a safety-net emergency department. *Health Service Research, 37*(2), 361–376.

Epping-Jordan, J., Bengoa, R., Kawar, R., & Sabate, E. (2001). The challenge of chronic conditions: WHO responds. *British Medical Journal, 323*(7319), 947–948.

Frojdh, K., Hakansson, A., Karlsson, I., & Molarius, A. (2003). Deceased, disabled or depressed – A population-based 6-year follow-up study of elderly people with depression. *Social Psychiatry and Psychiatric Epidemiology, 38*(10), 557–562.

Fry, P. S. (1993). Mediators of depression in community-based elders. In P. Cappeliez & R. J. Flynn (Eds.), *Depression and social environment: Research and intervention with neglected populations*. Montreal: McGill-Queen's University Press.

Garrett, L. (2003). Gaps between the rich and the poor. The widening differences in wealth, life expectancy, public health infrastructure and perception of threats, and the consequences for global security. *EMBO Reports, 4 Spec No*, S15–S19.

Gerber, L., Nguyen, Q., & Bounkeua, P. (1999). Working with Southeast Asian people who have migrate to the United States. In K. Nader, N. Dubrow, & B.H. Stamm (Eds.), *Honoring differences: Cultural issues in the treatment of trauma and loss* (pp. 98–118). Philadelphia: Brunner/Mazel.

Gilman, S. E., Kawachi, I., Fitzmaurice, G. M., & Buka, L. (2003). Socio-economic status, family disruption and residential stability in childhood: Relation to onset, recurrence and remission of major depression. *Psychological Medicine, 33*(8), 1341–1355.

Gilmer, W. S., Trivedi, M. H., Rush, A. J., Wisniewski, S. R., Luther, J., Howland, R. H., et al. (2005). Factors associated with chronic depressive episodes: A preliminary report from the STAR-D project. *Acta Psychiatrica Scandinavica, 112*(6), 425–433.

von Goeler, D. S., Rosal, M. C., Ockene, J. K., Scavron, J., & De Torrijos, F. (2003). Self-management of type 2 diabetes: A survey of low-income urban Puerto Ricans. *The Diabetes Educator, 29*(4), 663–672.

Himes, C. L. (2001). Elderly Americans. *Population Bulletin, 56*(4).

Hirschfeld, R. M., Russell, J. M., Delgado, P. L., Fawcett, J., Friedman, R. A., Harrison, W. M., et al. (1998). Predictors of response to acute treatment of chronic and double depression with sertraline or imipramine. *Journal of Clinical Psychiatry, 59*(12), 669–675.

Hoe, J. (2001). An investigation to identify the needs of those older people with mental health problems who have attended an old age psychiatry day hospital for one year or more. *Aging and Mental Health.*

Jagust, W. J., Eberling, J. L., Wu, C. C., Finkbeiner, A., Mungas, D., Valk, P. E., et al. (2002). Brain function and cognition in a community sample of elderly Latinos. *Neurology, 59*(3), 378–383.

Johnsen, M., Samberg, L., Calsyn, R., Blasinsky, M., Landow, W., Goldman, H. (1999). Case management models for persons who are homeless and mentally ill: The ACCESS demonstration project. *Community Mental Health Journal, 35*, 325–346.

Kanter, J. (1989). Clinical case management: Definition, principles, components. *Hospital and Community Psychiatry, 40*(4), 361–368.

Kraaij, V., Arensman, E., & Spinhoven, P. (2002). Negative life events and depression in elderly persons: A meta-analysis. *The Journals of Gerontology. Series B, Psychological Sciences and Social Sciences, 57*(1), P87–P94.

Krause, N. (1987). Life stress, social support, and self-esteem in an elderly population. *Psychology and Aging, 2*(4), 349–356.

Krause, N. (1993). Neighborhood deterioration and social isolation in later life. *International Journal of Aging and Human Development, 36*(1), 9–38.

Krause, N. (2005). Negative interaction and heart disease in late life: Exploring variations by socioeconomic status. *Journal of Aging and Health, 17*(1), 28–55.

Lawson, H., Kahn, M., & Heiman, E. (1982). Psychopathology, treatment outcome, and attitude toward mental illness in Mexican American and European patients. *International Journal of Social Psychiatry, 28*, 20–26.

Leong, F. T. (1986). Counseling and psychotherapy with Asian-Americans: Review of the literature. *Journal of Counseling Psychology, 33*, 196–206.

Lin, H. F., Lai, C. L., Tai, C. T., Lin, R. T., & Liu, C. K. (2004). Apolipoprotein E polymorphism in ischemic cerebrovascular diseases and vascular dementia patients in Taiwan. *Neuroepidemiology, 23*(3), 129–134.

Lopez, O. L., Jagust, W. J., Dulberg, C., Becker, J. T., DeKosky, S. T., Fitzpatrick, A., et al. (2003). Risk factors for mild cognitive impairment in the Cardiovascular Health Study Cognition Study: Part 2. *Archives of Neurology, 60*(10), 1394–1399.

Lorant, V., Deliege, D., Eaton, W., Robert, A., Philippot, P., & Ansseau, M. (2003). Socioeconomic inequalities in depression: A meta-analysis. *American Journal of Epidemiology, 157*(2), 98–112.

Mellor, J. M., & Milyo, J. (2003). Is exposure to income inequality a public health concern? Lagged effects of income inequality on individual and population health. *Health Services Research, 38*(1 Pt 1), 137–151.

Mio, J. S., Nagata, D. K., Tsai, A. H., & Tewari, N. (2006). Racism against Asia/Pacific Island Americans. In F. T. Leong, A. G. Inman, A. Ebreo, L. H. Yang, L. Kinoshita, & M. Fu (Eds.), *Handbook of Asian American psychology* (2nd ed., pp. 341–361). Thousand Oaks, CA: Sage.

Miranda, J., Azocar, F., Organista, K., Dwyer, E., & Areán, P. (2003). Treatment of depression among impoverished primary care patients from ethnic minority groups. *Psychiatric Services, 54*, 219–225.

Moos, R. H., Brennan, P. L., Schutte, K. K., & Moos, B. S. (2006). Older adults' coping with negative life events: Common processes of managing health, interpersonal, and financial/work stressors. *International Journal of Aging and Human Development, 62*(1), 39–59.

Mueser, K. T., Bond, G. R., Drake, R. E., & Resnick, S. G. (1998). Models of community care for severe mental illness: A review of research on case management. *Schizophrenia Bulletin, 24*(1), 37–74.

Mui, A. C. (1993). Self-reported depressive symptoms among Black and Hispanic frail elders: A sociocultural perspective. *Journal of Applied Gerontology, 12*(2), 170–187.

Nunez, D. E., Armbruster, C., Phillips, W. T., & Gale, B. J. (2003). Community-based senior health promotion program using a collaborative practice model: The Escalante Health Partnerships. *Public Health Nursing, 20*(1), 25–32.

Nyborg, V. M., & Curry, J. F. (2003). The impact of perceived racism: Psychological symptoms among African American boys. *Journal of Clinical Child and Adolescent Psychology, 32*, 258–266.

Ostir, G. V., Eschbach, K., Markides, K. S., & Goodwin, J. S. (2003). Neighbourhood composition and depressive symptoms among older Mexican Americans. *Journal of Epidemiology and Community Health, 57*(12), 987–992.

Phillips, L. H., Smith, L., & Gilhooly, K. J. (2002). The effects of adult aging and induced positive and negative mood on planning. *Emotion, 2*(3), 263–272.

Ruiz, P. (2002). Hispanic access to health/mental health services. *Psychiatry Quarterly, 73*, 85–91.

Sanchez-Lacay, J. A., Lewis-Fernandez, R., Goetz, D., Blanco, C., Salman, E., Davies, S., et al. (2001). Open trial of nefazodone among Hispanics with major depression: Efficacy, tolerability, and adherence issues. *Depression and Anxiety, 13*, 118–124.

Simpson, A., Miller, C., & Bowers, L. (2003). Case management models and the care programme approach: How to make the CPA effective and credible. *Journal of Psychiatric and Mental Health Nursing, 10*(4), 472–483.

Sirey, J. A., Bruce, M. L., Alexapoulos, G. S., Perlick, D., Raue, P., & Friedman, S. J. (2001). Perceived stigma as a predictor of treatment discontinuation in young and older outpatients with depression. *American Journal of Psychiatry, 15*, 479–481.

Spillmann, M., Borus, J. S., Davidson, K. G., Worthington, J. J., III, Tedlow, J. R., & Fava, M. (1997). Sociodemographic predictors of response to antidepressant treatment. *International Journal of Psychiatry and Medicine, 27*(2), 129–136.

US Bureau of the Census. (2000). Population projections of the United States by age, sex, race, and Hispanic origin: 1995–2050. *Current Population Reports*, P25–P1130.

Whaley, A. L. (2001). Cultural mistrust: An important psychological construct for diagnosis and treatment of African Americans. *Professional Psychology: Research and Practice, 32*(6), 555–562.

Woloshin, S., Bickell, N. A., Schwartz, L. M., Gany, F., & Welch, H. G. (1995). Language barriers in medicine in the United States. *JAMA, 273*, 724–728.

Wong, E. C., Kim, B. S. K., Zanc, N. W. S., Kim, I. J., & Huang, J. S. (2003). Examining culturally based variables associated with ethnicity: Influences on credibility perceptions of empirically supported interventions. *Culturally Diversity and Ethnic Minority Psychology, 9*, 88-96.

Zane, N., Hall, GC.N., Sue, S., Young, K., & Nunez, J. (2004). Research on psychotherapy with culturally diverse populations. In M. J. Lambert (Ed.), *Handbook of psychotherapy and behavior change* (5th ed., pp. 767–804. New York: Wiley).

Zane, N., & Yeh, M. (2002). The use of culturally-based variables in assessment: Studies on loss of face. In K. S. Kurasaki, S. Okazaki, & S. Sue (Eds.), *Asian American mental health: Assessment theories and methods* (pp.123–138). New York: Kluwer Academic/ Plenum.

15
Post-Stroke Depression and CBT with Older People

Ken Laidlaw

Understanding the Context of CBT for Post-Stroke Depression

The empirical evidence as it pertains to Stroke and CBT is at an early stage. There are many issues that remain unresolved and there remains a great deal of psychological research that requires to be undertaken, especially as recent meta-analyses have concluded that while CBT may improve mood post-stroke, there is insufficient evidence for the routine use of psychotherapy for the treatment of post-stroke depression (PSD) (Anderson, Hackett, & House, 2004; Hackett, Anderson, & House, 2004).

While the research evidence in favour of CBT as a treatment for PSD is limited, this should not deter further research into the application of CBT as it seems an especially good fit to meet the needs of people who have become depressed after a stroke. CBT is based on a here and now conceptualisation model that perfectly matches the immediate nature of stroke survivors' concerns. CBT adopts a skills enhancing, problem-solving focus that fits with the aims and needs of people who have survived a stroke when learning to manage the personal consequences of their stroke. CBT monitors and evaluates the cognitions of stroke survivors, many of whom have 'objectively' made a good recovery but whose subjective appraisal is to see only failure. Finally, the primary aim of CBT is symptom reduction, with an emphasis on reducing symptoms of depression such as apathy, hopelessness and low mood. These symptoms of depression are likely to result in excess disability, i.e. the impact of a stroke is magnified by depressive symptoms, resulting in decreased levels of functioning. For these reasons, CBT may be considered to be of great potential as a treatment for PSD.

Stroke can be seen to be a post-acute acquired impairment with chronic consequences for an individual's functioning resulting in disability and handicap. CBT interventions are based on the simple premise that behaviour can change, and in working with individuals who have survived a stroke it is important to conceptualise the nature of an individual's problem within a behavioural frame of reference (Laidlaw, Thompson, Dick-Siskin, & Gallagher-Thompson, 2003).

CBT also has utility in the overall rehabilitation of an individual after a stroke, as it provides a means of reducing depression after stroke and this can be very important for the individual's post-stroke recovery generally (Hibbard, Grober, Gordon, Aletta, & Freeman, 1990).

Rather than acting as a deterrent, the relative lack of empirical evidence for CBT of PSD ought to motivate researchers and clinicians to provide definitive answers to the questions of efficacy and the application of psychotherapy in an area of high unmet-need. Thus, this chapter sets out to provide a summary of the empirical evidence for CBT as a treatment for PSD, but equally, this chapter provides clinical suggestions for therapists working with, or planning to work with this population.

Stroke

In the western world, stroke is the third leading cause of death in older people and the main cause of disability in later life (Kinsella & Velkoff, 2001). On average in the US, someone has a stroke every 45 s (American Heart Association, 2003). In the UK stroke affects approximately 130,000 people a year and in the US, 600,000 people a year are affected. At any one time, there are 4.5 million stroke survivors in the US (Casper et al., 2003) and in the UK stroke care directly costs the National Health Service £2.8 billion per year, with informal care costs (that provided by families) estimated to cost an additional £2.4 billion per year (Source: National Audit Office, 2005).

Women typically tend to live longer than men, thus they are more likely to die as a result of a stroke with 61% of all stroke deaths occurring in women (American Heart Association, 2003). The effect of increased numbers of older people surviving a first stroke is likely to be amplified by demographic change as the developed and developing world is on the threshold of a substantial increase in the relative numbers of older people (United Nations, 2003). As stroke is primarily but not exclusively a condition found in old age, and given that mortality rates after stroke have declined in recent years this will result in increased numbers of older people in need of stroke rehabilitation and psychosocial interventions to deal with the emotional consequences of experiencing a stroke. Stroke therefore is likely to become an important issue for therapists working with older people.

Post-Stroke Depression

Depression following a stroke is a common complication that can inflate mortality and morbidity levels and may impair an individual's ability to participate fully in rehabilitation treatment (Turner-Stokes & Hassan, 2002). The prevalence of depression can change depending upon the methodological differences in detecting and measuring depression in stroke, with House (1987) reporting prevalence rates for PSD 18–61% and Turner-Stokes and Hassan reporting prevalence rates from 0 to 55%. Hackett, Yapa, Parag, and Anderson (2005) used different sources to calculate prevalence rates for depression following a stroke. Using hospital-based evidence, the frequency of depression following a stroke ranged from 6 to 55%, whereas for population-based studies, the frequency of depression following a stroke ranged from 1 to 47%, and for rehabilitation studies, the frequency of depression following a stroke ranged from 4 to 63%. Overall, Hackett et al. calculated that 33% of people experience a stroke develop depression. Thus it is fair to say that stroke results in an increased risk of developing depression.

Treatment of depression after a stroke is important for two main reasons, the distress that results and is caused to the individual and their carers and the association between depression and poor outcome following a stroke (Turner-Stokes & Hassan, 2002; Visser-Meily, van Heugten, Post, Schepers, & Lindeman, 2005). With stroke, poor outcome for depression may mean increased risk of mortality (Carney & Freedland, 2002; House, Knapp, Bamford, & Vail, 2001). While reduction of mortality post-stroke is an important aim of any service, for individuals who survive after a stroke, reduction of morbidity is of primary importance. It is in the reduction of morbidity that CBT can have an important role to play. Reducing depression is likely to result in better outcome for the person with a stroke (Turner-Stokes & Hassan; Thomas & Lincoln, 2006).

Research has focused on the possibility that depression arises as a result of cortical damage following a stroke, with Narushima, Koster, and Robinson (2003), hypothesising that lesion location is associated with the development of PSD. However, a recent meta-analysis looking at lesion location and depression concluded the evidence was inconclusive (Carson et al., 2000). However, for psychologists wishing to use CBT for depression after stroke this is really a non-issue as CBT has always adopted a here and now focus for its interventions (Beck, 1995; Thompson, 1996). In CBT, interventions are targeted at the factors that maintain low mood not necessarily intervening at the level of causation (Laidlaw, 2006). Thus when someone has a stroke, it is important that CBT interventions are targeted at the consequences for the individual, such as hopelessness about change, and isolation as a result of depressive apathy and anhedonia as impairments experienced by the individual following their stroke cannot be reversed by CBT. Thus, it is immaterial whether depression has been caused by cortical damage or is a consequence of it. Stroke can be understood in terms of three components: impairment, disability and handicap (WHO, 1980). For an explanation of the utility of the WHO model for psychotherapists working with individuals with depression comorbid with physical illness see Laidlaw, Thompson, and Gallagher-Thompson (2004).

For example, in stroke, impairment refers to the any loss/abnormality of body structure, appearance, organ or system such as damage caused to neural tissue. This is where the biological nature of stroke is important, it is the cause of the difficulties experienced by the individual. Disability is the impact of the impairment on the individual's ability to carry out 'normal' activities, for example when one cannot walk without assistance or with the use of mobility aids. However, handicap is more socially mediated. Consequences of handicap are reflected in the disadvantages experienced by an individual interacting with, and adapting to, the environment. One who has mobility problems following a stroke may encounter difficulties in accessing environments and may end up feeling excluded from normal communications and social interactions. With a loss of opportunity to participate in society at a general level and a loss of opportunity to participate at the level one normally operated at prior to stroke, the individuals idiosyncratic appraisal of the meaning of these events can become important and a CBT intervention becomes potentially effective. The usefulness of the WHO (1980) system to psychotherapists is that it allows one to consider the consequences of impairment or disease for an individual at a level of subjective personal meaning rather than at a level of disease.

Assessment of Depression Following a Stroke

A particular issue when working with older people with PSD is assessment. Often the presence of a medical condition has the effect of reducing the rate of detection of depression in older people (Bruce, 1999). Many methods of diagnosing depression rely on somatic symptoms that may complicate the diagnostic issues in PSD. The same could also be said of standardised questionnaires that also often measure somatic symptoms in depression. The clinician could therefore reasonably ask, what measures of depression may be valid for use in therapy with patients with PSD? Aben, Verhey, Lousberg, Lodder, and Honig (2002) reviewed the performance of common measures such as the Hamilton Rating Scale for Depression, Hospital Anxiety and Depression Scale and the Beck Depression Inventory and concluded that these could be used to screen for PSD with no changes to standard cut-offs. However a gender difference in the assessment of depression was detected where the aforementioned measures performed more poorly for women than men as women tended to record less specific complaints of distress than men (Aben et al.). In a separate but equally thorough review of assessment issues in PSD, Lincoln, Nicholl, Flannaghan, Leonard, and van der Gucht (2003) examined the validity of a number of standardised questionnaires such as the Beck Depression Inventory, Wakefield Depression Inventory, and General Health Questionnaire (GHQ28) to screen and diagnose depression in comparison to structured psychiatric diagnostic interviews. The results were somewhat disappointing as agreement between psychiatric diagnosis and questionnaires was poor, and while the questionnaires showed high sensitivity in detecting depression symptoms there was poor specificity for detection of depression using standardised questionnaires. Lincoln et al. calculated that using the BDI-II (Beck, Steer, & Brown, 1996) using a cut-off of 15/16 (as compared to the usual cut-off of 13) had a sensitivity of 91% and a specificity of 56%. Lincoln et al. note that measures considered to be good diagnostically are expected to achieve a sensitivity of at least 80% and a specificity of at least 60%. It is of course important to remember that the BDI-II (Beck et al.) was developed as a measure of symptom change rather than as a diagnostic instrument. Bearing this in mind it may be more sensible to consider the conclusion of Lincoln et al. that while standardised measures may be useful for screening of depression these need to be supplemented with a psychiatric diagnostic interview such as the DSM-IV (American Psychiatric Association, 2000) before confirming a diagnosis of PSD. Turner-Stokes and Hassan (2002) conclude that a number of assessment measures have some validity and utility in assessing depressive symptoms in people with a stroke such as the Geriatric Depression Scale, Hamilton Rating Scale for Depression and the Center for Epidemiological Studies Depression Scale (CES-D) however Turner-Stokes and Hassan recommend using the Beck Depression Inventory. Standardised questionnaires need to be used with caution by clinicians as they have not been primarily constructed with thought to detecting depression comorbid with a physical impairment and therefore may not demonstrate validity. Lincoln et al. recommend using specifically developed scales for detecting PSD.

The Efficacy of CBT as a Treatment for Post-Stroke Depression

Kneebone and Dunmore (2000) in a review of studies looking at the psychological management of PSD conclude that current interventions are lacking in sophistication and the evidence to date is limited. Hackett et al. (2005) note that very few stroke patients receive adequate management of their depression. Stroke is an area much neglected by psychological therapists (Kneebone & Dunmore) with very few interventions published to date empirically evaluating CBT as a treatment for PSD (Anderson et al., 2004).

While not strictly a study looking at CBT for PSD, Kemp, Corgiat, and Gill (1992) showed that a 12-week course of group CBT was effective for individuals diagnosed with a range of chronic disabling physical conditions including stroke. In this study 41 participants received group-based CBT treatment for depression. Eighteen of the participants had chronic disabling illnesses. Following treatment, both groups of participants (those with disabling conditions and those without) achieved similar improvement in depression symptoms at the end of treatment. However while the non-disabled group continued to show improvement at 6 months follow-up, the disabled group maintained treatment gains but did not achieve further improvement at 6 months follow-up. As the authors note, both the disabled and non-disabled participants increased the frequency of tasks they could perform and this study demonstrates that disabled older adults can benefit from psychotherapy. Kemp et al. provide a rationale for the use of CBT as a treatment for depression co-morbid with a medical condition such that "Disabling conditions cause psychological distress, decrements in enjoyable activities, loss of independence and lowered self-esteem. These factors are often the critical elements leading to a depressive disorder and they can be aided through psychotherapy." This rationale fits perfectly for CBT for people with PSD.

The first study to evaluate CBT as a treatment for post-stroke depression was a small pilot study by Lincoln, Flannaghan, Sutcliffe, and Rother (1997) Treatment was delivered to 19 participants out of a possible 155 identified as depressed from hospital records. All participants in this study were diagnosed with major depression following a stroke. In this study each participant acted as their own control during baseline and within-treatment assessments as treatment was delivered using standard AB design methods. Participants received an average of eight sessions over a 3-month course of treatment, with three participants withdrawing after a few sessions and five participants completing treatment after five sessions. Over the course of treatment, a significant decrease in depression severity, as measured by the BDI, was observed overall, although there were no changes in levels of functional disability. Of the 19 participants receiving CBT, four participants showed consistent benefits, six showed some benefits and nine showed little if any benefit. Lincoln et al. note that those participants who experienced recurrent illnesses and strokes, and those with cognitive impairment tended to benefit less from CBT treatment. In this study, however, there are a number of limitations, such as the small number of participants and a very high level of people refusing to participate in the study. There was also a lack of experience in the delivery of CBT, and it is unclear as to whether treatment was manualised. Nonetheless, results indicated that CBT may show some promise as a treatment for PSD.

Lincoln and Flannaghan (2003) followed up the earlier work by Lincoln et al. (1997) and carried out a randomised controlled trial of CBT for PSD. Participants were randomly allocated to receive one of three options, ten sessions of CBT, ten sessions of attention control interviews with no therapeutic intervention or usual care (no treatment). In total 123 participants were recruited into this study 1, 3 and 6 months after experiencing a stroke. Only 60 participants met diagnostic criteria for depression although all participants scored above cut-off for depressive symptoms on the Beck Depression Inventory. The results were disappointing in that there were no differences between the three conditions. Lincoln and Flannaghan note a number of possible limitations to their study. The CBT therapy component may not have been optimally delivered as the therapist was inexperienced and apparently did not receive training or supervision to the level usually found in such studies. There is undoubtedly a need for further more rigorously controlled studies examining the efficacy of CBT with people with PSD.

Overall it is clear that the evidence base for the efficacy of CBT for PSD is currently limited and in need of further work (Laidlaw et al., 2003; Nicholl, Lincoln, Muncaster, & Thomas, 2002). It is clear that when one looks at the potential for CBT it seems obvious that a present-oriented, skills-enhancing, problem-solving therapy holds much promise in the amelioration of depression following a stroke and yet the evidence does not at the moment support this. Primarily the evidence base is too thin in terms of the quantity and quality of studies that have been carried out. The pioneering work by Lincoln and colleagues is very difficult to carry out and such studies that have been executed have provided some indications as to what types of interventions may hold promise. Further research is urgently needed to answer the important question of what types of interventions are most effective with what types of stroke impairments. While the need for more evidence is strong, as yet all that can be said is that CBT *should* work rather than CBT *does* work for PSD.

The Application of CBT for Post-Stroke Depression

Characteristics of CBT for PSD

CBT is particularly appropriate as an intervention for individuals with PSD because it is skills enhancing, present-oriented, problem-focused and takes into account the reality of the situation. A skills enhancing approach combined with a problem focus is beneficial because it looks at the gaps in the abilities of the individual to perform activities following a stroke and considers the strategies utilised by the individual to deal with their problems.

CBT for PSD has a number of characteristics that emphasise the problem-solving, here and now, focus with stroke survivors. It is characterised by an orientation towards:

(i) *Psychoeducation* to ensure that patients are provided with information that can help orient them to focus on realistic goals for therapy and longer term expectations for recovery.
(ii) *Collaborative Empiricism* is adopted as a method throughout therapy in order to collaboratively test out the limits of what the patient can achieve. As such CBT recognises the idiosyncratic nature of any intervention with someone with PSD.
(iii) Active *problem-solving* to deal with disabilities and to manage associated cognitive distortions, emotional responses and behavioural disturbances.
(iv) Evaluating the *nature and quality of supports* available to the person and to consider the pre-morbid nature of the primary caregiving relationship.
(v) Enhancement of the *adjustment to the new reality* of life after a stroke and coming to terms with loss whilst striving towards a realistic appraisal of the impact of the stroke and achievement of the highest possible level of functioning. (Source: Laidlaw et al. 2003).

The Application of CBT for PSD

The acute nature of the onset of stroke has consequences for the individual and for those who care for them (Hibbard et al., 1990; Laidlaw et al., 2003). Stroke is an extremely distressing and frightening experience. Unlike many other illnesses that occur in later life that result in impairment and disability, stroke often occurs out of the blue with little warning. Therefore the individual and their carers have little time to adjust to the change that stroke brings, with the consequence that there can be elevated levels of anxiety after a stroke leaving the person fearful about the future and feeling vulnerable about their increased risk of developing a further stroke (Morrison, Pollard, Johnston, & MacWalter, 2005). The increased risk of developing anxiety problems after a stroke can be understood from a psychological standpoint. It is estimated that 25% of people experiencing a first stroke will have a second stroke within 5 years (American Heart Association, 2003). When people experience a stroke they are often supplied with educational leaflets that emphasise the increased risk of developing subsequent strokes albeit with information about what an individual can do to lessen their risk, nonetheless this elevates a person's level of fearfulness and sense of vulnerability.

From the perspective of a clinician discussing this issue with many patients, it appears that after discharge from hospital, at least in the short term, fear can also be elevated in survivors of a stroke. The first night after discharge can be an uncomfortable experience for the stroke survivor as they are no longer close to the security of life saving personnel and medical equipment. Following a stroke, because they fear a further stroke is imminent, the person seeks reassurance by having a person close by to call in the event that a catastrophe happens. The net effect here is that the person with the stroke is perpetually in fear of having a further stroke, and generally overestimates the likelihood of risk. As a consequence, sleep very often becomes disturbed for both the person and their carer. The carer can very often become demoralised at the level of contact and dependency of the person with a stroke. Here, graded behavioural strategies can also be useful in helping the person with the stroke extend the time on their own by a specified number of minutes each day or week, until they become more comfortable about being on their own.

Cognitive restructuring using diary forms such as dysfunctional thought records often helps the individual to make sense of their over-estimation of the likely danger. While someone who has survived a stroke may be left with cognitive impairment, it may be that the CBT therapist may need to simplify the thought diaries used for the purpose of monitoring and evaluating thoughts. In all cases when working with someone with PSD, it is important *not* to assume that people are impaired and collaborative empiricism here would suggest that the therapist ought to adopt an open mind about what the patient can achieve.

Cognitive restructuring is a potentially important intervention as it helps individuals to develop a sense of control over their depression and an enhanced sense of recovery from their stroke (Nicholl et al., 2002). PSD patients with a perceived lack of control over their recovery had a poorer outcome and remained depressed at 6 months follow-up after a stroke (Thomas & Lincoln, 2006).

It is common for people experiencing a stroke to have a lowered sense of personal agency. People lose confidence after experiencing a stroke and may be more likely to reduce their engagement in activities pursued prior to the onset of a stroke. By reducing their activity level in this way people are more likely to become isolated and sedentary with the consequence that their depression symptoms are more likely to increase, and their stroke recovery rate slows down creating a vicious circle. Thus, depression can be an important determinant of level of morbidity after a stroke as it has the potential to interfere with an individual's attempts to benefit from physical rehabilitation (Hibbard et al., 1990). As physical rehabilitation after a stroke is physically demanding and often painful, depression symptoms like apathy and hopelessness may result in a stroke survivor prematurely giving up on rehabilitation. A common negative thought in these circumstances is 'Nothing is going to change, so why bother?' The concept of 'excess disability' may be important here. Excess disability is the proportion of disability a person has which is not determined by the actual physical or cognitive impairment of a stroke. When examined in more detail, there are very often psychological rather than physical factors that limit recovery of function.

In essence thoughts biased by depression result in the individual assuming that things will not work out for them and hopelessness can set in. Thoughts about the future are erroneous as it is evident that nobody knows the future. Acting as if one does know the future generally results in increased anxiety and increased restriction of activities. Knight (2004) notes that therapists unused to working with older people often believe that older people are unlikely to benefit from psychotherapy. Therapists unused to working with older people with physical impairments are much more likely to be seduced into thinking that there is very little that can be done to overcome the disability and handicap faced by the patient. Thus, therapists are invited to engage in CBT themselves to ensure that their clinical decision-making is not biased by therapeutic nihilism. In essence, there may be large individual variation in stroke survivors, and clinical judgement needs to be exercised at all times when working with this population in the selection of behavioural versus cognitive interventions.

CBT can be used to examine and understand the discrepancy between actuality of loss following a stroke and the idiosyncratic perception of loss (Hibbard et al., 1990). It is here that active problem-solving techniques can be useful to examine what strategies the patient use to deal with their problems or even to attempt to overcome their problems. It is not uncommon for people to feel so overwhelmed by the sense of loss that they underestimate their ability to cope, and loss characterised in this way is a cognitive distortion (Grant & Casey, 1995). This is particularly acute when people compare their current level of functioning with that prior to the onset of their stroke. This is termed a 'baseline distortion' (see Laidlaw et al., 2003), where patients fail to use a realistic baseline in assessing current level of functioning, i.e. comparing themselves now to how they were the day *after* their stroke, rather than the day *before*. In the clinical experience of this author, it has been found to be extremely useful to discuss with people with PSD the consequences of seeing their current progress as inadequate; it often results in hopelessness, apathy and frustration. It is useful to discuss with the individual what would be the effect of comparing their current level of functioning with that of their level of functioning the day after a stroke. This is more helpful if the therapist draws a graph as illustrated below in Fig. 1. Here it can be seen that outcome in this case is the same, i.e. recovery is about 75% of level of functioning. If one compares oneself to the day before a stroke, which is the trajectory of the solid line, then naturally the level of functioning is poorer. However, this is not a fair comparison to make, the person has had a stroke now, and logically the day before they did not. A fairer and more realistic comparison to make is to compare the situation now, with how they were the day after they experienced a stroke, which is reflected in the dashed line. In this way, the progress that has been made can be evaluated fairly and realistically.

Homework is an important part of any CBT treatment program, as it is how patients become their own therapists (Beck, 1995). However, given the

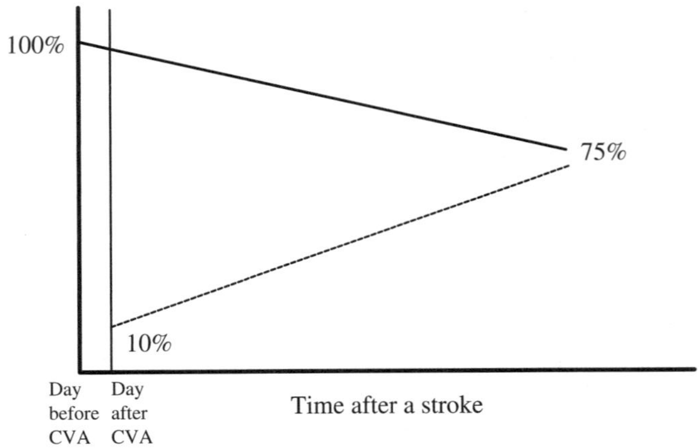

FIGURE 1. Baseline distortions after a stroke.

possibility for restriction of function following a stroke, the therapist may have to be creative in developing homework tasks so that they are short and more focused. Given also the possibility of cognitive impairment, the understanding of the patient in terms of what they are being asked to do needs to be explored in advance of the patient participating in any new activity. Patients are very often tempted to push themselves the 'extra mile' when they feel well but this results in exhaustion with the result that patients sabotage their own progress. It is important when designing homework tasks with patients that behavioural experiments are set up in a 'no-lose' way. If a patient is successful in completing a task, then this is good as it indicates how much they can do, it can be enlightening. It is also enlightening when the homework has been set at a level too difficult for the patient. In such instances, the therapist should take responsibility for this simple error. It may be that a number of elements of the task may not have been foreseen in advance of completion of the task and only by engaging in the task this can be learned. Therefore, the patient has nothing to lose from homework and everything to gain. In this regard, homework with individuals with PSD is an essential part of CBT.

Assessment and Therapy

When working with older people who have experienced a stroke there is also the possibility that they are cognitively impaired (Kneebone & Dumore; Lincoln et al., 1997). In some cases, a full neuropsychological assessment may be helpful in assessing the needs of the patient and assessing their ability to engage in CBT. There is a balance to be achieved here however as depressed individuals may underperform in cognitive testing and become more demoralised by perception of the loss of intellectual capacity.

As a way of gaining some insight into the abilities of the patient it may be useful to adopt a more ecological approach to testing a person's ability as a temporary way of assessing the need for assessment. It is not advocated that formal cognitive assessment is replaced by these procedures. It can however, be useful to ask the person with a stroke to read a newspaper. You may wish to ask the person to read out aloud the headlines of the newspaper that day, and to tell you a little bit about the story. This helps to ascertain whether the person has any visual impairment that may prevent them from reading and it also assists in assessing comprehension. Clearly these competences are necessary if you are going to be able to ask your patient to complete cognitive therapy diaries. This technique can also be used therapeutically.

Case Examples of CBT for PSD

First Case Mr. C

Mr. C was a 68-year-old retired information systems manager. He had experienced a stroke that had left him with a left sided weakness. He used to enjoy doing crosswords and now that he had experienced a stroke he assumed that he would no longer be able to complete crosswords and so had stopped attempting them as he commented that he did not want to find out how badly his brain had been affected by his stroke. No new activity had replaced this and Mr. C spent most of his time dozing. By asking Mr. C to read the newspaper and by getting him to read out aloud successively smaller size of text the therapist was able to ascertain that Mr. C was able to read the clues to the crossword. Mr. C was then encouraged to engage in this activity again. It was a surprise to Mr. C that he could read the text with no difficulties. This example illustrates the fact that in many cases using CBT with older people does mean the adoption of new hobbies and activities but merely the encouragement of the resumption of activities (Thompson, 1996).

The addition of Selection, Optimisation and Compensation (SOC) within CBT as a specific treatment component may assist an individual to successfully adjust to changed circumstances. The theory of *selective optimisation with compensation* (Baltes & Smith, 2002; Freund & Baltes, 1998) focuses on maintaining functioning in later life in the faces of challenges experienced when aging. People cope with restrictions in capacity by compensating for these losses using the strategy of optimisation and selection (Baltes & Smith, 2002; Freund & Baltes, 1998). In SOC for PSD, patients are encouraged to identify a particular problem and to reduce the repertoire of a larger set of possibilities (selection). Optimisation is the increase in skill of the completion of a predetermined activity or the greater ease at which activities are completed. Compensation results in the modification of activities in order to permit functioning to be maintained at its optimal level.

An Example Illustrating the Use of SOC in CBT for Post-Stroke Depression

Mr. S was a 71-year-old retired mechanical engineer referred by his Consultant Geriatrician because of a sense that psychological issues were preventing him making the most of his recovery from a stroke. Mr. S had in fact experienced two strokes. His first had been quite mild and he had made a very good and very quick recovery. As Mr. S had been relatively young at the time of his first stroke he had initiated a fitness regime that included going to the gym and he had gained a sense of achievement at starting exercising for the first time in his life. Mr. S had reformulated his first stroke as an event that had helped him to adopt a healthier lifestyle.

Unfortunately, Mr. S had a second stroke and his response in this instance was markedly different. Mr. S experienced his second stroke while undertaking a strenuous activity. He had been doing some heavy lifting in his garden at home when he had a stroke. Following this event, Mr. S admitted to being very fearful of exerting himself. In effect Mr. S would not do anything that raised his heart rate, so when he went into town for a shopping trip and on returning to the bus station noticed that his bus was about to leave the terminus he did not increase his rate of walking and missed his bus. Likewise when changing a light bulb, Mr. S spent the whole day completing this activity where before his second stroke he would have completed this in a matter of minutes. To achieve some resolution of this problem, the therapist used SOC to help Mr. S increase his activity level and thereby reduce his excess disability. Mr. S and the therapist targeted housework as an activity to increase.

Mr. S was very proud of his neat and tidy bungalow. As such, the therapist and Mr. S chose to use house cleaning as an activity to increase gradually. Using the SOC model, there are a number of aspects in one's repertoire of house-cleaning activities. Mr. S elected to focus on vacuum cleaning as an activity. Prior to his stroke he would clean all rooms in the bungalow that he shared with his wife in addition to also dusting and tidying up. In this scenario, the therapist and patient selected to focus on vacuum cleaning only, rather than all cleaning activities. This is an example of selection as the repertoire of cleaning activity was reduced to one element. Mr. S compensated for the fact that he had experienced a stroke and therefore had reduced physical capacity by vacuum cleaning one room only as his target. The room chosen was purposefully free from a lot of furniture to navigate around, thus he compensated by choosing a simplified version of this type of task. Mr. S optimised functioning by vacuum cleaning more frequently over time. Prior to engaging in this activity, Mr. S had reduced his cleaning activities to simple activities that did not result in any physical exertion, and introducing optimisation here meant that after one week Mr. S was encouraged to gradually build up the number of times he vacuum cleaned one room in his house. Within a few weeks, Mr. S extended his optimisation of this activity by extending his cleaning to include rooms other than the living room and eventually Mr. S was able to return to a level of activity that was compatible with a more independent level of functioning. By breaking down the strategy into the component elements of SOC, Mr. S increased his activity level in a paced and gradual way that took account of the reality of his stroke experiences and afforded him a rationale for the graded increase in activity level. Mr. S was also able to challenge his feelings that stroke had taken away his life and that he needed to conserve his activity to remain safe. It is important when engaging in any intervention that may require physical exertion that the patient has had a recent physical examination from a qualified and experienced medical practitioner. In the case of Mr. S he had been examined on a number of occasions by the Consultant Geriatrician who had originally referred him for psychological intervention.

By breaking tasks into specific components and utilising a model such as SOC, the rationale for the intervention can be shared easily with the patients and their agreement sought in advance. The situation regarding Mr. S was his level of anxiety regarding physical exertion had left him excessively disabled and isolated, and depressed. Hobbies he had enjoyed prior to his stroke such as going for walks with his dog along the seashore were now off limits because of his fear that he may 'cause' himself to have another stroke by increasing his heart rate above resting level. While it is possible that Mr. S could have another stroke when he was engaging in a physical activity, it was also possible he could experience another stroke when he was sitting in a chair at home alone. Indeed as Mr. S exhibited a number of signs and symptoms of depression at the start of treatment, this may have elevated his risk of mortality (see House et al., 2001). This case example emphasises the need to exercise clinical judgement and to seek the opinion of medical colleagues and members of multi-disciplinary teams when working with older people with PSD. It also emphasises that CBT can improve a person's mood and introduce change that individuals would find difficult to implement on their own.

Summary

The structured, present-oriented and problem-focused, nature of CBT suggests it as an obvious psychological treatment option for PSD, especially as CBT is a very adaptable form of psychotherapy for use with older people (Satre, Knight, & David, 2006). To adapt following a stroke, there are certain strategies that ought to be considered in individuals dealing with adjustment to life after stroke, these can be enhanced by the use of compensation, optimisation and selection. The needs of people with PSD have been neglected by psychotherapists and researchers and so there is now an urgent need to provide evidence of the potential effectiveness of this treatment for PSD. With conditions where there remains doubt as to whether depression is as a consequence of organic impairment, this may leave some psychotherapists feeling in doubt about the usefulness of their interventions, but when one focuses on the behavioural consequences of stroke for the individuals and their carer there is much that psychological therapy has to offer. While there is still a lot that can be understood about assessment as well as treatment for PSD, it is nonetheless important to maintain a focus on creativity in developing interventions and the use of assessment strategies. As Grant and Casey (1995) state, 'In order for CBT to be successful in treating frail elderly patients, therapists must feel free to creatively individualize the therapy.' Psychotherapy can and should be focused upon meeting the needs of depressed and disabled older people so that every older person is given the opportunity to live as independently as possible and to benefit as much as possible from rehabilitation interventions.

References

Aben, I., Verhey, F., Lousberg, R., Lodder, J., & Honig, A. (2002). Validity of the beck depression inventory, hospital anxiety and depression scale, SCL-90, and Hamilton depression rating scale as screening instruments for depression in stroke patients. *Psychosomatics, 43*, 386–393.

American Heart Association. (2003). *Heart disease and stroke statistics – 2004 update.* Dallas, TX: American Heart Association.

American Psychiatric Association. (2000). *Diagnostic and statistical manual of mental disorders*(4th ed.), Text revision (DSM-IV-TR). Washington, DC: American Psychiatric Association.

Anderson, C. S., Hackett, M. L., & House, A. O. (2004). Interventions for preventing depression after stroke. *Cochrane Database of Systematic Reviews, 2*. Art. No.: CD003689. doi:10.1002/14651858.CD003689.pub2

Baltes, P. B., & Smith, J. (2002). *New frontiers in the future of aging: From successful ageing of the young old to the dilemmas of the fourth age.* Paper presented at the Valencia Forum, Researchers, Educators and Providers contribution to the Second World Assembly on Ageing Valencia, Spain.

Beck, J. S. (1995). *Cognitive therapy: Basics and beyond.* New York: Guilford Press.

Beck, A. T., Steer, R. A., & Brown, G. K. (1996). *The Beck depression inventory-II.* San Antonio, TX: Psychological Corporation.

Bruce, M. L. (1999). The association between depression and disability. *American Journal of Geriatric Psychiatry, 7*, 8–11.

Carney, R. M., & Freedland, K. E. (2002). Psychological distress as a risk factor for stroke-related mortality. *Stroke, 33*, 5–6.

Carson, A., MacHale, S., Allen, K., Lawrie, S., Dennis, M., House, A., et al. (2000). Depression after stroke and lesion location: A systematic review. *Lancet, 356*, 122–126.

Casper, M. L., Barnett, E., Williams, G. I., Jr., Halverson, J. A., Braham, V. E., & Greenlund, K. J. (2003). *Atlas of stroke mortality: Racial, ethnic, and geographic disparities in the United States.* Atlanta, GA: Department of Health and Human Services, Center for Disease Control and Prevention.

Freund, A. M., & Baltes, P. B. (1998). Selection, optimization, and compensation as strategies of life management: Correlations with subjective indicators of successful aging. *Psychology and Aging, 13*, 531–543.

Grant, R. W., & Casey, D. A. (1995). Adapting cognitive behavioral therapy for the frail elderly. *International Psychogeriatrics, 7*, 561–571.

Hackett, M. L., Anderson, C. S., & House, A. O. (2004). Interventions for treating depression after stroke. *Cochrane Database of Systematic Reviews, 3*. Art. No.: CD003437. doi:10.1002/14651858.CD003437.pub2

Hackett, M. L., Yapa, C., Parag, V., & Anderson, C. S. (2005). Frequency of depression after stroke: A systematic review of observational studies. *Stroke, 36*, 1330–1340.

Hibbard, M. R., Grober, S. E., Gordon, W. A., Aletta, E. G., & Freeman, A. (1990). Cognitive therapy and the treatment of poststroke depression. *Topics in Geriatric Rehabilitation, 5*, 43–55.

House, A. (1987). Mood disorders after stroke: A review of the evidence. *International Journal of Geriatric Psychiatry, 2*, 211–221.

House, A., Knapp, P., Bamford, J., & Vail, A. (2001). Mortality at 12 and 24 months after stroke may be associated with depressive symptoms at 1 month. *Stroke, 32*, 696–701.

Kemp, B. J., Corgiat, M., & Gill, C. (1991/1992). Effects of brief cognitive-behavioral group psychotherapy on older persons with and without disabling illness. *Behavior, Health and Aging, 2*, 21–28.

Kinsella, K., & Velkoff, V. A. (2001). *US Census Bureau, Series P95/01-1, an aging world: 2001*. Washington, DC: US Government Printing Office.

Kneebone, I. H., & Dunmore, E. (2000). Psychological management of post-stroke depression. *British Journal of Clinical Psychology, 39*, 53–65.

Knight, B. G. (2004). *Psychotherapy with older adults* (3rd ed.). Thousand Oaks: Sage Publications.

Laidlaw, K. (2006). Psychological treatment for depression and anxiety in older adults. In C. P. Freeman & M. J. Power. (Eds.), *The handbook of evidence-based psychotherapy*. Chichester: John Wiley & Sons Ltd.

Laidlaw, K., Thompson, L. W., Dick-Siskin, L., & Gallagher-Thompson, D. (2003). *Cognitive behaviour therapy with older people*. Chichester: John Wiley & Sons Ltd.

Laidlaw, K., Thompson, L., & Gallagher-Thompson, D. (2004). Comprehensive conceptualisation of cognitive behaviour therapy for late life depression. *Behavioural and Cognitive Psychotherapy, 32*, 1–8.

Lincoln, N. B., & Flanagan, T. (2003). Cognitive behavioral psychotherapy for depression following stroke: A randomized controlled trial. *Stroke, 34*, 111–115.

Lincoln, N. B., Flannaghan, T., Sutcliffe, L., & Rother, L. (1997). Evaluation of cognitive behavioral treatment for depression after stroke: A pilot study. *Clinical Rehabilitation, 11*, 114–122.

Lincoln, N. B., Nicholl, C. R., Flannaghan, T., Leonard, M., & Van der Gucht, E. (2003). The validity of questionnaire measures for assessing depression after stroke. *Clinical Rehabilitation, 17*, 840–846.

Morrison, V., Pollard, B., Johnston, M., & MacWalter, R. (2005). Anxiety and depression 3 years following stroke: Demographic, clinical and psychological factors. *Journal of Psychosomatic Research, 59*, 209–213.

Narushima, K., Kosier, J. T., & Robinson, R. G. (2003). A reappraisal of poststroke depression, intra- and inter-hemispheric lesion location using meta-analysis. *Journal of Neuropsychiatry and Clinical Neuroscience, 15*, 422–430.

National Audit Office. (2005). *Reducing brain damage: Faster access to better stroke care*. Report by the Comptroller and Auditor General, Department of Health (UK) HC 452.

Nicholl, C. R., Lincoln, N. B., Muncaster, K., & Thomas, S. (2002). Cognitions and post-stroke depression. *British Journal of Clinical Psychology, 41*, 221–231.

Satre, D., Knight, B. G., & David, S. (2006). Cognitive-behavioral interventions with older adults: Integrating clinical and gerontological research. *Professional Psychology: Research and Practice, 37*, 489–498.

Thomas, S. A., & Lincoln, N. B. (2006). Factors relating to depression after stroke. *British Journal of Clinical Psychology, 45*, 49–61.

Thompson, L. W. (1996). Cognitive-behavioral therapy and treatment for late life depression. *Journal of Clinical Psychiatry, 57*(Suppl. 5), 29–37.

Turner-Stokes, L., & Hassan, N. (2002). Depression after stroke: A review of the evidence base to inform the development of an integrated care pathway. Part 1. Diagnosis, frequency and impact. *Clinical Rehabilitation, 16*, 231–247.

United Nations. (2003). *World population prospects: The 2002 revision*. New York: United Nations Population Division.

Visser-Meily, A., Van Heugten, C., Post, M., Schepers, V., & Lindeman, E. (2005). Intervention studies for caregivers of stroke survivors: A critical review. *Patient Education and Counseling, 56*, 257–267.

WHO. (1980). *International classification of impairments, disabilities and handicaps: A manual of classification relating to the consequences of disease*. Geneva: World Health Organisation.

16
Cognitive Behavioral Therapy for Older Adults with Bipolar Disorder

Robert Reiser, Diana Truong, Tam Nguyen, Wendi Wachsmuth,
Rene Marquett, Andrea Feit, and Larry W. Thompson

Bipolar disorder, a chronic and often debilitating mental health illness, can involve recurrent episodes of either mania or depression throughout the lifespan (Goodwin & Jamison, 1990). A large percentage of patients with bipolar disorder, have been hospitalized psychiatrically at least once; more than half have been hospitalized twice or more (Lish et al., 1994). The financial burden of this disorder for patients, their families, and society at-large, is extremely high, estimated by some to exceed $45 billion a year (Wyatt & Henter, 1995). Despite extensive research on bipolar disorder in younger individuals, there is limited empirical data pertaining to symptomatology, course, risk factors, and treatment in older adults (Bartels, Forester, Miles, & Joyce, 2000).

The lifetime prevalence of bipolar I disorder is 0.4–1.6% (American Psychiatric Association (APA), 2000), and increases roughly to 7%, when other bipolar spectrum disorders are considered (Rihmer & Angst, 2005). In the elderly, the prevalence of bipolar I disorder is 0.1% (Weissman et al., 1988), and 5–19% of older adults with a mood disorder have bipolar I disorder (Cassano, McElroy, Brady, Nolen, & Placidi, 2000). The lowered prevalence rate of bipolar I disorder in older adults, compared to younger age groups, may be explained by mortality, particularly increased suicide rates (Aizenberg, Olmer, & Barak, 2006; Dhingra & Rabins, 1991; Frank & Thase, 1999), and misdiagnosis (Judd & Kunovac, 1998; Umpathy, Mulsant, & Pollock, 2000). While there are reports that roughly 10% of all individuals with bipolar disorder develop it after age 50 (Sajatovic, 2002), this is likely to be an underestimate, given that it is often confused with dementia, delirium, and agitated depression in older individuals (Charney et al., 2003; Judd & Kunovac; McDonald, 2000). The gender ratio for younger adults diagnosed with bipolar I disorder is approximately equal; however, among older bipolar inpatients, the ratio of women to men is approximately two to one, which again may be due to increased mortality in the males. The rate of bipolar I disorder is fairly constant across ethnicities and cultures (APA).

Clinical Presentation in Older Adults

Bipolar I disorder is classified as "the occurrence of one or more Manic Episodes or Mixed Episodes (APA, 2000, p. 382)." There are currently four types of bipolar disorders classified in the *DSM-IV-TR*: Bipolar I Disorder; Bipolar II Disorder; Cyclothymic Disorder; and Bipolar Disorder Not Otherwise Specified (APA). In this chapter we will focus on bipolar I disorder – the most severe form, which most often results in hospitalization and severe functional impairments. While the diagnosis for bipolar I disorder can be made with the occurrence of one manic episode without any depressive episodes, it is much more common that individuals have at least one Major Depressive Episode (APA).

Bipolar disorder may present as early as the second decade of life; or, it may be identified in patients whose first episode of depression at a younger age is followed by a late life occurrence of mania, (Umpathy et al., 2000). Bipolar disorder can also be experienced secondary to neurological disorders. In such cases, mania usually accompanies or closely follows a first episode of depression or constitutes a patient's first affective episode altogether (Shulman, Tohen, Satlin, Mallya, & Kalunian, 1992).

Although depressive episodes appear more similar across the life span, the manic phase of bipolar disorder often presents atypically in adults over age 60 (Young & Falk, 1989). Some atypical signs of mania include mixed states, consisting of concurrent manic and dysphoric symptoms, and less intense hyperactivity and sexual drive, along with greater subsyndromal symptomatology. These variations in later-life mania make it more difficult to rely upon the DSM-IV-TR criteria for the diagnosis of mania in the older adult population (Forester, Antognini, & Stoll, 2006). Manic episodes in patients 60 years and older are manifested more often by irritability and agitated mood than euphoria or grandiosity, as well as featuring more paranoia, cognitive impairment, and frequent mixed states. Confusion, increased apathy, agitation, disorientation, and distractibility are often apparent (McDonald & Nemeroff, 1998), thus making it difficult to distinguish mania from organic conditions and schizophrenia (Charney et al., 2003; Judd & Kunovac, 1998).

Most older adults with bipolar I disorder first experience mania between 15 and 45 years of age (Almeida & Fenner, 2002), with only 10% having their first manic episode after the age of 50 (Yassa, Nair, & Iskandar, 1988). Earlier-onset bipolar I disorder is less likely to have medical and neurological comorbidity, less likely to have psychotic features, and more likely to have a family history of bipolar disorder. The late-onset group is more likely to have psychotic features and greater mortality rate in the decade following the episode, which suggests that earlier-onset and later-onset bipolar I disorder in older adults may have different etiologies (Cassidy & Carroll, 2002; Moorhead & Young, 2003; Sajatovic, Blow, Ignacio, & Kales, 2005). Older adults are particularly at increased risk for late-onset mania when there is an accompanying medical illness (Bartelset al., 2000), which can include delirium and dementia (Forester et al., 2006).

Assessment

Depression

The assessment of depression is covered in other chapters (cf. Moss & Scogin, Chap. 1) and therefore will not be discussed here.

Mania

The presence and severity of manic symptoms can be assessed by both clinician-rated and patient self-report scales noted below. One of the most widely used methods of assessing level of manic symptoms is the *Young Mania Rating Scale* (YMRS), which is a clinician-rated scale consisting of 11 items (Young, Biggs, Ziegler, & Meyer, 1978). Although not reported in detail here, psychometric analysis of the scales mentioned here suggest adequate reliability and validity. The *Altman Self Rating Mania Scale* (ASRM) was developed as a self-report rating for inpatient populations, designed to be brief and in accordance with the DSM-IV criteria of mania (Altman, Hedeker, Peterson, & Davis, 1997). The *Bech–Rafaelsen Mania Scale* is a clinician-rated scale consisting of 11 items, each scored on a 5 point Likert scale (Bech, 2002; Bech, Baastrup, de Bleeker, & Ropert, 2001). The *Clinical Global Impressions scale-BP* (CGI-BP) is a version of the CGI that was created to address criticisms of the CGI in order to assess bipolar illness, specifically (Spearing, Post, & Leverich, 1997).

There are also measures that focus on a review of the history of this disorder, such as the National Institute of Mental Health-Life Chart Method (NIMH-LCM: Denicoff et al., 2002), and the MOODS-SR, a self-report version of the Structured Clinical Interview for mood disorders (SCI-MOODS). The NIMH-LCM has several versions, including a retrospective life chart, utilizing monthly charting for retrospective analysis, and a prospective version using daily charting. Because it gives the clinician an overall picture of mood fluctuation in bipolar disorder, it is particularly well suited to the monitoring of patients throughout treatment. The SCI-MOODS interview captures the full spectrum of bipolar illness symptoms more successfully (Cassano, Dell'Osso, & Frank, 1999). A self-report version (MOODS-SR) has good agreement with the SCI-MOODS with correlation coefficients ranging from 0.88 to 0.97 (Dell'Osso, Armani, & Rucci, 2002).

Also critical to the assessment of an older adult with a bipolar spectrum disorder is the administration of additional tests, which address specifically the cognitive impairment that may be present in older adults experiencing a manic or hypomanic episode. In a recent review, Young, Heo, Schulberg, & Alexopolous (2006) reported that older adults with bipolar illness have lower scores on the MMSE and the Mattis Dementia Rating Scale (DRS; Mattis, 1988) reference. Concurrent alcohol or drug abuse with an older adult who has a bipolar spectrum disorder is also something that clinicians must screen for. A useful instrument for this is the Michigan Alcoholism Screening Test-Geriatric (MAST-G: Beullens & Aertgeerts, 2004).

Treatment Approach

While psychopharmacological interventions are the first line of accepted treatment, ongoing psychotherapy and/or psychosocial rehabilitation is now considered an equally essential component in a comprehensive treatment model for individuals with bipolar disorder (Agronin & Maletta, 2006; Swartz & Frank, 2001). Even with optimized psychopharmacological intervention alone, individuals with bipolar disorder may experience relapse at rates of 50% at 1 year and 70% at 5 years after a manic episode (Perry, Tarrier, Morriss, McCarthy, & Limb, 1999).

Pharmacotherapy

To date, there are no known randomized controlled studies examining the treatment of geriatric mania (Agronin & Maletta, 2006). Results from younger population studies suggest that the first choice of medication for mania is lithium with anticonvulsants, and antipsychotics being a second line treatment option (Gildengers et al, 2005; Sajatovic, 2002). A detailed discussion of medication use is beyond the scope of this chapter. The reader is referred to the references above, along with a detailed review by Post and Altshuler (2005).

Psychosocial Treatment

The course of bipolar disorder is influenced by a variety of psychosocial factors including family communication patterns, negative life events, dysfunctional attitudes, and disruptions in social rhythms (Otto, Reilly-Harrington, & Sachs, 2003). Psychosocial treatments that are reasonably well structured with a strong psychoeducational component, such as cognitive-behavior therapy (CBT) and interpersonal and social rhythm therapy (IPSRT) are likely to be well suited for the treatment of individuals with bipolar disorder. While there is considerable evidence to support their use with adult bipolar patients, to our knowledge, there are no randomized clinical trials evaluating the effectiveness of psychosocial treatments with older adults. Treatment of bipolar disorder in the geriatric population must be largely extrapolated from current interventions employed for the adult population. Clinicians must rely on mixed aged studies, uncontrolled clinical studies, and case reports for making treatment decisions with older adults (Sajatovic et al, 2005).

Psychosocial interventions for individuals with bipolar disorder are aimed at decreasing both syndromal and subsyndromal symptoms and increasing quality of life. Recent studies of psychosocial treatments for bipolar disorder have focused on teaching individuals specific behavioral and cognitive strategies to decrease recurrent, acute episodes of depression and mania, and increase treatment compliance (Lam et al., 2003). Colom et al. (2003) integrated early detection of prodromal symptoms, the enhancement of treatment compliance, and the induction of lifestyle regularity in a group-based psychoeducational treatment program. Teaching individuals to identify and cope with their idiosyncratic prodromes in cognitive-behavioral therapy has become an important focus in randomized clinical trials (e.g., Lam et al., 2003). Lam et al. (2003) developed a comprehensive individual treatment which focused on how to identify prodromes, develop specific coping strategies, and maintain social rhythm stability. Currently, both individual and group-based psychosocial interventions for individuals with bipolar disorder integrate the psychosocial techniques described above.

The Role of Social Rhythm Stability in Reducing Episodes

Studies have also shown that the course of bipolar disorder is sensitive to changes in circadian rhythm cycles, including long distance traveling and jet lag (Lam & Wong, 1997). Malkoff-Schwartz et al. (1998) found that among individuals with Bipolar I, changes in daily routines or sleep/wake cycles have the potential to bring about manic (but not depressive) episodes. In order to address this vulnerability, Frank, Swartz, & Kupfer (2000) developed IPSRT, a treatment which focuses on teaching individuals about the relationship between stress, the environmental context, and the onset of mood disorder symptoms and encourages them to maintain stable sleep/wake cycles and daily routines.

A Conceptual Model for Psychosocial Treatment of Older Adults

CBT with modifications tailored for the elderly is effective in treating unipolar depression in this age group (cf. Laidlaw, Thompson, Dick-Siskin, & Gallagher-Thompson for a summary, 2003). Very little work has been done using this or other psychosocial models with older bipolar patients. However, as noted above, the effectiveness of a CBT model for treating young adult bipolar patients is well established, and it is reasonable to assume that with appropriate age-sensitive adjustments CBT would also be effective with elderly bipolar patients. Preliminary clinical work in our laboratory provides strong support for this assumption (Nguyen, Truong, Feit, Marquett, & Reiser, 2006). Several specific elements, which warrant consideration when treating older patients in general, are outlined in a conceptual model that influenced the development of our CBT protocol and manual for treating older depressed patients (Laidlaw, Thompson, & Gallagher-Thompson, 2003).

While the basic CBT model emphasizing the interrelationship of thoughts, mood fluctuations, behavioral reactions, and physiological responses is applicable, the following factors should also receive emphasis (a) *Cohort Beliefs* – the unique cultural and sociohistorical perspective concerning mental illness, roles, and expectations related to treatment and developing acceptable treatment goals; (b) *Transition in Role Investments* – healthy aging is marked by the need to successfully adapt to changing social roles and navigate transition and loss related to aging; (c) *Intergenerational Linkages* – intergenerational relationships can be conflict-ridden with tensions between generations centered on autonomy vs. continuity and the transmission of values; (d) *Sociocultural Context* – negative attitudes toward aging conceived as increasing vulnerability to depression and reinforcing stereotyped views of the limitations of older adults; and

(e) *Physical Health* – The impact of chronic medical conditions and comorbid medical illness and understanding the three independent components of health status: impairment, disability, and handicap.

This model for treating depression in older adults can be further adapted to bipolar disorder by considering the following age-specific factors known to be relevant to the psychosocial treatment of the disorder. (a) The impact of social role transitions and interpersonal losses, (Frank et al., 2000) that contribute to mood instability. Specifically, older adults experience higher levels of stress related to role transitions and interpersonal losses and this has been demonstrated to increase vulnerability to bipolar episodes in other populations; (b) The role of disruptions in social rhythms as posited by Frank et al. (2000) in inducing new episodes of illness applies uniquely to older adults in term of the loss of structure associated with retirement and changes in activity levels and consequent changes in sleep–wake routines; and (c) The need to "grieve for the healthy self" (Frank et al., 2000). Initially conceived of as primarily related to the impairment, loss of function and limitations imposed by the disorder itself, this concept expanded can include the impact of other physical health and medical problems due to the known medical comorbidity of older adults with bipolar disorder.

Socializing Older Adults to Cognitive Behavioral Therapy

As noted above, traditional CBT modalities need to be modified to address the specific needs of older adults with bipolar disorder. In the formulation stage, clinicians should be aware of the unique issues older bipolar adults may present such as medical comorbidity, cognitive deficits, physical difficulties, decreased social support, social isolation, social rhythm disruption due to loss of structured activities, and transition in role function. Older adults may present to therapy with cohort beliefs about the formal nature of the doctor–patient relationship that are unhelpful in supporting the collaborative nature of CBT. This more traditional view of the doctor–patient relationship emphasizes the authority of the doctor over a collaborative working alliance in which the client is an active participant. The older adult client may hesitate to actively raise questions or discuss side effects of medication or disappointment over the progress of therapy thereby inadvertently limiting useful feedback to the therapist. It will be especially important in the initial sessions to focus on "resocializing" older adults to a more collaborative role in treatment and to address any problematic cohort beliefs about therapy, getting help, and the ability to make changes as an older adult (Laidlaw et al., 2003).

Adapting Cognitive Therapy to Older Adults with Physical and Cognitive Limitations

Cognitive behavioral strategies can be adapted to help older adults with bipolar disorder cope with mild cognitive deficits that may interfere with recall and retention of material in treatment. For example, the therapy agenda can be simplified to cover a limited number and range of topics, focusing on one essential problem at a time and deliberately paced to ensure that coping strategies are being fully incorporated into the older client's lifestyle.

It may also be useful to consider multimodel approaches to delivering information that incorporate several different types of processing: use of a whiteboard, use of written materials, therapy notes, and audio tapes summarizing key points. The therapist will want to ask the client to summarize key points and key understandings a few times in each session to make sure that information is delivered in a way that maximizes cognitive processing. Where memory problems or poor recall may be a significant factor, assignments can be supplemented with the use of written reminders, alarm clocks, or a detailed activity schedule or check-off list.

Declining physical health may result in decreased motor skills, fatigue, decreased pain tolerance, and visual and hearing impairments. Clients with physical impairments can benefit from a modified approach to thought records, evidence records, positive data logs, and activity schedules that are completed during the therapy session. For example, clients can fill out daily mood rating logs and other structured checklists that require a limited amount of writing. As an alternative to requiring extensive writing, tape recorders can be used to complete journal or log-based homework.

Treating Bipolar Depression

Assessing depressed moods in older adults may be complex, as some of the symptoms common to depression such as poor sleep, poor appetite, and decreased energy can be related to physical disabilities or the result of pharmaceutical or medical treatments (Winell & Roth, 2005). Laidlaw et al. (2003) review some of the myths and stereotypes and the impact of cohort specific beliefs that can complicate the treatment of cognitive symptoms of bipolar depression in older adults. The therapist is cautioned to become sensitized to counter-transference feelings about aging and the value of psychotherapy. Cognitive symptoms of depression including feelings of hopelessness, worthlessness, guilt, and suicidal ideation should be carefully reviewed and differentiated from the process of "expected normal" development in aging. The major treatment approach to treating bipolar depression is coping skills enhancement involving assisting patients in monitoring mood, identifying prodromes, and developing coping plans. As noted above, the impact of social role transitions and interpersonal losses should be considered early on in the formulation of bipolar depression.

Behavioral Strategies for Treating Bipolar Depression in Older Adults

Elderly clients often feel increasingly frustrated with the lack of accessibility and ability to do previously enjoyed activities. Specific behavioral approaches can be most useful in addressing increasing limitation and restricted activities. For example, behavioral activation combined with activity scheduling is helpful in developing structured routines for the client. This strategy is consistent with Laidlaw et al. (2003) who recommend a behaviorally oriented first line of treatment for more severe depression in older adults. For bipolar depression, the clinical presentation can have very clinically challenging vegetative and atypical features such as psychomotor retardation, agitation, insomnia, lethargy, fatigue, and overall loss of energy. Initial attempts at examining problematic cognitions may not be very successful especially when the individual experiences a high level of hopelessness and a low level of energy available for reflection and introspection. Furthermore, negative thoughts may be hard to challenge and are likely to be experienced as "valid" and truthful reflections of a very diminished sense of capability and competence associated with increasing inactivity, social isolation, and withdrawal.

In order to address this negative cycle, the therapist should consider initial use of behaviorally oriented activation strategies such as activity scheduling to increase overall activity and address decreases in pleasant events associated with role disruptions, transitions, and interpersonal losses. Reductions in activity lead to associated reductions in positive reinforcers with a "downward spiral" effect in which increasing withdrawal and isolation increasingly limit the availability of positive reinforcements in the individual's life. There is also a complex reciprocal interaction between diminished activity and a consequent diminished sense of mastery and pleasure that can activate maladaptive core beliefs about ability, competence, and sense of autonomy, etc. Often engaging in new prescribed behaviors can result in powerful behavioral tests that serve to challenge these underlying ideas and add momentum to the progress of the therapy.

Treating Mania in Older Adults

The clinical presentation of mania in older adults may incorporate restlessness, irritability, and agitation without the characteristic euphoria and grandiose thinking seen in other populations. The treatment of acute mania will be primarily addressed through the use of medications as described above to rapidly target overactivity, agitation, restlessness, and insomnia. However, once the client can be managed in an outpatient treatment setting, CBT can be rapidly initiated even if there are ongoing problems with agitation, insomnia, restlessness, "hyperpositive" thinking, or residual delusions.

The treatment of hypomania and mania in older adults incorporates many of the strategies identified above emphasizing training in mood monitoring, identification of hypomanic prodromes, and the development of coping strategies to decrease the likelihood of a severe episode with significant consequences. Again, a behavioral approach which targets reducing overstimulation and limiting maladaptive coping strategies is most likely to be effective at least initially. This can be approached in a simple fashion by asking the patient "What makes things worse when you are beginning to feel agitated, restless or overenergetic?" and "How can you best calm yourself down and relax based upon past experience?" Coping plans that are developed in collaboration with the patient should be as simple as possible, typically reducible to a three by five index card. As in the case examples below, it will be noted that sometimes, rather simple adjustments and changes can have a significant prophylactic effect in terms of preventing a full blown episode.

Case Example – Coping with Hypomania: "I'm a Superwoman"

Sally is in her early 60s with a 20-year history of bipolar disorder and a chronic physical condition that has worsened and is presenting increasing challenges in her daily life. She has enjoyed a very independent lifestyle managing her own business and likes taking control of situations and being "in charge." Over the past several years she has been experiencing increasing levels of pain and disability from her chronic inflammatory condition. During periods of overactivation associated with hypomania she is more likely to "overdo" it physically with the result that she often suffers from increased pain and depression over a period of several days or weeks as her physical condition deteriorates.

Sally reported that over the past weekend she had decided to go on a long hike with her partner. Half way through the hike, she started to develop a lot of pain. Knowing that her physical condition was causing her pain, her partner asked to stop and take a rest. Sally got upset and angry with her partner insisting that she is not lazy and that she can carry on. She reported that she kept going despite the pain and just kept ignoring the pain. She says that she wanted to be a "superwoman" and not "wimpy" and that she wanted to complete this steep hike despite very negative physical consequences. She felt that if she does not finish the hike, she would see herself as lazy, "wimpy," and incompetent. This is a problem that seemed to repeat itself in different contexts. Typically, although she recognizes intellectually that she is exceeding her limits; she ignores warning signs and ends up having a painful relapse.

Her core problem was formulated as follows: her problematic behavior (overdoing it) was associated with a belief system that prized her independence, rejected depending on others ("I have to push myself and ignore the pain"), and insisted that only complete autonomy was acceptable ("I can't be a wimp") with the outcome of not asking for help and painful overexertion. One of the key features here, ignoring or minimizing painful consequences, is also a common feature of manic/hypomanic thinking that leads to significant financial, interpersonal, and social problems. The therapists used cognitive restructuring techniques including a standard thought record to help Sally challenge her unhelpful beliefs about her invulnerability and powerfulness. A cognitive continuum intervention (ranging from "I'm a superwoman to "I'm a complete wimp") was used to help illustrate the exaggerated all or nothing features in her thinking process. As homework, Sally was asked to write down all the characteristics of a "superwoman" and a "wimp" and then to rate herself independently on each component or characteristic.

Finally, the therapists collaborated with Sally to put together a plan as to how she would address this situation differently in the future. The most salient features of the plan involved paying attention to physical warning signs of overexertion, communicating to her significant other, and asking for help; challenging her unrealistic beliefs; and, reminding herself of the negative painful consequences of ignoring warning signs of potential problems.

Case Example: Mrs. M

Mrs. M, a 60-year-old, Caucasian, divorced female, diagnosed with Bipolar I Disorder with psychotic features, was referred to our study from a mental health county clinic where she has been seen over several years for weekly outpatient individual and group psychotherapy, monthly medication, and case management. Initially high functioning, she had received a postgraduate degree in applied social research but has been unemployed for the last 12 years and on SSI. Presently, she lives independently at a rooming house and has regular contact with her daughters. At intake, she reported relative mood stability with no psychiatric hospitalizations over the past 10 years. Her other medical conditions include congestive heart failure, sleep apnea, hypertension, diabetes mellitus type 2, and a history of substance/alcohol abuse now in remission.

History

Mrs. M has an extensive and complex psychiatric history beginning in her early teens. She was initially diagnosed with schizoaffective disorder after a hospitalization for attempted suicide where she presented as severely depressed with prominent auditory hallucinations. In her early 50s, when Mrs. M was first diagnosed with Bipolar I Disorder in addition to ongoing problems with substance and alcohol abuse/dependence, she began receiving regular mental health services that included psychotherapy, case management, and pharmacological treatment. She was placed on a mood stabilizer in addition to her antidepressant. During this time, she also joined AA/NA to address her substance-related problems. Although she has been compliant with treatment, she has experienced a number of relapses in between periods of improvement. Mrs. M indicated that her manic episodes were typically precipitated by seasonal changes whereas the depressive episodes commonly followed extreme stressors in her life such as educational/occupational difficulties and relational conflicts with family and friends. Mrs. M also complained that side effects from medication significantly impacted her ability to function in her daily activities.

Current Family and Social Context

Mrs. M's primary support systems are her daughters, but she feels that she relies too heavily on them for assistance and support. As a result, she often isolates herself rather than turning to her family during times when she needs support the most. Although Mrs. M has casual friends she does not feel comfortable asking them for help. As for romantic relationships, Mrs. M has not dated since her divorce.

Specific Age-Related Issues

Mrs. M had specific age-related medical issues that further complicated her clinical picture including diabetes and hypertension that put her at higher risk for developing cerebrovascular disease. Often older adults may have a number of health problems that can make depression more difficult to treat. It is also common for older people to hold the belief that with a chronic medical condition it is "only natural to be depressed." However, depression is not an automatic consequence of physical illness. Therefore, an important part of Mrs. M's treatment plan was to inquire about her understanding of her health problems and to examine catastrophic thoughts about the consequences of her chronic conditions.

In one particular session, Mrs. M shared how she discovered that the source of her most current depressive episode was related to her sleep apnea. She reported that she held the belief, "I'm a failure" because she did not use her sleep apnea machine. However, after she reviewed the reasons for why she did not use the machine she realized that she felt claustrophobic when she wore the machine. Once she understood the reason behind her experienced difficulty in using the machine, she developed a plan in which she wore the machine for a few minutes longer each night to get use to the feeling of having something on her face while decreasing her level of anxiety. As a result, she no longer believed that she was a failure and experienced a positive mood shift.

It was also important to address possible age-related changes in her cognitive capacity. Mrs. M had several medical problems that further increased her risk of experiencing changes in her cognitive abilities. As such, it was important to discuss with her how these cognitive changes might impact her daily functioning. For example, Mrs. M was highly invested in returning to school despite several unsuccessful attempts. Upon entry into our study, one of her primary goals was to return to school and successfully obtain a degree. Consequently, a discussion around determining reasonable goals and understanding her limitations was initiated. It was helpful to learn more about her individual belief system and coping strategies as it appeared that Mrs. M had strongly adhered to the cohort belief that "if you set out to do something, you need to complete it otherwise you are a failure." Therefore, part of the treatment goal was to normalize Mrs. M's belief by explaining that her generational cohort may have shared similar beliefs, but that such perspective may no longer be shared by the general society, particularly given changes in knowledge about mental illness.

Mrs. M also struggled with the changes in her occupational and relational status. She was once an independent and highly functional individual that held a professional position, who was now dependent on SSI and family members to meet her basic needs. It was difficult for Mrs. M to understand and accept the changes in her level of functioning due to her psychiatric illness. With her children, she was embarrassed to share her difficulties and to ask for support and as a result, she was often lonely and depressed. Thus, an important element of therapy was finding ways Mrs. M could still feel important and valuable. It was pointed out to Mrs. M by a fellow group member that she consistently participated during group and often provided useful suggestions and feedback to other group members. Specifically, group members indicated that they appreciated hearing about her struggle with bipolar disorder and learned from her experiences. This helped Mrs. M begin to regain her sense of importance to her daughters, her friends, and to society.

Summary

While evidence-based approaches to the treatment of older adults with bipolar disorder are still in a developmental stage, best practices in the treatment of this disorder in older adults can clearly be discerned. Adapting well-established cognitive behavioral approaches to bipolar disorder in other age groups to older adults requires some sensitivity to the specific social and cultural factors relevant to this group. We have presented some promising case data (Nguyen, et al., 2006) and there is reason to be hopeful that targeted psychosocial treatments can reduce suffering and disability. It is our hope that investments in further research into psychosocial treatments for older adults with bipolar disorder can have a salutary effect in further expanding the therapeutic armamentarium of effective treatments and reducing stigma and barriers to treatment for this vulnerable and underserved population.

Acknowledgments We would like to express our profound gratitude for the continuing support of The Health Trust in San Jose, California. We also want to recognize the Santa Clara County Mental Health Department and the Alliance for Community Care, our community partners in a multiyear effort to improve mental health services to seriously mentally ill clients. Finally, we want to thank our clients who have taught us so much.

References

Agronin, M. E., & Maletta, G. (2006). *Principles and Practice of Geriatric Psychiatry.* Philadelphia: Lippencott, Williams, & Wilkens.

Aizenberg, D., Olmer, A., & Barak, Y. (2006). Suicide attempts amongst elderly bipolar patients. *Journal of Affective Disorders, 91*(1), 91–94.

Almeida, O. P., & Fenner, S. (2002). Bipolar disorder: Similarities and differences between patients with illness onset before and after 65 years of age. *International Psychogeriatrics, 14*, 311–322.

Altman, E. G., Hedeker, D., Peterson J. L., & Davis J. M. (1997). The Altman Self-rating Mania Scale. *Biological Psychiatry, 42*(10), 948–955.

American Psychiatric Association. (2000). *Diagnostic and Statistical Manual of mental disorders* (4th ed., text revision). Washington, DC: Author.

Bartels, S. J., Forester, B., Miles, K. M., & Joyce, T. (2000). Mental health service use by elderly patients with bipolar disorder and unipolar major depression. *American Journal of Geriatric Psychiatry, 8*(2), 160–166.

Bech, P. (2002). The Bech–Rafaelsen Mania Scale in clinical trials of therapies for bipolar disorders: A 20-year review of its use as an outcome measure. *CNS Drugs, 16*(1), 47–63.

Bech, P., Baastrup, P. C., de Bleeker, E., & Ropert, R. (2001). Dimensionality, responsiveness, and standardization of the Bech–Rafaelsen Mania Scale in the ultra-short therapy with antipsychotics in patients with severe manic episodes. *Acta Psychiatrica Scandinavica, 104,* 25–30.

Beullens, J., & Aertgeerts, B. (2004). Screening for alcohol abuse and dependence in older people using DSM criteria: A review. *Aging & Mental Health, 8*(1), 76–82.

Cassano, G. B., Dell'Osso, L., & Frank, E. (1999). A clinical reality in search of diagnostic criteria and an assessment methodology. *Journal of Affective Disorders, 54*(3), 319–328.

Cassano, G. B., McElroy, S. L., Brady, K., Nolen, W. A., & Placidi, G. F. (2000). Current issues in the identification and management of bipolar spectrum disorders in 'special populations'. *Journal of Affective Disorders, 59*(Suppl. 1), S69–S79.

Cassidy, F., & Carroll, B. J. (2002). Vascular risk factors in late-onset mania. *Psychological Medicine*, *32*, 359–362.

Charney, D., Reynolds, C., Lewis, L., Lebowitz, B., Sunderland, T., et al. (2003). Depression and bipolar support alliance consensus statement on the unmet needs in diagnosis and treatment of mood disorders in late life. *Archives of General Psychiatry, 60*, 664–672.

Colom, F., Vieta, E., Sanchez-Moreno, J., Martinez-Aran, A., Torrent, C., Reinares, M., Goikolea, J. M., Benabarre, M., & Comes, R. (2003). Psychoeducation in bipolar patients with comorbid personality disorders. *Bipolar Disorders: An International Journal of Psychiatry & Neuroscience, 6*, 294–298.

Dell'Osso, L., Armani, A., & Rucci, P. (2002). Comparison of interview (SCI-MOODS) and self-report (MOODS-SR) instruments. *Comprehensive Psychiatry, 43*(1), 69–73.

Denicoff, K. D., Ali, S. O., Sollinger, A. B., Smith-Jackson, E. E., Leverich, G. S., & Post, R. M. (2002). Utility of the daily prospective National Institute of Mental Health Life-Chart Method (NIMH-LCM-P) ratings in clinical trials of bipolar disorder. *Depression and Anxiety, 15*, 1–9.

Dhingra, U., & Rabins, P. V. (1991). Mania in the elderly: a 5–7 follow-up. *Journal of American Geriatric Society, 39*, 531–539.

Forester, B., Antognini, F. C., & Stoll, A. (2006). Geriatric bipolar disorder. In M. Agronin & G. Maletta (Eds.), *Principles and practice of geriatric psychiatry* (pp. 369–391). Philadelphia: Lippincott Williams & Wilkens.

Frank, A., & Thase, M. E. (1999). Treatment Outcome Studies. In S. L. Johnson & R. L. Leahy (Eds), *Psychological Treatment of Bipolar Disorder* (pp. 226–241). New York, NY: Guiford Press.

Frank, E., Swartz, H., & Kuper, D. (2000). Interpersonal and social rhythms therapy: managing the chaos of bipolar disorder. *Biological Psychiatry, 48*, 234–246.

Frank, E., Swartz, H. A., & Kupfer, D. J. (2000). Interpersonal and social rhythm therapy: managing the chaos of bipolar disorder <http://linkinghub.elsevier.com/retrieve/pii/S0006322300009690>. *Biological Psychiatry*, *48*, 593–604.

Gildengers, A., Mulsant, B., Begley, A., McShea, M., Stack, J., Miller, M., et al. (2005). A pilot study of standardized treatment in geriatric bipolar disorder. *American Journal of Geriatric Psychiatry, 134*, 319–323.

Goodwin, F. K., & Jamison, K. R. (1990). *Manic-depressive illness*. New York: Oxford University Press.

Judd, L. L., & Kunovac, J. L. (1998). Bipolar and unipolar depressive disorders in geriatric patients. *Mental Disorders in the Elderly: New Therapeutic Approaches, 13*, 1–10.

Lam, D., & Wong, G. (1997). Prodromes, coping strategies, insight and social functioning in bipolar affective disorders. *Psychological Medicine, 27*, 1091–1100.

Lam, D. H., Watkins, E. R., Hayward, P., et al. (2003). Psychoeducation and cognitive behavioral therapy for bipolar disorder. *Archives of General Psychiatry, 60*, 145–152.

Laidlaw, K., Thompson, L. W., Dick-Siskin, L., & Gallagher-Thompson, D. (2003). *Cognitive Behavioral Therapy with Older Adults,* Chichester: John Wiley & Sons.

Lish, J., Dime-Meenan, S., Whybrow, P. C., Price, R. A., & Hirschfeld, M. (1994). The National Depressive and Manic-depressive Association survey of bipolar members. *Journal of Affective Disorders, 31*, 281–294.

Malkoff-Schwartz, S., Frank, E., Anderson, B. P., Hilastala, S. A., Luther, J. F., Sherrill, J. R., Houck, P. R., & Kupfer, D. J. (1998). Social rhythm disruption and stressful life events in the onset of bipolar and unipolar episodes. *Psychological Medicine, 30*, 1005–1016.

Mattis, S. (1988). *Dementia Rating Scale (DRS). A professional manual.* Odessa, FL: Psychological Assessment Resources, Inc.

McDonald, W. M., & Nemeroff, K. R. (1998). Practice guidelines for diagnosing and treating mania and bipolar disorder in the elderly. *Medscape Psychiatry & Mental Health Column ejournal.*

McDonald, W. M. (2000). Epidemiology, etiology, and treatment of geriatric mania. *Journal of Clinical Psychiatry, 61*(Suppl. 13), 3–11.

Moorhead, S. R., & Young, A. H. (2003). Evidence for a late-onset bipolar I disorder subgroup from 50 years. *Journal of Affective Disorders, 73*, 271–277.

Nguyen, T., Truong, D., Feit, A., Marquett, R., & Reiser, R. (2006). Response to group-based cognitive behavioral therapy for older adults with bipolar disorder. *The Clinical Gerontologist, 30*(2), 103–110.

Otto, M., Reilly-Harrington, N., & Sachs, G. S. (2003). Psychoeducational and cognitive-behavioral strategies in the management of bipolar disorder. *Journal of Affective Disorder, 73*, 171–181.

Perry, A., Tarrier, N., Morriss, R., McCarthy, E., & Limb, K.(1999). Randomized controlled trial of efficacy of teaching patients with bipolar disorder to identify early symptoms of relapse and obtain treatment. *British Medical Journal, 318*, 149–153.

Post, R. A., & Altshuler, M. D. (2005). Risk of switch in mood polarity to hypomania or mania in patients with bipolar depression during acute and continuation trials of venlafaxine, sertraline, and bupropion as adjuncts to mood stabilizers. *The American Journal of Psychiatry, 163*, 232–239.

Rihmer, Z., & Angst, J. (2005). Mood disorders: Epidemiology. In B. J. Sadock, & A. J Sadock (Eds.), *Kaplan & Sadock's comprehensive textbook of Psychiatry* (Vol. 8). New York: Lippincott, Williams & Wilkins.

Sajatovic, M. (2002). Treatment of bipolar disorder in older adults. *International Journal of Geriatric Psychiatry, 17*, 865–873.

Sajatovic, M., Blow, F. C., Ignacio, R. V., & Kales, H. C. (2005). New-onset bipolar disorder in later life. *American Journal of Geriatric Psychiatry, 13*(4), 282–289.

Sajatovic, M., Gyulai, L., Calabrese, J., Thompson, T. R., Wilson, B. G., White, R., et al. (2005). Maintenance treatment outcomes in older patients with bipolar I disorder. *American Journal of Geriatric Psychiatry, 134*, 305–311.

Shulman, K., Tohen, M., Satlin, A., Mallya, G., & Kalunian, D. (1992). First-episode mania in late life. *The American Journal of Psychiatry, 151*, 130–132.

Spearing, M. K., Post, R. M., & Leverich, G. S. (1997). Modifications of the Clinical Global Impressions (CGI) Scale for use in bipolar illness (BP): The CGI-BP. *Psychiatry Research, 73*(3), 159–171.

Swartz, H. A., & Frank, E. (2001). Psychotherapy for bipolar depression: a phase-specific treatment strategy. *Bipolar Disorder: An International Journal of Psychiatry & Neuroscience, 3*, 11–22.

Umpathy, C., Mulsant, B. H., & Pollock, B. G. (2000). Bipolar disorder in the elderly. *Psychiatric Annals, 30*(7), 473–480.

Weissman, M. M., Leaf, P. J., Tischler, G. L., Blazer, D. G., Karno, M., Bruce, M. L., et al., (1988). Affective disorders in five United States communities. *Psychological Medicine, 18*(1), 141–153.

Winell, J., & Roth, J. A. (2005). Psychiatric assessment and symptom management in elderly cancer patients. *Oncology, 19*, 1479–1507.

Wyatt, R. J., & Hunter, J. (1995). An economic evaluation of manic-depressive illness. *Social Psychiatry and Psychiatric Epidemiology, 30*, 213–219.

Yassa, R., Nair, N. P. V., & Iskandar, H. (1988). Late-onset bipolar disorder. *Psychaitric Clinics of North America, 11*, 117–131.

Young, R. C., Biggs, J. T., Ziegler, V. E., & Meyer, D. A. (1978). A rating scale for mania: reliability, validity, and sensitivity. *British Journal of Psychiatry, 133*, 429–435.

Young, R. C., & Falk, J. R. (1989). Age, manic psychopathology and treatment response. *International Journal of Geriatric Psychiatry, 4*, 73–78.

Young, R. C., Heo, M., Schulberg, H. C., & Alexopolous, G. S. (2006). Cognitive impairment in bipolar disorder in old age: literature review and findings in manic patients. *Journal of Affective Disorder, 92*, 125–131.

17 Meaning Reconstruction in Later Life: Toward a Cognitive-Constructivist Approach to Grief Therapy

Robert A. Neimeyer, Jason M. Holland, Joseph M. Currier, and Tara Mehta

Some 18 months after her husband's death, Martha, aged 63, describes herself as "drowning in a sea of grief." Far from moving toward some form of recovery, she experiences herself as "stuck" in a futile protest against the impossibility of living without John, who had been the "compass" for her life for the past two decades. Without the special caring, attunement, and structure he provided her, Martha feels "disoriented," "unreal," as if his death is "just some sort of terrible joke" on the part of a malicious God. John's fast demise from an aggressive cancer gave her little time to adapt to the harsh reality of his impending loss, but Martha confesses that she spent the majority of this "warning period" actively resisting the knowledge of his eventual death, just as she continues to resist the full emotional implications of his absence. Now, she feels deeply lonely and "cut-off" from others, with the exception of her concerned adult son and daughter, and is caught up in an angry "fight" with John's children by a previous marriage about the estate. Because the dispute arose partly from critical ambiguities in his will, Martha confesses that she also feels betrayed by John, and wonders whether his apparent love for her was really "a lie." She therefore feels as if his death not only deprived her of a hoped-for future with John in retirement, but also eroded a cherished view of their past. Tearfully, she describes how she has "no purpose for living" since John's death, and as her health has begun to suffer as a result of the stress of the loss, she finds herself wishing that it were she, rather than he, who had died.

In many respects Martha exemplifies the diagnostic category of *complicated grief*, a condition whose coherence, correlates, and consequences have received increasing scrutiny over the past decade in both the psychological and psychiatric literatures (Prigerson & Jacobs, 2001).[1] Our goals in the present chapter are to orient the reader to the data on the incidence of such complications in bereavement,

[1] Note that this diagnosis, currently being reviewed for inclusion in the DSM V, is in the process of being relabeled "prolonged grief disorder" to emphasize its persistence beyond the period of normal grief. However, we have retained the term "complicated grief" to maintain continuity with the present published literature.

especially among older adults, and to sketch how they might be assessed and treated from a broadly cognitive-constructivist viewpoint, one centered in the human need to reestablish continuity and meaning in a life story that has been challenged by loss (Neimeyer, 2001).

Background: Pathways Through Bereavement

If bereavement were itself a disorder, then its lifetime incidence would approach 100%. It is sobering to reflect that ultimately, every person, every place, every project, and every possession that we love, we will someday lose – at least in an earthly sense. The accumulating succession of such losses throughout life, culminating in the loss of all that remains at the point of our own death, repeatedly challenges us as we age, often in the form of the normative losses of loved ones through natural, anticipated causes. At any time, there are over 12 million widowed persons in the US, with 800,000 new cases added annually. By the age of 65, one half of all women and 10% of all men will have been widowed at least once, and these figures will rise to 80 and 40%, respectively, by age 85 (Rosenzweig, Prigerson, Miller, & Reynolds, 1997). As striking as these numbers are, they dramatically underestimate the extent of bereavement as a psychosocial transition in later life, as the loss of siblings and friends exceeds spousal bereavement by a factor of three-to-one and nine-to-one, respectively (Hays, Gold, & Peiper, 1997). Other losses are less normative, and as evidence suggests, more difficult for survivors, such as the "off-time" death of an adult child suffered by 10% of elders (Moss, Moss, & Hansson, 2001), or the sudden or violent death of a loved one through accident, suicide, or homicide (Currier, Holland, Coleman, & Neimeyer, 2007). If only because such losses grow increasingly frequent as we age – and simultaneously deprive us of the support during times of adversity that the deceased might previously have provided – one might think that debilitating responses to bereavement like that suffered by Martha would be relatively common in later life.

And yet the best available evidence suggests that this is not the case (Wortman & Silver, 2001). Though the loss of a loved one in late life typically presents significant challenges of an instrumental, emotional, and social sort, recent findings from a unique prospective longitudinal study suggest that most bereaved elders ultimately adapt successfully to loss. Specifically, the Changing Lives of Older Couples study (CLOC; Carr, Nesse, & Wortman, 2006) involved over 1,500 baseline interviews with married couples from the Detroit metropolitan area who were diverse in SES and ethnicity, with the only criterion for participation being that the husband was age 65 or older. The research team then followed them closely over as many as 8 years, during which time 205 lost their spouses to death. Detailed follow-up assessments at 6 months, 18 months, and 4 years after the death measured not only survivors' bereavement-related distress, but also their marital satisfaction and coping styles prior to and after the loss, mental health histories, social support – in sum, a remarkably comprehensive range of mourner resources and liabilities for negotiating this profound psychosocial transition. Results identified a subset of bereaved older spouses who followed what was presumed to be a "common grief" trajectory, with depression rising substantially from a low preloss baseline over the first 6 months of bereavement, and then tapering off to baseline levels in the ensuing year (Bonanno, Wortman, & Nesse, 2004). Such spouses seemed to exemplify the expected pattern of adaptive grieving, finding comfort in positive memories of their partner and in many respects engaging in constructive forms of coping with the loss. Interestingly, however, this group accounted for a mere 11% of the sample, raising questions about the status of the 89% of bereaved elders who failed to follow this presumably typical trajectory. Remarkably, 45% of the widows and widowers in the study evidenced a "resilient" pattern, which was marked by only transient distress following the loss of their spouse, such that their level of depression approached their baseline 6 months after their bereavement and remained low throughout the study period. Interviews further indicated that these survivors displayed generally good adjustment across a host of measures, and were low in avoidance-oriented coping strategies, suggesting that they were not simply "denying" the reality of the loss. Another intriguing group, termed "depressed improved," were marked by extremely high levels of depression on average *three years prior* to the spouse's death, and experienced enduring *relief* following the loss, reflected in very low levels of distress in subsequent assessments. Results for the 10% of spouses who conformed to this trajectory suggested that they had experienced considerable distress during the spouse's life related to the quality of the marriage (perhaps due to an oppressive relationship or the demands of caregiving for a chronically ill partner). For them, death was less the problem than the solution to a problem, and they reported many surprising benefits of widowhood, which gave them a new lease on life.

The two remaining pathways were more worrying, however. These included a "chronic depression" subgroup that showed high distress preloss, often exhibiting long-standing marital problems and difficulties coping, and that continued to report high levels of distress 6 and 18 after the loss (Bonanno et al., 2004). For this group, accounting for about 8% of the sample, bereavement seemed to have exacerbated sustained struggles with adjustment characterized by avoidance coping, an inability to access comforting memories and other problems that could benefit less from grief therapy, per se, than from competent treatment targeting their depression (Neimeyer, 2005). Finally, 16% of widows and widowers were classified as suffering from "chronic grief." This subgroup not only failed to return to preloss levels of positive adjustment, but they also reported the highest levels of yearning and regret about the lost relationship, had the greatest likelihood of engaging in an intensive but unsuccessful attempt to find meaning in the loss, and were most likely to report a history of dependency in their attachment to their partner (Bonanno et al., 2002, 2004). These latter groups suggest a focus for clinical assessment, the concern of the next section.

Assessment

Even under normal circumstances, the loss of a loved one can represent a profound life transition that challenges the survivor's adaptation at many levels, yielding typical symptoms of shock and numbness, sadness, yearning for the deceased, anger, guilt, depression, anxiety, and a sense of purposelessness and futility about the future (Rosenzweig et al., 1997). In addition, it is worth recognizing that more intimate forms of loss, such as widowhood, imply sweeping social and behavioral adjustments that transcend these psychological symptoms. As Moss et al. (2001, p. 247) point out, "during a marriage of many years, there forms a system of roles, traditions, and interdependencies" that must be reworked in the wake of loss. In keeping with this argument, data from the CLOC study indicate that spousal loss in later life prompts a reorganization of relationships between surviving parents and adult children, decreasing the dependence of children on parents, and increasing the parents' reliance on their children, a pattern that is more accentuated for Caucasian than African-American families (Ha, Carr, Utz, & Nesse, 2006). Finally, several studies suggest that bereavement perturbs basic physiological systems, compromising sleep, neuroendocrine, and immune function for adults of all ages (Hall & Irwin, 2001; Hardison, Neimeyer, & Lichstein, 2005), although the causal pathways accounting for these effects require further research (Center for the Advancement of Health, 2004).

Beyond these general biopsychosocial challenges associated with bereavement, there is clear evidence that some survivors will experience ongoing disruption of their mood and social or occupational functioning to a degree that clinical assessment and intervention is warranted. Some of these responses can be diagnosed in terms of familiar psychiatric syndromes, such as major depression, the assessment of which in older adults has been exhaustively reviewed by Edelstein and his colleagues (Edelstein, Kalish, Drozdick, & McKee, 1999). Other responses, however, are sufficiently distinct from bereavement-related depression (with its symptoms of sad mood, anhedonia, worthlessness, and psychomotor retardation) and anxiety (marked by nervousness, sweating, and palpitations) to merit separate diagnosis. Across the last decade, evidence has built for considering complicated grief (CG) an entity in its own right, whose diagnostic criteria have been evaluated in a *DSM V* field trial providing strong evidence of their reliability and validity (Prigerson & Maciejewski, 2006). Table 1 lists the revised criteria that have survived empirical tests to determine the items yielding the most information with respect to the underlying attribute of CG and whose high performance was invariant across subgroups defined by age, sex, race, and kinship relationship to deceased.

As a comparison of these criteria with the problems displayed by Martha in the opening case study will suggest, such symptoms are readily observed in clinical practice, and can be assessed through both a structured diagnostic interview (Silverman et al., 2000) and the self-report Inventory of Complicated Grief (Prigerson & Jacobs, 2001). A large body of research indicates that bereaved persons who satisfy criteria for complicated grief are at higher risk of subsequent

TABLE 1. Criteria for complicated grief proposed for DSM-V adapted from Prigerson and Maciejewski (2006).

Criterion A. Chronic and disruptive yearning, pining, longing for the deceased
Criteria B. The person must have four of the following eight remaining symptoms at least several times a day or to a degree intense enough to be distressing and disruptive:

1. Trouble accepting the death
2. Inability to trust others
3. Excessive bitterness or anger related to the death
4. Uneasiness about moving on (e.g., making new friends, pursuing new interests)
5. Numbness/detachment (e.g., feeling disconnected from others)
6. Feeling that life is empty or meaningless without deceased
7. Feeling bleak about the future
8. Agitation (e.g., jumpiness or edginess)

Criterion C. The above symptom disturbance causes marked and persistent dysfunction in social, occupational, or other important domains
Criterion D. The above symptom disturbance must last at least 6 months

psychiatric and physical morbidity (e.g., depression, suicide ideation, high blood pressure), adverse health behaviors (e.g., increased smoking, alcohol consumption, poor sleep), and quality of life impairments (Prigerson & Jacobs, 2001).

Precisely because bereavement can lead to enduring complications in multiple spheres of the survivor's life, assessment should be (a) multidimensional, (b) appropriately timed, and (c) responsive to the character of loss and the life circumstances of the client. The first of these guidelines is underscored by the finding that, at least in the case of spousal bereavement, depression and grief follow different courses, with the former typically subsiding within a period of months, whereas the separation distress entailed in grief can persist at subclinical levels for a period of years (Thompson, Gallagher-Thompson, Futterman, Gilewski, & Peterson, 1991). As the CLOC data also illustrate (and as additional evidence bearing on treatment response to be reviewed later further attests), these two responses therefore can be discriminated in terms of their phenomenology, their characteristic trajectories, their associated coping styles, and their implications for intervention, warranting separate assessment of them as potentially comorbid conditions. A similar argument can be made with reference to other disorders that can complicate adaptation to loss, such as generalized anxiety, posttraumatic stress responses, or panic disorder (Prigerson & Maciejewski, 2006). More generally, in view of the serious health consequences associated with grief, especially in its more complicated forms, a medical assessment of physiological functioning (e.g., immune response, cardiac health) and cognitive function (as memory deficits associated with bereavement-related depression can mimic dementia) is strongly advised (Edelstein et al., 1999). However, clinicians should be cautioned against regarding only pathological responses as worthy of attention. Indeed, a growing body of evidence indicates that many of the bereaved display resilience (Bonanno, 2004) or even personal growth (Neimeyer, Prigerson, & Davies, 2002) following loss. Fortunately, instruments like the Hogan Grief Reactions Checklist (Hogan, Greenfield, & Schmidt, 2001) provide valid assessments of these adaptive responses as well.

The timing of assessment also deserves consideration. Whereas some syndromes can be reliably assessed in the early weeks or months of loss (such as mood or panic disorders), complicated grief cannot be validly diagnosed until a marked pattern of impairment has been observed for at least 6 months (as the criteria dictate). This conservative guideline discourages over-diagnosis in the early aftermath of loss, when a high percentage of bereaved persons who will ultimately overcome their acute distress would endorse high levels of symptomatology. However, for the 10–15% of the bereaved who meet CG criteria for 6 months or longer, evidence suggests that the successful adaptation achieved in most uncomplicated cases is much less likely to occur over the succeeding 3 years (Ott, 2003).

Finally, assessment should be guided by the character of the loss itself as well of the client who suffers it. For example, violent losses (i.e., by suicide, homicide, and accident) have been found to generate more complicated grief symptoms than natural deaths, even when the latter are sudden and unexpected (Currier et al., 2007), and the gradual demise of a loved one in a state of progressive dementia poses unique demands upon caregivers long before the death itself occurs (Meuser & Marwit, 2001). These and other forms of bereavement might therefore call for focused assessment using any of a growing number of grief measures tailored to specific forms of loss, as extensively reviewed by Neimeyer and his colleagues (Neimeyer & Hogan, 2001; Neimeyer, Hogan, & Laurie, 2007). The increasing translation of these scales into Spanish and other common languages and greater attention to issues of culture is a further salutary trend reflected in these reviews. But even with a repertory of formal assessment tools (which can be complemented by a range of familiar measures of coping styles, social support, spiritual resources, and other relevant factors), there is clear compass for clinical acumen in evaluating distinctive themes pertaining to given losses and clients. For example, losses of siblings – the most common form of family loss for older adults – deprive survivors of their longest-lived relationships, which have often provided a sense of history and shared validation of identity across a lifetime (Hays et al., 1997). On the other hand, loss of a spouse or even an elderly parent ends the most security-enhancing relationships in adult life, posing potential problems for survivors with patterns of insecure attachment (Field, Gao, & Paderna, 2005; Neimeyer et al., 2002). Finally, death of an adult child – often by violent means – has been found to have a particularly enduring impact (Malkinson & Bar Tur, 2005) and to pose special risks to the elderly parent's mental health (Clarke & Wrigley, 2004). Delicate therapeutic inquiry into the significance of the loss and the function of the relationship to the deceased is therefore indicated. Such inquiry need not be limited to diagnostic interviews, of course, as illustrated by the "empty chair" monologues used by Field and Bonanno (2001) to investigate the adaptation of older bereaved spouses 6 months after the death of their partner. Giving each spouse 5 min of time alone to address the deceased as if he or she were present in the room, these researchers found that the degree of self-blame (e.g., about things done or not done) that could be coded in the dialogues predicted more acute grief symptomatology. In contrast, the amount of other-blame (e.g., about the death itself or treatment of the surviving spouse) was related not to grief, but to subsequent development of somatic and anxiety symptoms. These issues concerning the meaning of the loss and the nature of the (postmortem) relationship to the deceased receive further attention in the section to follow.

Conceptual Issues

The conceptual framework through which we approach the life disruptions occasioned by bereavement is provided by *constructivism*, a broadly postmodern approach to psychology that emphasizes the personal and social processes by which people impose meaning and orientation on the experiences of their lives (Neimeyer & Bridges, 2003; Neimeyer & Mahoney, 1995). From a constructivist standpoint, grief can be viewed as a struggle to reaffirm or reconstruct a world of meaning that has been challenged by loss (Neimeyer, 2001). Converging evidence from several investigators has linked an inability to do so with intense and protracted grieving (see Gillies & Neimeyer, 2006, for a review), whereas an ability to "make sense" of a loss in spiritual, philosophical, or practical terms predicts more favorable bereavement outcomes (Center for the Advancement of Health, 2004; Davis, Nolen-Hoeksema, & Larson, 1998). In fact, a recent study of an ethnically diverse sample over 1,000 adults bereaved by the natural or violent deaths of their loved ones demonstrates that sense-making served as a nearly "perfect" mediator of the impact of suicide, homicide, or accident on the complicated grief symptomatology of survivors, even when compared with unexpected natural deaths that shared the element of suddenness (Currier, Holland, & Neimeyer, 2006). Additional research suggests that for many mourners bereavement adaptation entails maintaining an adaptive continuing bond of attachment with the one who has died (Rubin, 1999), at least at later phases of bereavement (Field & Friedrichs, 2004). Thus, viewing complicated grief through the twin lenses of attachment theory (Bowlby, 1980) and meaning reconstruction (Neimeyer, 2001) could provide an integrative frame for understanding not only problematic adjustment to loss, but also resilience and growth (Neimeyer, 2006).

A particularly apropos theoretical contribution of constructivism to bereavement research and practice is the concept of *narrative*, the neurologically, psychologically, and socially instantiated process by which human beings "emplot" events, attributing to them organization and thematic meaning. Nowhere is this narrative activity more evident or important than in the formulation of a *self-narrative*, defined as "an overarching cognitive–affective–behavioral structure that organizes the 'micronarratives' of everyday life into a 'macronarrative' that consolidates our self-understanding, establishes our characteristic range of emotions and goals, and guides our performance on the stage of the social world" (Neimeyer, 2004, pp. 53–54). From this perspective, identity can be seen as a narrative achievement, as our sense of self is established through the stories that we tell about ourselves, the stories that relevant others tell about us, and the stories we enact in their presence. Profound "psychosocial transitions" such as bereavement (Parkes, 1993), however, perturb the coherence of this self-narrative, presenting survivors with difficult-to-integrate losses and sometimes invalidating the fundamental assumptions of predictability, fairness, and control on which our life stories rely (Janoff-Bulman, 1992). In this context, postloss resilience can be understood in terms of the successful *assimilation* of bereavement into still viable and essentially positive preloss narrative structures, as when the death of a spouse in late life is viewed as a sad, but anticipated expression of the natural order, or as an expression of "God's will" for a deeply devout survivor. Likewise, posttraumatic growth (Calhoun & Tedeschi, 2006) can be viewed as result of significant *accommodation* in the form of reorganization, deepening, or expansion of the self-narrative to be adequate to the changed life of the survivor. Alternatively, bereavement that merely exacerbates an existing depressive outlook can be construed in terms of assimilating the loss into a previous, but predominantly negative life story. Finally, failure to either assimilate or accommodate to the reality of the loss can leave the individual with only the remnants of a "fragmented story" that no longer makes sense in the present (Currier & Neimeyer, 2006). The conceptualization of these adverse outcomes accords with a recently advanced cognitive-behavioral model of complicated grief as an inability to integrate the loss into existing autobiographical memory structures (Boelen, van den Hout, & van den Bout, 2006), a circumstance that clearly calls for therapeutic intervention. It is to this topic that we now turn.

Treatment

Until recently, clinicians working with the bereaved faced a conundrum when they shifted from the assessment of life-limiting or perhaps even life-threatening responses to loss to attempts to provide psychotherapy to those who displayed them. Specifically, meta-analysis of controlled outcome studies suggested that existing treatments were manifestly ineffective for those to whom they were offered, displaying effect sizes of one-eighth the size of conventional psychotherapy for most conditions (Kato & Mann, 1999). Though few outcome studies exclusively focus on bereaved elders, those that do seem to follow a similar pattern of showing small to no effects (Lund & Caserta, 1992; Reich & Zautra, 1989) or in some cases negative effects (Sabatini, 1988–1989). Significantly, these therapies, typically offered in a group format to bereaved adults irrespective of their level of distress or symptomatology, consisted mainly of a psychoeducational grab-bag of ideas for coping with grief and opportunities for emotional disclosure and catharsis. It therefore seems possible that many of these studies yielded disappointing results because they (a) failed to use inclusion criteria to identify chronic and severe sufferers and (b) employed interventions that were not based on a coherent and empirically informed theory of grief (Jordan & Neimeyer, 2003).

Recent studies have attempted to overcome these limitations by using stricter inclusion criteria and developing theoretically relevant interventions, including a prominent narrative component. Along these lines, two studies deserve mention. First, in 2005, a 16-week therapy termed complicated grief treatment (CGT) was tested against interpersonal psychotherapy (IPT) in a randomized clinical trial with 95 Caucasian and African-American participants who met criteria for CG (Shear, Frank, Houck, & Reynolds, 2005), many of whom were older adults. Guided by the Dual Process Model of bereavement (Stroebe & Schut, 1999), therapists in the CGT condition promoted the dual goals of helping clients both process their loss and seek restoration in a changed world that required the development of new life goals. Key interventions included not only psychoeducation about oscillating attention to these two processes, but also distinctive therapeutic procedures such as "revisiting," in which the client retold, recorded, and reviewed the story of the loss, and "imaginal conversations," in which a renewed connection to the deceased was promoted through imaginary dialogue with the lost loved one. Further work was done to consolidate positive memories of the relationship and to project toward a changed future. IPT followed the usual procedures for this empirically supported treatment (Weissman, Markowitz, & Klerman, 2000), linking symptoms of grief to interpersonal problems and working toward a realistic view of the deceased and the development of satisfying relationships. Clients in both conditions showed improvement over time. Significantly, however, CGT was shown to be superior to IPT in reducing symptoms of CG and improving participants' work and social adjustment. CGT also produced a faster rate of improvement in CG symptoms, although the two treatments yielded comparable outcomes on measures of depression and anxiety. Though preliminary, these findings are encouraging, particularly in light of the fact that therapists in this study had a "strong allegiance" to IPT (Shear et al., p. 2603).

Not all older adults can conveniently avail themselves of face-to-face therapy, however, especially when problems with health or transportation militate against doing so. However, a recent survey of mostly retired, widowed people revealed that about half of participants used the Internet and e-mail, which was shown to be a valuable form of social support that might protect against secondary symptoms of CG (e.g., depression, posttraumatic stress; Vanderwerker & Prigerson, 2004). In keeping with this finding, Wagner and her colleagues (Wagner, Knaevelsrud, & Maercker, 2006) compared an individualized 5-week Internet-based cognitive-behavioral therapy (CBT) for CG to a wait-list control group with 55 clients exhibiting symptoms of CG. Specifically, treatment consisted of a series of carefully tailored writing assignments with therapist feedback, focusing on (a) recalling and recording the story of the loss in all of its most painful sensory and emotional details, (b) drafting a letter to an imaginal friend with a similar loss to foster perspective, compassion, and meaning attribution, (c) integrating positive memories of the loved one into ongoing life, (d) redefining identity, (e) sharing the story of the loss with a sympathetic other, and (f) projecting new goals. At posttreatment, clients who received these therapeutic interventions were substantially better off than those in the wait-list condition across a variety of dimensions, particularly on measures of intrusive thoughts and avoidance. Taken together, these two studies suggest that a systematic use of narrative procedures – whether oral or written – can help clients with complicated grief to confront and integrate the most painful aspects of their loss into their life stories while also consolidating sustaining memories of the relationship and reconstructing a changed sense of themselves in the future. Detailed semistructured interviews, loss characterizations, biographical methods, metaphoric and poetic techniques, guidelines for memory books, and numerous additional narrative procedures for promoting these goals have been described and illustrated elsewhere (Neimeyer, 1999, 2001).

Case Illustration

As conveyed in the vignette with which this chapter opened, Martha exemplified virtually all of the diagnostic criteria for CG, with the possible exception of agitation. Keenly aware of being "unable to move on with life," she sought therapy from the first author (RAN) for an existence that for her had become unlivable. Therapy began with a review of her symptoms and formulation of her loss through an extension of her own poignant metaphors of disorientation (e.g., being "at sea" without the "compass and anchor" that John had provided), as well as encouragement to consult a physician colleague in view of her considerable weight loss and depressive symptomatology, a referral she accepted. Martha then voiced her ambivalence about the advice she had received from friends that she had to "say goodbye" to John in order to move forward, and expressed relief and intrigue at the therapist's suggestion that grieving might be less a process of "letting go" than of finding ways psychologically to continue the connection with the loved one. When she spontaneously referred to the comfort she found in memories of special, and sometimes "hilarious" times with her husband, the therapist encouraged her to record these in the form of a written story of their relationship, something she might at some point share with her grown son and daughter. Martha enthusiastically accepted this assignment, and subsequently credited it with helping her "turn a corner," feeling that the writing helped her validate key themes in her past.

In subsequent weeks Martha's mood improved as she began a course of antidepressant medication, but several problems specific to her bereavement remained to be addressed. Among these were the "void" she faced in the wake of the loss of her intimacy with John, and her fear that she could never again find someone who could love and understand her as completely. Ironically, making progress on this issue by opening herself to the possibility that other relationships could someday prove rewarding in their own right led to an unexpected complication: a brief affair with an older man that ended in his withdrawal and her disappointment. Martha then began a more cautious and considered attempt to expand her social world, focusing primarily on lunches with women friends or group activities that allowed her to maintain greater control over potential relationships until she was ready for greater closeness.

A thornier problem concerned her simmering anger at people more central to her life, including John's children with whom she was embroiled in a legal dispute regarding his estate. It soon became apparent that this anger distracted her from both the intensity of her grief and her anger at John himself for having "abandoned" her and "left her at the mercy" of "unscrupulous manipulators." Reading this anger as a painfully discordant theme that resisted integration into Martha's narrative of her relationship, the therapist then suggested a few sessions of focused work on this issue, including several experientially vivid "two chair" dialogues with John centering around her hurt and rage at him for "betraying" her by dying and leaving so much incomplete. Eventually, alternating between John's chair and her own, Martha was able to move the dialogue toward mutual understanding and forgiveness.

As therapy moved toward its final stage some 6 months after it began, Martha was able to formulate a more complex understanding of her long marriage to John, and reach an acceptance of his human foibles as well as remarkable strengths. She was also able to place both their love and her loss in the broader context of a self-narrative that affirmed her considerable resources and resilience, and moved toward a renewed spiritual conviction that God still played a constructive role in her life, as did her children. At the point of this writing she has begun to consider new career opportunities, although she continues to acknowledge waves of sadness and uncertainty in a life that nonetheless feels worth living.

In addition to the psychological approaches emphasized above, a body of literature is emerging that suggests that pharmacological interventions might address some of the secondary symptoms of CG among older adults. For example, in a study of 80 bereaved elders Reynolds et al. (1999) found that nortriptyline and nortriptyline plus IPT successfully remitted symptoms of depression for 56 and 69% of participants, respectively, a superior outcome compared to placebo with either IPT or medication management. However, nortriptyline appeared to have little impact on symptoms of grief. Research on bereaved older adults also provides evidence supporting the use of tricyclic antidepressants for enhancing activities of daily living (Oakley, Khin, Parks, Bauer, & Sunderland, 2002). In addition, other studies point to the efficacy of newer medications such as paroxetine (Zygmont et al., 1998) and bupropion (Zisook, Shuchter, Pedrelli, Sable, & Deacine, 2001). Thus, although pharmacotherapy does not seem to address the core symptoms of separation distress and disruption of a world of meaning stemming from bereavement, it does appear to have an adjunctive role to play, especially when prominent symptoms of mood disorder complicate the clinical picture.

Conclusion

Viewed through a constructivist lens, later life is shaped less by the objective reality of cumulative loss than by the way such losses are integrated into the unique life stories of survivors. In this brief chapter we have tried to sketch the outlines of an empirically informed cognitive-constructivist view of grief, and to suggest its promise in guiding relevant assessment and intervention in cases of complication. We hope that readers will join us in exploring the hopeful implications of these concepts and methods for clinical research and practice, so that the variety of challenges that loss introduces in life might one day be matched by the richness of resources for meeting them.

References

Boelen, P. A., van den Hout, M. A., & van den Bout, J. (2006). A cognitive-behavioral conceptualization of complicated grief. *Clinical Psychology, 13*, 109–128.

Bonanno, G. (2004). Loss, trauma, and human resilience. *American Psychologist, 59*, 20–28.

Bonanno, G., Wortman, C. B., Lehman, D. R., Tweed, R. G., Haring, M., Sonnega, J., et al. (2002). Resilience to loss and chronic grief. *Journal of Personality and Social Psychology, 83*, 1150–1164.

Bonanno, G., Wortman, C. B., & Nesse, R. M. (2004). Prospective patterns of resilience and maladjustment during widowhood. *Psychology and Aging, 19*, 260–271

Bowlby, J. (1980). *Attachment and loss*. New York: Basic Books.

Calhoun, L. G., & Tedeschi, R. G. (2006). *Handbook of posttraumatic growth: Research & practice*. Mahwah, NJ: Erlbaum.

Carr, D., Nesse, R., & Wortman, C. B. (2006). *Spousal bereavement in late life*. New York: Springer.

Center for the Advancement of Health. (2004). Report on bereavement and grief research. *Death Studies, 28*, 489–575.

Clarke, C. S., & Wrigley, M. (2004). Suicide-related bereavement and psychiatric morbidity in the elderly. *Irish Journal of Psychological Medicine, 21*, 22–24.

Currier, J. M., Holland, J. M., Coleman, R., & Neimeyer, R. A. (2007). Bereavement following violent death: An assault on life and meaning. In R. Stevenson & G. Cox (Eds.), *Perspectives on violence and violent death*. Amityville, NY: Baywood.

Currier, J. M., Holland, J. M., & Neimeyer, R. A. (2006). Sense-making, grief, and the experience of violent loss: Toward a mediational model. *Death Studies, 30*, 403–428.

Currier, J. M., & Neimeyer, R. A. (2006). Fragmented stories. In E. K. Rynearson (Ed.), *Violent death*. New York: Brunner-Routledge.

Davis, C. G., Nolen-Hoeksema, S., & Larson, J. (1998). Making sense of loss and benefiting from experience. *Journal of Personality and Social Psychology, 75*, 561–574.

Edelstein, B., Kalish, K. D., Drozdick, L. W., & McKee, D. R. (1999). Assessment of depression and bereavement in older adults. In P. A. Lichtenberg (Ed.), *Handbook of assessment in clinical gerontology*. New York: John Wiley & Sons, Inc.

Field, N. P., & Bonanno, G. A. (2001). The role of blame in adaptation in the first five years following the death of a spouse. *American Behavioral Scientist, 44*, 764–781.

Field, N. P., & Friedrichs, M. (2004). Continuing bonds in coping with the death of a husband. *Death Studies, 28*, 597–620.

Field, N. P., Gao, B., & Paderna, L. (2005). Continuing bonds in bereavement: An attachment theory based perspective. *Death Studies, 29*, 277–299.

Gillies, J., & Neimeyer, R. A. (2006). Loss, grief, and the search for significance. *Journal of Constructivist Psychology, 19*, 31–65.

Ha, J. H., Carr, D., Utz, R. L., & Nesse, R. (2006). Older adults' perceptions of intergenerational support after widowhood? *Journal of Family Issues, 27*, 3–30.

Hall, M., & Irwin, M. (2001). Physiological indices of functioning in bereavement. In M. S. Stroebe, R. O. Hansson, W. Stroebe, & H. Schut (Eds.), *Handbook of bereavement research* (pp. 473–492). Washington, DC: American Psychological Association.

Hardison, H. G., Neimeyer, R.A., & Lichstein, K.L. (2005). Insomnia and complicated grief symptoms in bereaved college students. *Behavioral Sleep Medicine, 3*, 99–111.
Hays, J., Gold, D., & Peiper, C. (1997). Sibling bereavement in late life. *Omega, 35*, 25–42.
Hogan, N. S., Greenfield, D. B., & Schmidt, L. A. (2001). Development and validation of the Hogan Grief Reaction Checklist. *Death Studies, 25*, 1–32.
Janoff-Bulman, R. (1992). *Shattered assumptions*. New York: Free Press.
Jordan, J. R., & Neimeyer, R. A. (2003). Does grief counseling work? *Death Studies, 27*, 765–786.
Kato, P. M., & Mann, T. (1999). A synthesis of psychological interventions for the bereaved. *Clinical Psychology Review, 19*, 275–296.
Lund, D. A., & Caserta, M. S. (1992). Older bereaved spouses' participation in self-help groups. *Omega, 25*, 47–61.
Malkinson, R., & Bar Tur, L. (2005). Long term bereavement processes of older parents: The three phases of grief. *Omega, 50*, 103–129.
Meuser, T. M., & Marwit, S. J. (2001). A comprehensive, stage-sensitive model of grief in dementia caregiving. *Gerontologist, 41*, 658–670.
Moss, M. S., Moss, S. Z., & Hansson, R. O. (2001). Bereavement and old age. In M. S. Stroebe, R. O. Hansson, W. Stroebe, & H. Schut (Eds.), *Handbook of bereavement research* (pp. 241–260). Washington, DC: American Psychological Association.
Neimeyer, R. A. (1999). Narrative strategies in grief therapy. *Journal of Constructivist Psychology, 12*, 65–85.
Neimeyer, R. A. (Ed.). (2001). *Meaning reconstruction and the experience of loss*. Washington, DC: American Psychological Association.
Neimeyer, R. A. (2004). Fostering posttraumatic growth: A narrative contribution. *Psychological Inquiry, 15*, 53–59.
Neimeyer, R. A. (2005). Widowhood, grief, and the quest for meaning: A narrative perspective on resilience. In D. Carr, R. M. Nesse, & C. B. Wortman (Eds.), *Late life widowhood in the United States* (pp. 227–252). New York: Springer.
Neimeyer, R. A. (2006). Re-storying loss: Fostering growth in the posttraumatic narrative. In L. Calhoun & R. Tedeschi (Eds.), *Handbook of posttraumatic growth: Research and practice* (pp. 68–80). Mahwah, NJ: Lawrence Erlbaum.
Neimeyer, R. A. & Bridges, S. K. (2003). Postmodern approaches to psychotherapy. In A. S. Gurman & S. B. Messer (Eds.), *Essential psychotherapies* (2nd ed., pp. 272–316). New York: Guilford.
Neimeyer, R. A., & Hogan, N. S. (2001). Quantitative or qualitative? Measurement issues in the study of grief. In M. S. Stroebe, R. O. Hansson, W. Stroebe, & H. Schut (Eds.), *Handbook of bereavement research* (pp. 89–118). Washington, DC: American Psychological Association.
Neimeyer, R. A., Hogan, N. S., & Laurie, A. (2007). The measurement of mourning. In M. Stroebe, R. O. Hansson, H. Schut, & W. Stroebe (Eds.), *Handbook of bereavement research and practice*. Cambridge Books: New York.
Neimeyer, R. A., & Mahoney, M. J. (Eds.). (1995). *Constructivism in psychotherapy*. Washington, DC: American Psychological Association.
Neimeyer, R. A., Prigerson, H. G., & Davies, B. (2002). Mourning and meaning. *American Behavioral Scientist, 46*, 235–251.
Oakley, F., Khin, N. A., Parks, R., Bauer, L., & Sunderland, T. (2002). Improvement in activities of daily living in elderly following treatment for post-bereavement depression. *Acta Psychiatrica Scandinavica, 105*, 231–234.
Ott, C. H. (2003). The impact of complicated grief on mental and physical health at various points in the bereavement process. *Death Studies, 27*, 249–272.

Parkes, C. M. (1993). Bereavement as a psychosocial transition. In M. S. Stroebe, W. Stroebe, & R. O. Hansson (Eds.), *Handbook of bereavement* (pp. 91–101). New York: Cambridge.

Prigerson, H. G., & Jacobs, S. C. (2001). Traumatic grief as a distinct disorder. In M. S. Stroebe, R. O. Hansson, W. Stroebe, & H. Schut (Eds.), *Handbook of bereavement research* (pp. 613–645). Washington, DC: American Psychological Association.

Prigerson, H. G., & Maciejewski, P. K. (2006). A call for sound empirical testing and evaluation of criteria for complicated grief proposed for DSM-V. *Omega, 52*, 9–19.

Reich, J. W., & Zautra, A. J. (1989). A perceived control intervention for at-risk older adults. *Psychology and Aging, 4*, 415–424.

Reynolds, C. F., Miller, M. D., Pasternak, R. E., Frank, E., Perel, J. M., Cornes, C., et al. (1999). Treatment of bereavement-related depression in later life. *American Journal of Psychiatry, 156*, 202–208.

Rosenzweig, A., Prigerson, H., Miller, M. D., & Reynolds, C. F. (1997). Bereavement and late-life depression. *Annual Review of Medicine, 48*, 421–428.

Rubin, S. S. (1999). The two-track model of bereavement. *Death Studies, 23*, 681–714.

Sabatini, L. (1988–1989). Evaluating a treatment program for newly widowed people. *Omega, 19*, 229–236.

Shear, K., Frank, E., Houck, P. R., & Reynolds, C. F. (2005). Treatment of complicated grief: A randomized controlled trial. *Journal of the American Medical Association, 293*, 2601–2608.

Silverman, G. K., Jacobs, S. C., Kasl, S. V., Shear, K., Maciejewski, P. K., Noaghiul, F. S., et al. (2000). Quality of life impairments associated with diagnostic criteria for traumatic grief. *Psychological Medicine, 30*, 857–862.

Stroebe, M., & Schut, H. (1999). The dual process model of coping with bereavement. *Death Studies, 23*, 197–224.

Thompson, L. W., Gallagher-Thompson, D., Futterman, A., Gilewski, M. J., & Peterson, J. (1991). The effects of late-life spousal bereavement over a 30-month interval. *Psychology and Aging, 6*, 434–441.

Vanderwerker, L. C., & Prigerson, H. G. (2004). Social support and technological connectedness as protective factors in bereavement. *Journal of Loss and Trauma, 9*, 45–57.

Wagner, B., Knaevelsrud, C., & Maercker, A. (2006). Internet-based cognitive-behavioral therapy for complicated grief: A randomized controlled trial. *Death Studies, 30*, 429–453.

Weissman, M. M., Markowitz, J. C., & Klerman, G. L. (2000). *Comprehensive guide to interpersonal psychotherapy*. New York: Basic Books.

Wortman, C. B., & Silver, R. C. (2001). The myths of coping with loss revisited. In M. S. Stroebe, R. O. Hansson, W. Stroebe, & H. Schut (Eds.), *Handbook of bereavement research* (pp. 405–429). Washington, DC: American Psychological Association.

Zisook, S., Shuchter, S. R., Pedrelli, P., Sable, J., & Deaciuc, S. C. (2001). Bupropion sustained release for bereavement. *Journal of Clinical Psychiatry, 62*, 227–230.

Zygmont, M., Prigerson, H. G., Houck, P. R., Miller, M. D., Shear, K., Jacobs, S., et al. (1998). A post hoc comparison of paroxetine and nortriptyline for symptoms of traumatic grief. *Journal of Clinical Psychiatry, 59*, 241–245.

18 PTSD (Post-Traumatic Stress Disorder) in Later Life

Lee Hyer and Amanda Sacks

Posttraumatic stress disorder (PTSD) is a stress reaction characterized by symptoms of reexperiencing, avoidance/numbing, and hyperarousal following 1 month or more an exposure to extreme trauma. In recent years, there has been a spate of studies and meta-analyses on the assessment and treatment of this disorder. However, the variable of age in the context of treatment has not been adequately addressed. In addition, little data exist regarding the role of aging as a mediating or moderating influence in the expression of trauma. We do not yet know the importance of a life lived, one's life history, and how this might affect any current trauma. Importantly too, clinical data of client-treatment factors with elders are lacking. That said, we have apt models of treatment for other age groups and can reasonably export empirically supported principles for older adults.

As a general position, older victims tend to be no more vulnerable or reactive than younger individuals despite increasing stress rates (Hyer & Sohnle, 2001). Current rates of PTSD in people of all ages exposed to traumatic events range from about 5 to 11 percent (Kessler, Sonnega, Bromet, Hughes, & Nelson, 1995) and about the same for older adults (Hyer & Sohnle, 2001). Lifetime exposure to at least one major stressor event sums to 69%, slightly higher for older adults (Norris, 1992). In later life, too PTSD has been linked to suicide, adverse reactions to health care, smoking and alcohol abuse, more outpatient visits and primary care use, and behavioral disturbances in long-term care (Cook, O'Donnell, Moltzen, Ruzek, & Skeith, 2005; Davidson, Hughes, Blazer, & George, 1991; Frueh et al., 2003; Hyer & Stanger, 1997; Schnurr & Spiro, 1999; Spiro, Schnurr, & Aldwin, 1994).

Empirically Supported Treatment and Principles

The traditional view is that people develop PTSD in the wake of exposure to a traumatic event, as the imprint of that trauma comes to dominate how they organize their way in the world (van der Kolk, 2002). From a biological perspective, it is argued that when trauma occurs, a distinct neural circuit develops that is easily triggered by even minimal imaginal or other sensory processes, thereby creating the perception that it seems as if the traumatic event were occurring again. Biologically, the formation of PTSD symptoms results from a complex relationship between cental nervous system and immune system that is mediated through a disregulated hypothalamic pituitary adrenal axis. At some point the shear act of imagination powered by the human cognitive machine (or a simple stimulus) creates havoc in day-to-day behavior and interaction. Given the engrained nature of trauma imprints, effective therapy needs to help survivors tolerate the sensory reminders of the trauma, and physically experience efficacy and purpose in response to stimuli that once triggered feelings of helplessness and dependence.

Most studies on treatment outcome for PTSD have used variants of CBT programs, which include mild exposure therapy, anxiety management, or cognitive therapy (see Hyer & Sohnle, 2001). Outcome studies with older adults have been modest, small studies or case studies (Boehnlein & Sparr, 1993; Hyer et al., 1990; Lipton & Schaffer, 1986; Molinari & Williams, 1995; Snell & Padin-Rivera, 1997). Cook et al. (2005) applied a CBT treatment group to eight older veterans and found changes in clinician-rated measures in severity of PTSD. The methods applied reflect those of Boehnlein and Sparr (1993), Hyer et al. (1990), and Thorp and Stein (2005) using anxiety management, rational thinking, self-reward, mastery, and relaxation, as well as psychoeducation. Trauma processing via exposure was absent in all cases. Thorp and Stein (2005) was one of the few studies that applied exposure, adapted for late-life to an older group of adults with PTSD. This consisted of 12 sessions over 6 weeks of exposure. Results were favorable indicating that older adults can tolerate these conditions. In one of the few case studies, on an older patient, Russo, Hersen, and Van Hasselt (2001) carried out a single-case analysis to assess the effects of imaginal exposure in a 57-year-old female suffering from current and reactivated PTSD following a transient ischemic attack. A positive outcome was noted.

Four therapies have special merit for the treatment of PTSD symptoms: (1) anxiety management training (AMT, see below); (2) stress inoculation therapy, representing a coping model; (3) EMDR, providing a dosed exposure that targets state-specific information related to the trauma and has applicability with older victims (Hyer & Kushner, in press); and (4) cognitive processing therapy, applying gradual and multiple exposure to the trauma, and consisting largely of rescripting and altering distortions (Resnick, 2003). A related treatment that may be more applicable to older adults who have intense and continued trauma is multimodal therapy (Falsetti, Erwin, Resnick, Davis, & Combs-Lane, 2003; Hyer & Sohnle, 2001). Reminiscence therapy, the act of recollecting a memory in one's past for varied purposes, has been used with some efficacy with older adults (Hyer & Sohnle, 2001). It has been applied for purposes of a positive self-focus (reminiscence for identity, death preparation, and problem solving) and a negative self-focus (reminiscence for boredom reduction, bitterness revival, and intimacy maintenance) effects with older adults (Cappeliez & O'Rourke, 2006). Additionally, there have been a few combined studies with medication and psychotherapy, but again these have not specifically assessed older adults (Engels and Verney, 1997).

The positive news for the treatment of PTSD (International Society for Traumatic Stress Studies; Foa, Keane, & Friedman, 2000) is that exposure therapy remains the best empirically supported intervention, followed by cognitive restructuring (CR) and SSRIs. These studies have not been evaluated systematically in older trauma victims, however. Older victims find the exposure treatments distressing (Hyer & Sohnle, 2001). While there are many benefits to the use of prolonged exposure in treatment of young adults, older patients generally respond poorly. In contrast, acute stress disorder (ASD), an immediate variant of PTSD, has been better evaluated with older adults than PTSD (Blanchard & Hickling, 1997; Bryant, Moulds, & Nixon, 2003). In this case, the modified exposure models have merit.

Acute and Chronic Trauma at Late life

Acute trauma, involving both ASD and early PTSD, is prevalent in older adults. A limited number of studies have attempted to differentiate PTSD responses or symptoms according to age (Averill & Beck, 2000). The most studied acute trauma at late life involves motor vehicle accidents (MVAs) (Norris, 1992). Given this trauma, estimates of PTSD range from 10 to 46% for all ages (Blanchard & Hicking, 1997), as well as other psychiatric distress at 1 year follow-up (Mayou, Bryant, & Duthie, 1993). MVAs are the second leading cause of injury in geriatric patients, resulting in the highest crash fatality rate of any age group, with most accidents related to driver error (Harvey, Bryant, Dang, Sackville, & Basten, 1998). Additionally, MVAs cause a high mortality rate in older pedestrian accidents due to inability to hear traffic, inability to attain speeds, and poor judgment. Harvey and Bryant (1998) have shown that, for ASD participants, a high percentage later qualify for a PTSD diagnosis 6 months later. Additionally, when elders are victimized, they may not only sustain greater physical injuries than other age groups but also have a decreased ability to heal after injury and trauma (Brown et al. 2004).

Several other forms of trauma specific to older adults have been noted. The incidence of PTSD following the completion of cancer treatment ranges from 0 to 32% (Kangas et al., 2005). Additionally, loss of spouse, natural disaster, other health problems, and victimization by crime have been cited as severe stressors during late life. Elder abuse is descriptive of many types of maltreatment, including neglect, physical, emotional, financial, and sexual abuse (Benbow & Haddad, 1993; McCaertney & Severson, 1997; Newton, 2001).

Typically, chronic or delayed PTSD presents as a subclinical state of psychiatric symptoms that worsens over time and becomes manifest as PTSD. Factors that initiate psychopathology are generally not the same as those that maintain its persistence into older age (Rutter & Maughan, 1997). A myriad of factors related to abuse (nature of the trauma), the balancing of risk and resilient components, extant cognitive schemata, social network, and predisposing factors interact to "produce" a response to trauma, most often one that results in adequate adjustment (McFarlane, 1992; Paris, 1997).

Modally, a poor trauma reaction probably begins with an acute reaction. Although this may remit, individuals rework the original trauma at each subsequent developmental stage, as new cues related to the trauma become salient. Salamon (2001), for example, discussed how the terrorist attacks in the US on September 11, 2001, acted as a second episode of PTSD for many Holocaust survivors from World War II. Symptoms do not usually become worse over time, they become more rigid. In addition, over time personality may be altered as an accommodation to changes in life. While perhaps not the best model for PTSD, for elders who have experienced combat trauma during their lifetime, roughly 20% are continuously troubled, 20% are symptom free totally, and the remainder experience intermittent symptoms (Garb, Bleich, & Lerer, 1987; Hyer & Summers, 1995). Finally, symptoms of trauma often alter at later life, as aging-related variables assert a greater influence. This has been labeled heterotypic continuity (Hyer & Sohnle, 2001).

Aging Issues of Trauma: Vulnerability and Stress Inoculation Hypotheses

Given that PTSD is a diathesis disorder and that aging is itself varied and complex, there are many variables of note that mediate/moderate the unfolding of this disorder. We address two. An essential tension exists as to whether older adults are more vulnerable or have increased inoculation as a result of a stressor. On the one hand, due to reduced adaptive capacities, coping resources external resources (e.g., low income and lack of social support), and increased exposure to traumatic events (e.g., health-related events such as strokes or elder abuse in family) through aging, older adults may be more likely to develop psychiatric problems (including PTSD) or problems with psychosocial functioning than younger people. This has been labeled the vulnerability or the additive burden hypothesis (Dohrenwend & Dohrenwend, 1981). This hypothesis is bolstered by detrimental mental health problems among older people (e.g., Logue, Melick, & Struening, 1981; Melick & Logue, 1985) and the link between increased PTSD symptoms following traumatic events and older age (Ullman & Siegel, 1994). Additionally, older people have been found to report greater distress immediately following traumatic events (e.g., Bolin & Klenow, 1983) as well as 3 years later (Kilijanek & Drabek, 1979), and to perceive themselves at greater risk of extended disaster-related traumatic stress (Melick & Logue, 1985). Not surprisingly, older people have a slower period to remission of PTSD symptoms and depression than younger people (Cwikel & Rozovski, 1998), and often utilize poorer coping strategies (Charlton & Thompson, 1996).

The stress inoculation hypothesis, on the other hand, argues that age is considered a protective factor against the development of PTSD. This occurs particularly when older people have dealt successfully with previous trauma (Gibbs, 1989), or were able to reexperience a trauma with a high degree of control, to find meaning in some outcome of the trauma, and find social support

(Lyons, 1991). Considerable evidence exists for this view. Older people exposed to flood disasters reported fewer symptoms (e.g., depression, anxiety, somatic symptoms) than younger and middle-aged people (Gleser, Green, & Winglet, 1981; Phifer, 1990; Phifer & Norris, 1989) and those who had previously experienced floods were less distressed following subsequent flooding (Norris, 1992). To have experienced previous crises could enable older people to develop coping strategies necessary to adapt successfully to the traumatic event at hand (Elder & Clipp, 1989; Norris, 1992) or to increase their mastery skills and knowledge of themselves, by changing values (Tedeschi & Calhoun, 1995).

Assessment

Despite inadequate psychometric properties for older adults, we believe that the PTSD scales applicable to younger groups have merit at late life. Table 1 provides several measures that may be utilized for assessment of symptoms in the elderly (for a more detailed description and exhaustive list of measures see Hyer & Sohnle, 2001). For Criteria A of PTSD, Carlson (1999) advocates for three conditions: (1) perception of the event as highly negative valence, (2) the suddenness of the experience, and (3) the inability to control the event leading to fear of safety and security. This is adequately addressed by the trauma history questionnaire (THQ; Green, 1996). For PTSD assessment, the most used scale is the Clinician Administered PTSD Scale (CAPS; Blake et al., 1995).

TABLE 1. Objective measures of PTSD.

Psychosocial domain	Measures
Trauma exposure	Clinician Administered PTSD Scale
Interviewer scale	(CAPS)
Self-report	Trauma History Questionnaire (THQ)
	PTSD Check List (PCL)
	Impact of Events (IES)
Other psychiatric symptoms	Beck Depression Inventory
	Beck Anxiety Scale
	Penn State Worry Scale
Functional status and quality of life	ADL/IADL
	Brief Quality of Life Interview (QOLI)
Personality	Personality Assessment Inventory (PAI)
	Millon Clinical Multiaxial Inventory-III
Social supports/psychiatric comorbidities	Interview
Health/medical comorbidities	SF-36 and interview
	Contact with primary care physician
Distorted beliefs	Posttraumatic Cognitions Inventory
Therapy alliance	Working Alliance Inventory (WAI)
Cognitive processing abilities	Verbal intelligence, sustained attention, working memory, inhibition errors, new learning (acquisition and retention)

Self-report PTSD measures (PCL-C; Weathers, Litz, Herman, Huska, & Keane, 1993), along with other self-report psychiatric scales (BAI; Beck & Steer, 1988; BDI-II; Beck, Steer, & Brown, 1996; and PSWS; Meyer, Miller, Metzger, & Borkevec, 1990; O'Hare et al., in press; Ware, Kosinski, & Keller, 1994), are important for the establishment of the diagnosis and for measurement of change in therapy. The ADL/IADL (Katz, Ford, Moskowitz, Jackson, & Jaffee, 1963; Lawton & Brody, 1969) scales, as well as cognitive processing measures, are requisite for determining level of care and mediating problems in adjustment. Examination of health issues and social support are significant. Personality scales (MCMI-III; Millon, 1984, 1994 and PAI; Morey, 1991) can be important, as each addresses both PTSD and related comorbidities, specifically Axis II traits. The Working Alliance Inventory (WAI; Horvath & Greenburg, 1989) is an important measure in the therapy process, as it probably mediates change in PTSD. Finally, in therapy, the client is asked their subjective units of distress on a numerical scale, e.g., 0–100 (SUDS), PCL-C, BDI, additionally, the Impact of Event Scale (IES; Horowitz, Wilner, & Alvarez, 1979) scale can be informative about substantive changes and their possible causes.

The neuropsychological status of older trauma victims is a special concern. It is now well documented that neuropsychological deficits occur in PTSD (Bremner, 2002). Reviews on the structural and functional anatonomy of patients with PTSD (e.g., Bremner, 2002; Elzinga & Bremner, 2002; Hull, 2002) note compromised neurocognitive functioning. These are thought to impede a person's ability to cope with traumatic stressors (Brewin, Andrews, & Valentine, 2000). On the one hand, there appears to be a preferential bias or hypervigilance associated with the presentation of threat-related stimuli (Hyer & Sohnle, 2001). On the other hand, cognitive deficits in several domains are due to PTSD; memory (Bremner, Southwick, Johnson, Yehuda, & Charney, 1993; Elzinger & Bremner, 2002; Sutker et al., 1991; Vasterling, Brailey, Constans, Borges, & Sutker, 1997), attention (Vasterling et al., 2002), executive functioning (Beckham, Crawford, & Feldman, 1998), and global intellectual functioning (Sutker, Uddo-Carne, & Allain, 1991; Vasterling et al.). Cognitive deficits in PTSD are not well understood, however, and may reflect baseline cognitive abilities, learning disabilities, various occult neurological problems, and mood problems. The coexistence of neuorpsychological impairment and PTSD symptoms suggests a diminished prognosis, and are reflective of problems regarding PTSD change (Vasterling & Braily, 2005). Findings suggest that cognitive deficits are not global but are limited to specific domains.

Treatment Model

Hyer and Sohnle (2001) provided a treatment model of PTSD based on CBT components (Table 2). They provided a six-step model of treatment. In general, the more the intensity of the perturbation of the victim, the more therapeutic tasks relate to the beginning parts of this treatment model; the less the victim is impacted

TABLE 2. PTSD treatment model.

Goals	Methods
Stabilize symptoms	Treat comorbid disorders or stressors (including health)
Relationship building	Build trust, the ability to confront trauma with a trusted therapist. Apply the client-rated WAI
Attend to necessary developmental, treatment, and education (normalization) factors	Assure social supports, daily coping, social skills, and treatment compliance
Apply cognitive restructuring (CR)	Teach the five-step program of CR and consider tactics of personality style as therapy-guided treatment
Elicit positive core memories (PCM)	Foster a renarration of self with core memories that can generalize to current life situations
Decondition trauma memories	Apply AMT (see description in chapter)

by the trauma, the more the therapist can address the latter parts of the model and achieve lasting change. The rationale for the ordering of these tasks is to keep people committed, encourage experimentation, and create cognitive change.

First, the task is to stabilize symptoms, including the treatment of comorbid disorders and current stressors (including health). Some symptoms are "more equal" than others. Up front, the therapist is especially wise who addresses sleep problems, panic anxiety, depression, substance abuse, and psychotic symptoms. Eighty percent of PTSD patients (of a chronic nature) receive an additional diagnosis (Hyer & Sohnle, 2001). PTSD also involves abnormalities in a number of domains within the individual, such as memory, mood, bodily experience, interpersonal relatedness, emotion, and behavior. In addition, medical issues are a problem that requires an interface with the primary care physician. The therapist's role then is as a "therapy manager," initially and foremost with the symptoms (the trauma). Often the operative tasks of the therapy are basic: keeping the trauma victim in treatment, being supportive during difficult periods, maintaining appropriate arousal levels, and in general assuring commitment to the goals of therapy.

Second, the working alliance between client and therapist is critical. For some, the relationship is the therapy. The therapist must do three relationship tasks: (1) be liked/respected, (2) be believable (with a plan), and (3) initiate intervention strategies that result in some successes. We recommend that the WAI be used as a feedback marker in the therapy, since it has been shown to mediate success in therapy (Hyer, 2005).

The third task is to attend to necessary developmental and treatment factors, assure social supports, daily coping, and social skills and treatment compliance. Self-management components must be in place for the therapy to be successful. The clinician must be assured that the life style is reasonably healthy and not interfering with life or (later) narrative repair. This includes, sleep, eating, exercise, treating physical illnesses, and use of medications. Behavioral interventions include standard CBT interventions (Martell, Addis, & Jacobson, 2001), especially scheduling pleasant events, social interaction, activity scheduling, assertiveness training, and bibliotherapy.

Fourth, the therapeutic use of CR is warranted. This is the main therapeutic element that has applicability to PTSD for older adults. CR is critical at enhancing the efficacy of prolonged exposure, and in fact has been effective in trauma even without the use of exposure therapy (Mueser et al., 2007). CR may involve multiple curative factors including habituation anxiety, discrimination between remembering and reencountering, differentiation of trauma from similar but safe events, mastery experience, and assisting in the organization of trauma narratives. We apply a five-step CR process. This is a variant of the DTR used in CBT (Beck, 1995; Hamlin, Mueser, & Rosenberg, 2004) in which the victim is requested to apply five steps in everyday challenges related to trauma as well as other issues.

The fifth component involves the self-representational stories, positive core memories (PCM). They include written or taped autobiographies with feedback that may involve listening to tapes, making pilgrimages or reunions, developing a genealogy, doing a scrapbook or photo albums, looking at old letters and memorabilia, summarizing one's life work, and exercising the preservation of ethnic identity (Lewis & Butler, 1974). Rybarczk and Bellg (1997) and Hyer et al. (1990) applied PCMs alone with older adults to good effect. The application of PCMs involves a simple recounting of a positive event, reinforcing the reasonable positive interpretations, reengaging hope, reframing better world views, and supporting competence (see Hyer & Sohnle, 2001).

The very last component involves the trauma memory. The therapist takes time to assess whether the client is a candidate for this stage in the therapy, whether they will allow access to the trauma memory, and have the requisite skills for an exposure and unearthing of the content (see Hyer & Sohnle, 2001). The therapeutic task is to bring this implicit memory (easily triggered by emotional states, interpersonal contexts, external stimuli, and language cues) into awareness as self-as-past, i.e., into language in the form of a narrative. The more the trauma memory is organized, narrativized, and placed into explicit memory, as elaborated, disclosed, and validated, the more general improvement occurs (Foa et al., 1999; Pennebaker, 1989). These memories can now be retrieved less as summary and more positive.

AMT is recommended for this task. This is a soft exposure technique that involves a variety of procedures, including biofeedback, relaxation, and CR. Conceptually, trauma psychotherapy is a continual retelling of self, shifting from previous negative narratives to more positive ones (van der Kolk, 2002). With the combination then of assimilation (CBT and narrative work) and exposure work (AMT + exposure), trauma codes can become habituated (desensitized) *and* assimilated (corrective cognitive codes).

With AMT, use of exposure is done empathically and with relaxation as a lead-in (Hyer & Sohnle, 2001). Procedurally, the sequence of memory work is straightforward (Table 2). After the therapist has assessed for client skills, willingness, and needed safety issues, the therapist obtains the facts of the memory, a "clean" rendition of events. Confusion, misattribution, fears, and gaps are highlighted and gently challenged. The objective is not an exacting account of the trauma, as there is little evidence that this helps in therapy. Careful scaffolding is also required. Normalization and education regarding the reasons for unearthing the memory are critical. A model of PTSD can be provided, as well as the reasons for relaxation and exposure interchange (the therapy). Listening for fears is therapeutically important. Use of handouts and obtaining data (e.g., IES, BDI) is helpful. Normal safety and suicidal considerations are always performed.

Relaxation is applied first for eventual use of self-controlled desensitization (SCD; Borkovec, 2005) – a gentle calming down procedure according to the therapist's taste. This is performed according to AMT principles (hierarchal, gradual, slow, paired with relaxation, and persistent until desensitization). Typically this is done for several sessions. Once the memory has been obtained, two added features are employed: (1) a simple fear hierarchy is obtained for the use of the AMT exposure, often bits and pieces of the memory and (2) the memory itself from beginning to end.

The therapist then applies AMT procedures (see Hyer & Sohnle, 2001). When the AMT exposure is performed, it may be preceded by interoceptive exposure (identifying sensations that are a problem) and instructions on coping or use of imagery. Regardless, a simple AMT procedure SCD is applied, interchanging relaxation and fear initially at SUDS 4–6 for several minutes until remission (0–2). Upon reaching this goal, the next fear is applied. This is done until the complete memory is desensitized.

Finally, at therapy end, relapse awareness is helpful. The therapist anticipates the need for continuance of care, discusses issues of a change in treatment, and the importance of booster sessions. The client is taught how to recognize early warning signs, anticipate high risk situations, and develop an emergency plan for relapse. This includes a self-help strategy and involves significant others, as well as the consolidation of a relapse philosophy.

Case Example

An 83-year-old white, widowed female resident of an assisted living facility was assessed by a geropsychiatrist due to presenting symptoms of increased irritability, poor sleep, and hyperarousal. Prior to the manifestation of these symptoms, she had been described by staff as friendly and interactive with other residents. Medical history included glaucoma, arthritis, history of cancer treatment and hypertension, and MMSE score of 27, suggesting mild cognitive impairment. Additionally, she had a psychiatric history, including generalized anxiety disorder for which she had been treated pharmacologically and occasional psychotherapy. She had been residing in the assisted living facility for 4 years. During the past year, a male resident moved into the apartment next to her. Over the course of the year, he displayed harassing behaviors toward her including asking her repetitive questions during meals in the dining room, following her down the hallway, and at times entering her apartment. During the evening, she entered her apartment where she found the man on her bed where he assaulted her. She was able to obtain help through pressing her panic button and potential harm was avoided; however, after the incident she displayed symptoms of withdrawal, isolation, aimless pacing, constricted affect, and difficulty in maintaining concentration and attention. Additionally, she refused to return to her room after which the staff moved her to another floor. It was clear that this change in behavior occurred after the attempted assault. Despite the staff moving the male to a locked part of the facility, she continued to display these symptoms 3 months after the incident. When asked about specific details of the event, she was unable to provide specific details; however, expressed that every time she currently enters into her apartment she imagines him on her bed and subsequently is unable to enter the room without being accompanied by staff or family member.

Assessment

On interview with the THQ, Mrs. M fulfilled the three conditions for criteria A for PTSD. It is important to note further that on the THQ additional information was revealed that she had been mugged as a younger adult and had since avoided that area of the neighborhood. Additionally, she became a widow 2 years ago after having been the caretaker of her husband who suffered from cancer. Currently, her daughter is suffering from a terminal pulmonary illness and is in the hospital the majority of the time. On the CAPS, Mrs. M showed a moderate intensity for the criteria of most PTSD symptoms (total score 55), including recurrent and intrusive recollections and dreams of the event, physiological reactivity to cues that remind her of the event, and persistent efforts to avoid thinking about the event. On the Beck Anxiety Scale, she was in the clinically significant range (22), specifically endorsing items such as being terrified, unable to relax, and feeling shaky and unsteady. On the Beck Depression Scale, she was in the mild range of symptomatology (18). She was also given a personality scale, the PAI (personality assessment inventory). She was elevated on the anxiety subscale consisting of cognitive, affective, and physiological components, as well as the traumatic stress index.

According to staff, she appeared to display increased confusion since the incident and was therefore given a brief neuropsychological battery, which displayed impairments in attention, concentration, and processing speed and within normal limits scores in all other domains. Her functional status in terms of ADLs and IADLs was within normal limits; however, due to diminished concentration, there was reported forgetting of appointments and activities since the incident. On the Brief Quality of Life Interview, Mrs. M displayed a diminished quality of life as evidence by endorsement of symptoms regarding lack of deriving well-being from family relations, lack of engaging in pleasurable activities at her usual level, and diminished sleep.

Treatment

The geropsychiatrist administered a trial of an antidepressant medication to regain premorbid levels of functioning. Although there was a positive response in the first 2 weeks, Mrs. M continued to express having intrusive thoughts about the incident, feeling estranged from her family, and difficulty in sleeping and concentrating. Then treatment with the geropsychologist began with CBT.

1. *Stabilization*: After assessment, she was informed about PTSD and the importance of her commitment and probable targets in treatment. This was a collaborative endeavor that she was to be an active participant. Her response to the event was legitimated and her assurance that this treatment would make a difference was expressed. A model of PTSD and a simple contract was applied – nature of symptoms, her and therapist's role, crisis plan. In addition, she was assured that the depression and other anxiety disorders were complications, but ones of recent origin that would subside, given treatment.

2. *Relationship*: This component was fostered across sessions and her response on the WAI goals, bond, and tasks were all positively rated. She easily trusted the therapist and confided in him.

3. *Balance*: She was encouraged to seek out social supports, and to care for herself with adequate sleep, eating, exercise, and medication compliance. In fact, sleep was a problem (onset and maintenance) and one that required attention on each session. The therapist also requested that she indulge in several pleasant events, and to read up on PTSD. A self-help book was given. Relaxation exercises were also taught.

4. *Cognitive Restructuring*: She was taught the five-step CR skill, first in the session on non-PTSD areas and later on PTSD symptoms. She was also instructed to apply this in vivo which she did by Session 6. This skill allowed her to see how problems and symptoms could be handled by herself. She was also instructed to use this in the later exposure sessions. The personality scale revealed low dominance and high warmth, reflecting a more dependent stance on interpersonal issues.

5/6. *PCM/Exposure*: No effort was made to apply PCM, as she was willing and able to try exposure with the trauma memory. The components of imaginal exposure were assessed and assured (access to trauma memory, ego skills, psychologically minded, and self-soothing techniques), as well as preparation for the therapy (safety and crisis plan as well as instructions on coping). She had already had relaxation training. Mrs. M provided a short narrative of the trauma experience with sensory details and affect. This process occurred gradually, in short periods of time during five sessions, but was summarized in Session 7. Supportive therapy consisting of validation and reassurance of safety was practiced throughout the exposure process to ensure continued collaboration. She also provided a hierarchy of stimuli as well as coping images. No interoceptive exposure was used, as she was very willing to apply the exposure method.

Self-control desensitization was applied. She was given an SUDS level 5 on the trauma scale and this scene was held in memory for 30 s. followed by a coping image and relaxation. This was done eight times as the SUDS eventuated in a 0 rating. This leads to an SUDS level 8 image at the next session with similar results. After that (in the next session), she was willing to process the whole memory with a quick resolution. She became habituated to the exposure and displayed decreased anxiety. Throughout, the therapist used Mrs. M's cognitions regarding the trauma, herself and her future. These were also coping mechanisms for her to apply between sessions. CR took place during and after exposure therapy so that increased positive thoughts of self-mastery could be incorporated into her cognitions. Throughout the course of therapy then, she acquired the skill of becoming aware and controlling automatic thoughts as well as being able to elicit a relaxation response in the context of an anxiety provoking situation such as cues that reminded her of the traumatic event. She was also able to apply this in vivo.

Assessment at the end of 12 sessions demonstrated decreased anxiety symptoms on the BDI/BAI and decreased PTSD symptoms as measured by the PCL. Quality of life had also improved. Based on her anxiety, she may continually exhibit anxiety symptoms and continued psychotherapy may be beneficial.

Conclusion

Trauma at late life is plentiful and treatable. Data of the past two decades do not suggest that older adults are at greater risk than younger adults for poor psychosocial outcomes following exposure to trauma. There is simply insufficient evidence to conclude that trauma causes more negative psychosocial consequences in older adults. Empirically supported therapies that apply to younger groups have applicability to older groups as well.

The six-step model proposed here, which incorporates key elements of traditional cognitive and behavioral therapies but also expands them to include a trauma focus, is a comprehensive treatment for acute or chronic PTSD. The model treats the whole person with PTSD problems. Memories are modified or transformed directly only as a last resort. Throughout, factors related specifically to aging, mourning for losses, giving meaning to experiences, reestablishing self-coherence and self-continuity, achieving ego integration, and culture and social support are integrated.

References

Averill, A., & Beck, J. (2000). Posttraumatic stress disorder in older adults: A conceptual review. *Journal of Anxiety Disorders, 14*, 133–156.

Beck, J. (1995). *Cognitive therapy: Basics and beyond*. New York: Guilford.

Beck, A., & Steer, R. A. (1988). *Beck anxiety inventory*. San Antonio, TX: The Psychological Corporation.

Beck, A. T., Steer, R. A., & Brown, G. K. (1996). *Beck depression inventory manual* (2nd ed.). San Antonio, TX: Psychological Corporation.

Beckham, J. C., Crawford, A. L., & Feldman, M. E. (1998). Trail making test performance in Vietnam combat veterans with and without posttraumatic stress disorder. *Journal of Traumatic Stress, 11*, 811–819.

Benbow, S. M., & Haddad, P. M. (1993). Sexual abuse of the elderly mentally ill. Blake, D. D., Weathers, F. W., Nagy, L. M., Kaloupek, D. G., Gusman, F. D., Charney, D. S., et al. (1995). The development of a clinician-administered PTSD scale. *Journal of Traumatic Stress, 8*, 75–90.

Blanchard, E. B., & Hickling, E. J. (1997). *After the crash: Assessment and treatment of motor vehicle accident survivors*. Washington, DC: American Psychological Association.

Bremner, J. D. (2002). Neuroimaging studies in post-traumatic stress disorder. *Current Psychiatry Reports, 4*, 254–263.

Bremner, J. D., Southwick, S. M., Johnson, D. R., Yehuda, R., & Charney, D. S. (1993). Childhood physical abuse and combat-related posttraumatic stress disorder in Vietnam veterans. *American Journal of Psychiatry, 150*, 235–239.

Brewin, C. R., Andrews, B., & Valentine, J. D. (2000). Meta-analysis of risk factors for posttraumatic stress disorder in trauma-exposed adults. *Journal of Consulting and Clinical Psychology, 68*, 748–766.

Boehnlein, J., & Sparr, L. (1993).Group therapy with World War II ex-POWs: Longterm adjustment in a geriatric population. *American Journal of Psychotherapy, 47*, 273–282.

Bolin, R., & Klenow, D. (1983). Response of the elderly to disaster: An age-stratified analysis. *International Journal of Aging and Human Development, 16*, 283–296.

Borkovec (2005). *GAD in the Real World.* Symposium given at NJ-ACT, Westfield, NJ.

Brown, K., Streubert, G. E., Burgess, A. W. (2004). Effectively detect & manage elder abuse. *Nurse Practioner, 8*, 22–27.

Bryant, R. A., Moulds, M. L., & Nixon, R. V. D. (2003). Cognitive behaviour therapy of acute stress disorder: A four-year follow-up. *Behaviour Research and Therapy, 41*(4), 489–494.

Carlson, E. (1999). *Trauma assessments: A clinicians guide.* Towson, MD: Sidran Foundation and Press.

Cappeliez, P., & O'Rourke, N. (2006). Empirical validation of a model of reminiscence and health at later life. *Journal of Gerontology, 61*B, P237–P244.

Charlton, P. F. C., & Thompson, J. A. (1996). Ways of coping with psychological distress after trauma. *British Journal of Clinical Psychology, 35*, 517–530.

Cook, J., O'Donnell, C., Moltzen, J., Ruzek, J., & Sheikh, J. (2005). Clinical observations in the treatment of World War II and Korean War veterans with combat-related PTSD. *Clinical Geropsychologist, 29*(2), 82–94.

Cwikel, J., & Rozovski, U. (1998). Coping with the stress of immigration among new immigrants to Israel from Commonwealth of Independent States (CIS) who were exposed to Chernobyl: The effect of age. *International Journal of Aging and Human Development, 46*, 305–318.

Davidson, J. R., Hughes, D., Blazer, D. G., & George, L. (1991). PTSD in the community: An epidemiological study. *Psychological Medicine, 21*, 713–721.

Dohrenwend, B., & Dohrenwend, B. (1981). Sex differences and psychiatric disorders. *American Journal of Sociology, 81*, 1447–1454.

Elder, G., & Clipp, E. (1989). Combat experience and emotional health: Impairment and resilience in later life. *Journal of Personality, 57*, 311–341.

Engels, G. I., & Verney, M. (1997). Efficacy of nonmedical treatments of depression in elders: A quantitative analysis. *Journal of Clinical Geropsychology, 3*, 17–35.

Ensel, W. M. (1991). "Important" life events and depression among older adults: The role of psychological and social resources. *Journal of Aging and Health, 3*, 546–566.

Elzinga, B. M., & Bremner, J. D. (2002). Are the neural substrates of memory the final common pathway in posttraumatic stress disorder (PTSD)? *Journal of Affective Disorders, 70*, 1–17.

Falsetti, S. A., Erwin, B. A., Resnick, H. S., Davis, J., & Combs-Lane, A. M. (2003). Multiple channel exposure therapy of PTSD: Impact of treatment on functioning and resources. *Journal of Cognitive Psychotherapy, 17*, 133–147.

Foa, E. B., Dancu, C. N., Hembree, E. A., Jaycox, L. H., Meadows, E. A., & Street, G. P. (1999). A comparison of exposure therapy, stress inoculation training, and their combination for reducing posttraumatic stress disorder in female assault victims. *Journal of Consulting and Clinical Psychology, 67*, 194–200.

Foa, E., Keane, T., & Friedman, M. (2000). *Effective treatments for PTSD: Practice guidelines from the International Society for Traumatic Stress Studies*. New York: Giulford.

Frueh, C., Elhai, J., Gold, P., Monnier, J., Magruder, K., Keane, T., et al. (2003). Disability compensation seeking among veterans evaluated for posttraumatic stress disorder. *Psychiatric Services, 54*, 84–91.

Garb, R., Bleich, A., & Lerer, B. (1987). Bereavement in combat. *The psychiatric clinics of North America, 10*(3).

Gibbs, M. (1989). Factors in the victim that mediate between disaster and psychopathology: A review. *Journal of Traumatic Stress, 24*, 489–514.

Gleser, G. C., Green, B., & Winglet, C. (1981). *Prolonged psychosocial effects of disaster: A study of buffalo creek*. New York: Academic Press.

Green, B. L. (1996). Trauma History Questionnaire (self-report). In B. H. Stamm (Ed.), *Measurement of stress, trauma, and adaptation* (pp. 366–368). Lutherville, MD: Sidran.

Hamlin, J., Mueser, K., Rosenberg, S., & Rosenberg, H. (2004). *Brief cognitive therapy for PTSD*. New Hanover, NC: Dartmouth University Medical School Press.

Harvey, R., Bryant, A., Dang, S., Sackville, T., & Basten, C. (1998). Treatment of acute stress disorder: A comparison of cognitive-behavioral therapy and supportive counseling. *Journal of Consulting and Clinical Psychology, 66*(5), 862–856.

Harvey, A., & Bryant, R. (1998). The relationship between acute stress disorder and posttraumatic stress disorder; A prospective evaluation of motor vehicle accident survivors. *Journal of Counseling and Clinical Psychology, 66*(3), 507–512.

Horowitz, M., Wilner, N., & Alvarez, W. (1979). Impact of event scale: A measure of subjective stress. *Psychosomatic Medicine, 41*, 209–218.

Horvath, A. O., & Greenburg, L. S. (1989). Development and validation of the working alliance inventory. *Journal of Counseling Psychology, 36*, 223–233.

Hull, A. M. (2002). Neuroimaging findings in post-traumatic stress disorder. Systematic review. *British Journal of Psychiatry, 181*, 102–110.

Hyer, L. (2005). *Use of WAI in therapy with older adults*. Piscatway, NJ: Grand Rounds at Robert Wood Johnson.

Hyer, L., & Brandsma, J. (1997). EMDR minus eye movements equals good psychotherapy. *Journal of Traumatic Stress, 10*(3), 515–523.

Hyer, L., & Kushner, B. (in press). EMDR: A review. In P. Lehrer & R. Wolfolk (Eds.), *Handbook of Stress* (2nd ed.). New York: Guiford.

Hyer, L., & Sohnle, S. (2001). *Trauma among older people: Issues and treatment*. Philadelphia: Brunner-Routledge.

Hyer, L., & Stanger, E. (1997). The interaction of Posttraumatic Stress Disorder and depression among older combat veterans. *Psychological Reports, 80*, 785–786.

Hyer, L., & Summers, M. (1995). *An understanding of combat trauma at later life*. Augusta, GA: VA Merit Review.

Hyer, L., Swanson, G., Lefkowoitz, R., Hillesland, D., Davis, H., & Woods, M. (1990). The application of the cognitive behavioral model to older stressor groups. *Clinical Geropsychologist, 9*, 145–190.

Kangas, M., Henry, J. L., & Bryant, R. A. (2005). The relationship between acute stress disorder and posttraumatic stress disorder following cancer. *Journal of Consulting and Clinical Psychology, 73*(2), 360–364.

Katz, S., Ford, A. B., Moskowitz, R. W., Jackson, B. A., & Jaffee, M. W. (1963). Studies of illness in the aged. The index of ADL: A standardized measure of biological and psychosocial function. *Journal of the American Medical Association, 185*, 914–919.

Kessler, R., Sonnega, A., Bromet, E., Hughes, M., & Nelson, C. (1995). Posttrasuamatic stress disorder in the National Comorbidity Survey. *Archives of General Psychiatry, 52*, 1048–1060.

Kilijanek, T., & Drabek, T. (1979). Assessing long-term impacts of a natural disaster: Focus on the elderly. *Gerontologist, 19*, 555–566.

van der Kolk, B. A. (2002). Posttraumatic therapy in the age of neuroscience. *Psychoanalytic Dialogues, 12*(3), 381–392.

Lawton, M. P., & Brody, E. (1969). Assessment of older people: Self-maintaining and instrumental activities of daily living. *The Gerontologist, 9*, 179–186.

Lehman, A., F. (1996). Quality of life interview (QOLI). In L. I. Sederer & B. Dickey (Eds.), *Outcomes assessments in clinical practice* (pp. 117–119). Baltimore: Williams & Wilkins.

Lewis, M., & Butler, R. (1974). Life review therapy: Putting memories to work in individual and group therapy. *Geriatrics, 29*, 165–173.

Lipton, M., & Schaffer, W. (1986). Posttraumatic stress disorder in the older veteran. *Military Medicine, 151*, 522–524.

Logue, J. N., Melick, M. E., & Struening, E. L. (1981). A study of health and mental health status following a major natural disaster. *Research in Community and Mental Health, 2*, 217–274.

Lyons, J. A. (1991). Research in the trauma recovery program at the Jackson VA Medical Center. *PTSD Research Quarterly, 4*(3), 8.

Martell, C., Addis, M., & Jacobson, N. (2001). *Depression in context: Strategies for guided action*. New York: W. W. Norton.

Mayou, R., Bryant, B., & Duthie, R. (1993). Psychiatric consequences of road traffic accidents. *British Medical Journal, 307*(11), 647–651.

McCaertney, J. R., & Severson, K., (1997). Sexual violence, post-traumatic stress disorder, and dementia. *Journal of the American Geriatric Society, 45*(1), 76–78.

McFarlane, A. (1992). Avoidance and intrusion in posttraumatic stress disorder. *Journal of Nervous and Mental Disease, 180*(7), 439–445.

Melick, M., & Logue, J. (1985). The effect on the health and well-being of older women. *International Journal of Aging and Human Development, 21*, 227–238.

Meyer, T., Miller, M., Metzger, R., & Borkovec, T. D. (1990). Development and validation of the Penn State Worry Questionnaire. *Behaviour Research and Therapy, 28*, 487–495.

Millon, T. (1984). On the renaissance of personality assessment and personality theory. *Journal of Personality Assessment, 48*(5), 450–466.

Millon, T. (1994). *Millon clinical multiaxial inventory-III*. Minneapolis: National Computer Systems.

Molinari, V., & Williams, W. (1995). An analysis of aging World War II POWs with PTSD: Implication for practice and research. *Journal of Geriatric Psychiatry, 28*, 99–114.

Morey, L. C. (1991). *Personality assessment inventory*. Odessa, FL: Psychological Assessment Resources.

Mueser, K. T., Bolton, E., Carty, P. C., Bradley, M. J., Ahlgren, K. F., Distaso, D. R., Gilbride, A., Liddell, C. (2007). The trauma recovery group: a cognitive-behavioral program for post-traumatic stress disorder in persons with severe mental illness. *Community Mental Health Journal, 43*(3), 281–304.

Newton, K. (2001). Geriatric trauma. *Topics in Emergency Medicine, 23*(3):1–12.

Norris, F. H. (1992). Epidemiology of trauma: Frequency and impact of different potentially traumatic events on difficult demographic groups. *Journal of Consulting Clinical Psychology, 60*(3), 409–418.

O'Hare, T., Sherrer, M. V., Cutler, J., McCall, T. M., Dominique, K. N., & Garlick, K. (2002). Validating the psycho-social wellbeing scale with community clients. *Social Work in Mental Health, 1*(2), 15–30.

Paris, J. (1997). Introduction: Emotion and empiricism: Research on childhood trauma and adult psychopathology. *Journal of Personal Disorders, 11*, 1–4.

Pennebaker, J. W. (1989). Confession, inhibition, and disease. In L. Berkowitz (Ed.), *Advances in experimental social psychology* (Vol. 22). Orlando, FL: Academic Press.

Phifer, J. (1990). Psychological distress and somatic symptoms after natural disaster: Differential vulnerability among older adults. *Psychology and Aging, 5*, 412–420.

Phifer, J., & Norris, F. (1989). Psychological symptoms in older adults following natural disaster: Nature, timing, duration, and course. *Journal of Gerontology: Social Sciences, 44*, S207–S217.

Resnick, P. A. (2003). Cognitive therapy for posttraumatic stress disorder. *Journal of Cognitive Psychotherapy, 15*(4), 321–330.

Russo, S. A., Hersen, M., & Van Hasselt, V. B. (2001). Treatment of reactivated posttraumatic stress disorder: Imaginal exposure in an older adult with multiple traumas. *Behavior Modification, 25*(1), 94–115.

Rutter, M., & Maughan, B. (1997). Psychosocial adversities in childhood and adulthood. *Journal of Personal Disorders, 11*, 4–19.

Rybarczk, B., & Bellg, A. (1997). *Listening to life stories: A new approach to stress intervention in health care*. New York: Springer.

Salamon, M. (2001). Holocaust survivors and secondary PTSD. *Clinical Gerontologist, 24*(3–4), 183–185.

Schnurr, P., & Spiro, A. (1999). Combat exposure, posttraumatic stress disorder symptoms, and health behaviors as predictors of self-reported physical health in older veterans. *Journal of Nervous & Mental Disease, 187*(6), 353–359.

Spiro, R., Schnurr, P., & Aldwin, C. (1994). Combat related posttraumatic stress disorder symptoms in older men. *Psychology and Aging, 9*, 17–26.

Snell, F., & Padin-Rivera, E. (1997). Group treatment for older veterans with post-traumatic stress disorder. *Journal of Psychosocial Nursing, 35, 10–16.*

Sutker, P. B., Uddo-Carne, M., & Allain, A. N. (1991). Clinical and research assessment of posttraumatic stress disorder: A conceptual overview. *Psychological Assessment, 3*, 520–530.

Tedeschi, R. G., & Calhoun, L. G. (1995). *Trauma and transformation: Growing in the aftermath of suffering*. Thousand Oaks, CA: Sage.

Thorp, S., & Stein, M. (2005). PTSD and functioning. *PTSD Research Quarterly, 16*(3), 1–7.

Ullman, S. E., & Siegel, J. M. (1994). Predictors of exposure to traumatic events and posttraumatic stress sequelae. *Journal of Community Psychology, 22*(4), 328–338.

Vasterling, J., & Braily, K. (2005). Neuropsychological findings in adults with PTSD. In J. Vasterling & C. Brewin (Eds.), *Neuropsychology of PTSD: Biological, cognitive and clinical perspectives* (pp. 178–207). New York: Guilford.

Vasterling, J. J., Brailey, K., Constans, J. I., Borges, A., & Sutker, P. B. (1997). Assessment of intellectual resources in Gulf War veterans: Relationship to PTSD. *Assessment, 4*, 51–59.

Vasterling, J. J., Duke, L. M., Brailey, K., Constans, J. I., Allain, A. N., Jr., & Sutker, P. B. (2002). Attention, learning, and memory performances and intellectual resources in Vietnam veterans: PTSD and no disorder comparisons. *Neuropsychology, 16*, 5–14.

Ware, J. E., Kosinski, M., & Keller, S. D. (1994). *SF-36 physical and mental health summary scales: A user's manual.* Boston: Health Assessment Lab, New England Medical Center.

Weathers, F. W., Litz, B. T., Herman, D. S., Huska, J. A., & Keane, T. M. (1993). *The PTSD Checklist (PCL): Reliability, validity, and diagnostic utility.* Paper presented at the annual meeting of the International Society for Traumatic Stress Studies, Chicago.

19 Training of Geriatric Mental Health Providers in CBT Interventions for Older Adults

Nancy A. Pachana, Bob Knight, Michele J. Karel, and Judith S. Beck

With an increasingly wide range of mental health professionals working with older populations, the need for training in effective intervention strategies for this population is growing. While the literature describing theoretical developments in intervention strategies and controlled trials demonstrating efficacy with respect to older cohorts are on the rise, there is a lack of focus on the underpinnings of effective training of mental health professionals, including data on the efficacy of training methods and positive and negative outcomes of different training strategies and approaches. This issue may be of particular importance in work with older adults; for example, Pinquart and Sörensen's meta-analysis of treatment for depression in older adults showed an effect for training in work with older adults (Pinquart & Sörenson, 2001).

This chapter describes considerations in training students and clinicians to use cognitive-behavioral therapies (CBT) with older adults. Core competencies required for effective interaction with this population as well as the importance of incorporating core competencies into training programs for psychologists, psychiatrists, nurses, and other allied health professionals are described. Recent work on the development of therapist skills training models is presented. Assessing therapist skills training models is critical for providing students and professionals with optimum training experiences that effectively build expertise and enhance practice. Methods of teaching core competencies and therapist skills are inevitably affected, to some extent, by the quality of supervision, and ways to enhance supervision are also discussed. Finally, a resource guide including information on training opportunities, relevant professional societies, manuals, and publications of interest are included at the end of the chapter.

Core Competencies in Working with Older Adults

Over the last two decades, training programs in a variety of disciplines including psychology, medicine, and allied health have evolved from discipline-based apprenticeship models, with intensive curricula based on lectures and accumulating set units of knowledge and hours of experience, to a competency-based model. This shift has resulted in greater flexibility in the delivery of content (e.g., problem-based learning strategies), increased interdisciplinary interactions and greater attention paid to both the learning process (e.g., in terms of feedback given) as well as achievement of competencies that go beyond technical skills to include professional and reflective practice and strategies for continued professional development.

In order to guide curriculum development, several descriptions of professional competencies required to work effectively with older adults have been published. Such competency frameworks are important for training purposes in that they identify core knowledge and skill bases as well as strategies for continued development. Such guidelines are primarily discipline-based, and usually reflect differences in national and regional higher education structures. Key elements of a competency-based training structure include recognition of the trajectory from novice to competent performer to expert practitioner, gradual withdrawal of training structures over time to foster independent practice, and an emphasis on developing self-guided learning skills and reflection on performance.

Many professional organizations concerned with psychotherapy in general, and interventions for older adults in particular, have published guidelines or practice statements on these subjects (see Appendix for a listing of relevant professional societies). For example, the American Psychological Association (APA) developed its Guidelines for Psychological Practice with Older Adults in 2003 (American Psychological Association, 2004). The APA guidelines cover general knowledge about adult development and aging, attitudes toward older adults, clinical and educational issues, as well as assessment, intervention, and other service provision. APA recommends that clinicians acquire familiarity with intervention outcome literature for older adults, the ability to adapt skills in therapy to the needs of older adults, and situation-specific knowledge in settings where older adults may typically be seen (e.g., nursing homes). The guidelines also highlight the importance of working in interdisciplinary teams, and the necessity of appropriate referrals.

Clinicians working with older populations need to be cognizant of increasing diversity with increasing age, and the implications of this diversity in terms of research, assessment, and treatment in older populations. Country of birth, gender, ethnic identity, socioeconomic status, sexual orientation, disability status, and urban/rural residence may all influence the experience of aging in both positive and negative ways. Finally, particular ethical issues which may arise in working with older populations (e.g., competency assessment and guardianship issues) are discussed.

The 2006 National Conference on Training in Professional Geropsychology was convened to specify a model for training psychologists to develop competence in treating older adults. The "Pikes Peak" model elaborates geropsychology attitude, knowledge, and skill competencies, and recognizes that there are multiple training pathways to geropsychology competence, through a combination of graduate, internship, postdoctoral, and/or postlicensure training (Knight & Karel, 2006). The Pikes Peak model cites the importance of evidence-based psychological practice with older adults, but does not specifically address CBT competencies.

General guidelines for attaining core competencies in CBT have been developed (e.g., Sudak, Beck, & Wright, 2003), but CBT competencies specific to older adults have not yet been described. Among others, the CBT competencies described by Sudak et al. include the ability to formulate cases according to the cognitive model, to structure sessions to promote progress in therapy, and to monitor progress and outcomes. CBT competencies require elaboration when applied to specific populations such as older adults. For example, a significant age gap between therapist and client may impede the therapist's ability to form a sound therapeutic alliance, and may require additional attention in supervision. Some CBT competencies enumerated by Sudak and colleagues, such as achieving treatment compliance, including homework, have been applied to older adults (Kazantzis, Pachana, & Secker, 2003).

Despite the lack of articulated core competencies in CBT specifically aimed at older adults, outcome studies suggest that CBT are effective for a variety of presenting problems among this population. Gatz, Kaskie, Kasl-Godley, McCallum, and Wetherell (1998) reported in a review article that likely efficacious therapies for this population include cognitive-behavioral treatments for selected disorders (e.g., clinical depression, sleep disorders). Beck and Stanley (1997) and Wetherell, Lenze, and Stanley (2005) report similar efficacies for CBT and anxiety among older adults. More recently, Scogin, Welsh, Hanson, Stump, and Coates (2005) conducted an evidence-based review of psychological treatments for geriatric depression; while their article highlights the efficacy of psychological treatments, it also underscores the need to provide training in evidence-based treatments for the growing numbers of older adults seeking help for mental health problems.

Although the literature base of treatment studies specifically targeting older adults is limited, and thus the efficacy of psychotherapies, including CBT, for many disorders is still unspecified, there are currently several theoretical models of best practice in psychotherapy with older adults (e.g., Knight, 2004) as well as reviews of CBT for older adults that have found CBT to be efficacious among this population (e.g., Laidlaw, Thompson, Dick-Siskin, & Gallagher-Thompson, 2003). This growth in intervention theories and practice tailored to the needs of older adults mirrors the growth in assessment tools designed specifically for older populations such as the California Older Person's Pleasant Events Schedule (COPPES) (Gallagher & Thompson, 1981; Rider, Thompson, & Gallagher-Thompson, 2004 – revised computer version can be downloaded from the Older Adult and Family Center; http://oafc.stanford.edu).

Importance of CBT Competencies in Formal Training Programs as well as Continuing Education

Increasingly, the literature is addressing the specific training needs of students in geriatric mental health more broadly (Knight, Teri, Santos, & Wohlford, 1995; University of California, 2006) and also with particular respect to CBT training in specific disciplines such as psychiatry (Sudak et al., 2003), nursing (Freeman

& Freeman, 2004), and social work (Ronan & Freeman, 2006). Mental health professionals-in-training who aspire to work with older cohorts must learn the applicability of CBT conceptualizations and interventions across many clinical problems and across a range of care settings for older adults. It is important for training programs to help students develop competencies for: geropsychological assessment, often aided by CBT conceptualizations; awareness of special issues in the therapeutic relationship with older adult clients; identification of geriatric health, mental health, or behavioral problems that are amenable to CBT interventions; adaptation of interventions for older adults, their families, and professional caregivers; and modification of interventions in different care settings. These competencies can be attained only through clinical supervision with older adult clients who have a range of presenting problems. For professionals learning CBT skills or their applicability to older adult clients, models of clinical consultation with experienced practitioners may be particularly helpful after formal training has been completed (Pretorius, 2006).

Psychological problems commonly seen in geriatric health and mental health settings include not only depression and anxiety, but also a range of behavioral health issues (insomnia, pain, coping with disability, obesity, smoking), the interaction of medical and psychological issues (e.g., excess disability related to depression), substance abuse, sexual concerns, marital or other relationship concerns, behavioral problems in dementia, and caregiving stress. All of these problems can be addressed, at least in part, with CBT interventions, and all have at least some research demonstrating the benefit of CBT interventions with older adult clients (Gatz et al., 1998; Satre, Knight, & David, 2006). Within the context of a sound therapeutic relationship, a solid assessment, and a treatment plan that is based on a cognitive conceptualization of the case, students and clinicians can assemble a "toolbox" of CBT approaches to address a wide range of presenting problems in older adult clients.

Older adult clients often have multiple, and interacting, problems – medical, functional, psychiatric, social, interpersonal, and spiritual. Geriatric mental health training often focuses on addressing the needs of these complex patients: how to conceptualize the multiproblem patient, how and when to seek further expert consultation to aid assessment, who should be included in assessment and/or intervention (e.g., when to include family members), how to prioritize problems as foci for intervention (e.g., which problems may need to be deferred until others are addressed), whether the intervention should focus on the patient or the care system, whether the level of care is appropriate, whether multiple interventions are needed, and what frequency and duration of intervention(s) are indicated. Additionally, work with complex geriatric patients often challenges students to confront ethical dilemmas (e.g., regarding capacities for informed consent or independent living) and to work collaboratively with professionals from various disciplines (Zeiss & Steffan, 1996).

Training practitioners in geriatric mental health care must focus not only on evidence-based psychological interventions, but also on the ethical, interdisciplinary, and systemic contexts in which such interventions are provided. Students ideally develop competency to conduct assessment and intervention in a range of care settings – from outpatient mental health clinics to primary care (Haley, 1999), rehabilitation, and other inpatient medical settings (Lopez & Mermelstein, 1995), to long-term care settings (Hyer & Intrieri, 2006) and care in the home (Morgan & Weiman, 1990). It is highly desirable for students to gain experience with both healthy, independently functioning older adults, and with physically and/or cognitively frail older adults who depend upon others for some degree of care. Thus, training may include traditional CBT for depression through weekly outpatient psychotherapy visits, as well as brief interventions in medical settings, adaptations of therapy for nursing home residents, and training in behavioral principles for managing difficult behaviors of patients with dementia.

Those who teach or supervise students or clinicians who are learning to apply CBT interventions with older adult clients should have expertise in both gerontology and CBT. Currently, many trainers do not have expertise across both areas; some need further training in adult development, aging and geriatric mental health assessment and treatment and/or in cognitive-behavioral theory and interventions. However, resources are increasingly available for self-study, continuing education, and postgraduate training; see this chapter's appendix.

In brief, training students and clinicians to adapt CBT interventions for older adults includes helping them to understand how aging-related developmental changes (sensory, emotional, cognitive), cohort experiences, and specific late-life challenges and social contexts may affect the process, goals, and outcomes of psychotherapy (Knight & McCallum, 1998). A critical element of training is helping trainees to reflect on their own feelings about aging and disability, and how these attitudes may affect the therapeutic relationship and their own therapeutic optimism or pessimism. Trainees may also need guidance in helping older clients set realistic goals for therapy. Clients and trainees often need assistance in identifying room for improvement in a given situation, or assessing their reactions to it, even if the problem cannot be "cured."

Therapist Skills Training Models

Although theoretical models of CBT and CBT outcome studies have received a great deal of attention in the literature, evaluating the efficacy of training therapists in CBT has received far less empirical consideration. Although the number of mental health professionals seeking training in CBT has increased (e.g., Goisman, 2005) and the benefits of such training has been documented (e.g., Hafner, Crago, Christensen, Lia, & Scarborough, 1996), less attention has been paid to training methods themselves. In particular, theoretical underpinnings of successful training and specific training procedures have not been described and empirically evaluated. Various authors have described key frameworks and abilities that demonstrate therapist skill levels (e.g., Padesky, 1996; Liese & Beck, 1997). Four characteristics of a competent therapist are noted by Shaw and Dobson (1988): guidance by a theoretical or conceptual framework; knowledge and memory of the patient(s)' issues; adept and effective application of methods of intervention; and knowledge of when and when not to apply specific techniques. Other authors have cited therapist attitudes (e.g., Mahoney, 2000) as well as possession of technical and interpersonal skills (e.g., Trepka, Rees, Shapiro, Hardy, & Barkham, 2004) that contribute to successful therapy processes. Beck et al. (in press) describe the importance of teaching case conceptualization skills to therapists so that they can effectively plan treatment within and across sessions.

Recently, Bennett-Levy (2006) proposed a cognitive model for the acquisition and elaboration of therapist skills in both novice and expert practitioners. The model posits that successful application of therapy skills requires declarative or factual knowledge, procedural or skills-based knowledge, and the ability to reflect on one's implementation of this knowledge. Thus, the training of therapists should include means of teaching declarative and procedural knowledge through a variety of means: didactically (e.g., through lectures, observational learning, and supervision) and by practicing the application of these skills (e.g., through role-plays, practice in clinical settings, and supervision). The knowledge gained didactically is transferred and applied in real-world settings. Additionally, reflection on one's practice helps the clinician to gain therapeutic skill and confidence in treating a wide variety of individuals in complex situations.

Bennett-Levy proposes that two forms of reflection come into play in the training and continued evolution of therapist skills: reflection-on-action (consideration of one's actions after the fact) and reflection-in-action (consideration of one's actions while performing the actions). Reflection in treatment is fostered primarily through supervision for the novice therapist, but (ideally) becomes a more self-directed process, often spurred by problem-solving when a difficult case presents or when difficulties arise in the course of treatment. The capacity for self-reflection may be an important mechanism in the development of clinical wisdom (Skovholt & Ronnestad, 1992). The therapist's capacity for self-reflection results in continuous growth in knowledge and skills and ability to individualize treatment. Inadequate self-reflection may result in ineffectual applications of knowledge and skills, and/or poor outcomes in more complex or challenging cases. Skovholt, Ronnestad, and Jennings (1997) pose the question: "A therapist and a counselor can have 20 years of experience or one year of experience 20 times. What makes the difference? A key component is reflection." (p. 365)

Other researchers have developed assessment tools that measure therapist competence, in part to facilitate therapist training. The Cognitive Therapy Scale (Young & Beck, 1980) is a measure of therapist competence that has been used in numerous studies to measure the relationship between client outcomes and therapist outcomes (e.g., in therapy for depression, Trepka et al., 2004). Lambert's Outcomes Questionnaire (OQ-45) allows therapists to objectively monitor a patient's response to therapy with the aim of improving psychotherapy outcomes. Lambert has also proposed an objective system by which therapists can systematically address their therapy process in the case of a patient who is failing to make progress (Percevic, Lambert, & Kordy, 2006). Such a system can demonstrate, for example, that a therapist's assumption, such as early progress in therapy predicts later progress for the same patient, actually may not hold true for an individual patient, and that in fact the opposite may be more common (Percevic et al.); such data can be of great value to therapists-in-training. Innovations in IT have also facilitated the development of interactive problem-based learning paradigms, which are being incorporated into training programs (Weingardt, 2004). Therapist measures contribute to the training of novice therapists, as they allow objective measurement of efficacy which can then be incorporated into supervision, and promote self-reflection.

Training therapists to treat older populations raises particular challenges, including the need for specialized knowledge and practical skills in working with older clients, and the therapist's ability to reflect on how he or she is engaging the older client, as alluded to previously in this chapter. Another barrier to training may be a dearth of both older clients and appropriate supervisors in a training environment; in such circumstances novel use of technology (for example, in creating archives of case material) may be invaluable.

Supervision of Skills Training in Geropsychology

Increasing attention is being paid to the processes underlying effective supervision of psychotherapy skills. Recent publications by Pretorius (2006) and Newman and Beck (in press) provide an overview of recommended practice for CBT supervision, including effective structuring of supervision sessions, techniques for rating therapy sessions, and the nature of the supervisory relationship. It is important to assess the general level of expertise of supervisees in working with older adults and in conducting CBT as well as their specific expertise in using CBT with older adults. While a strong didactic foundation is a critical component of training for mental health professionals, supervised clinical experience is essential to putting the knowledge base into practice with older adults (e.g., Knight & Karel, 2006). Experience in training new therapists and consulting with seasoned therapists new to working with older adults suggests that while many therapists may know what interventions need to be implemented, some are inhibited from using that knowledge due to unexamined stereotypes or well-learned behaviors that interfere with putting knowledge into practice. At times, visual cues that the client is older or frail overpower well-established clinical skills that have been used with other clients.

Assessing older adults is more complex than assessing younger adults; it requires knowledge of neuropsychology and health psychology as well as competence in the assessment of physical functioning in the ADL/IADL sense (for more, see Lichtenberg, 1999). Behavioral approaches for assessing the functioning of older adults, such as direct observation by trained observers and structured reports by the psychologist, health care staff, and family members, can be particularly useful in public settings such as residential care facilities.

Assessment of older adults is enhanced by thoughtful exposure to and experience with a wide range of older adults, from the creatively aging to the extremely frail.

It is difficult to recognize psychopathology in later life if one's sense of normality is grounded solely in experience with younger adults. Clinicians need to distinguish, for example, the kinds of memory complaints that represent the normal aging process from those that signal subjective perception of early changes in a dementing illness. Doing so usually requires the guidance of a clinical supervisor in the same way that differentiating symptoms of paranoid psychosis and the preoccupations of a conspiracy hobbyist in younger patients require such guidance.

Conversely, clinicians may misattribute certain symptoms as stemming from the natural aging process. One of us (BGK) once helped a mental health nurse figure out the diagnosis for a puzzling patient in an inpatient psychiatric unit by asking her to watch him in the day room for a few minutes and pretend he was 35 rather than 75. She turned back quickly and said (correctly, as it turned out with more complete assessment), "He's an alcoholic." Again, experienced clinical guidance is often required to achieve accurate diagnosis in older adults.

A key issue in clinical work with older adults is recognizing and responding effectively to one's ageism in both its negative and positive valences. Supervisees are rarely aware of their ageism. Manifestations of ageism are prevalent and varied and can include (among others): uncritically accepting statements about forgetting to do therapy homework as reflecting age-related memory failure rather than investigating other factors (such as resistance); automatically siding with a younger family member instead of collecting data about a family conflict; discouraging clients from talking about current medical problems as they affect daily life; changing the subject when clients talk about death; refraining from appropriately guiding sessions; avoiding (gentle) confrontation; and failing to appropriately investigate the validity of the client's attribution of problems to aging rather than to other causes (Knight, 2004). In general, therapists do not think of themselves as ageist, and their supervisors need to question them to elicit and test their assumptions about their older adult clients. In this way, clinicians can develop more accurate assumptions and treat their clients more effectively. Clinical supervision requires observation (live or taped) of complete sessions by a supervisor who is competent in professional geropsychology.

Some therapists need to learn new ways of working with older clients. Many have been trained not to interrupt or disagree with older persons and assume that doing so will disturb them. They need to develop the skill of tactful interruption to appropriately guide the therapy session. When their clients express ideas that could be distorted, clinicians need to learn how to help their patients evaluate and respond to their cognitions. They need to learn the art of gentle persistence and motivational skills to persuade reluctant older adults to adopt more adaptive behaviors. They may also need to help avoidant clients develop more adaptive beliefs about experiencing negative emotions before their clients are willing to bring up problems they would rather avoid discussing.

How do these elements of supervision translate into training models? The precise nature of training is likely to vary depending on the resources available to the training program. Settings may differ; for example, older adults may be treated in clinical offices, their homes, in nursing homes, medical clinics or hospitals, or psychiatric inpatient units. The diagnoses of older patients are likely to vary, too – common psychiatric disorders include dementia, schizophrenia, depression, anxiety, and substance abuse. Clinical supervisors themselves should be competent in working with older adults (unfortunately, this is not always the case) and observation of clinical work should be an integral part of the supervisory program. Clinical training that includes supervised experience with older adults is becoming more common in graduate training programs, formal internships, and psychiatric residencies. Developing models for teaching cognitive behavior therapy for older adults, to mental health professionals who develop an interest in working with older adults after their formal training and licensure, is a major challenge for the future (Knight & Karel, 2006).

Conclusions

Training mental health professionals to work in a CBT framework requires a solid understanding of models that facilitate student learning, the provision of appropriate and graded practice experience with older adults in a wide variety of settings, and strong, supportive supervisory experiences. In addition, it is important to foster an awareness of the need for continuing education and professional development throughout the course of a career in mental health. It is therefore imperative that practitioners have access to such continuing educational opportunities, particularly in areas such as CBT for older adults, which have only recently been added to the curricula of many mental health training programs and for which there is a greater need for service provision than there are specialist providers. Such targeted training would allow generalists to specialize in this growing proportion of the population.

References

American Psychological Association. (2004). Guidelines for psychological practice with older adults. *American Psychologist, 59*, 236–260.

Beck, J. S., Sarnat, J., & Barenstein, V. (in press). Psychotherapy-based approaches to supervision. In C. Falender & E. Shafranske (Eds.), *Casebook for Clinical supervision: A competency-based approach*. Washington DC: American Psychological Association.

Beck, J. G., & Stanley, M. A. (1997). Anxiety disorders in the elderly: The emerging role of behavior therapy. *Behavior Therapy, 28*, 83–100.

Bennett-Levy, J. (2006). Therapist skills: A cognitive model of their acquisition and refinement. *Behavioural and Cognitive Psychotherapy, 34*, 57–78.

Freeman, S. M., & Freeman, A. (2004). *Cognitive behavior therapy in nursing practice*. New York: Springer.

Gallagher, D., & Thompson, L. W. (1981). *Depression in the elderly: A behavioural treatment manual*. Los Angeles: University of Southern California Press.

Gatz, M., Kaskie, B., Kasl-Godley, J. E., McCallum, T. J., & Wetherell, J. L. (1998). Empirically-validated psychological treatments for older adults. *Journal of Mental Health and Aging, 4*, 9–46.

Goisman, R. M. (2005). Medical residents seek CBT training. *Behavior Therapist, 28*, 36–46.

Hafner, R. J., Crago, A., Christensen D., Lia, B., & Scarborough, A. (1996). Training case managers in cognitive-behaviour therapy. *Australian and New Zealand Journal of Mental Health Nursing*, *54*, 163–170.

Haley, W. E. (1999). Psychotherapy with older adults in primary care medical settings. *Journal of Clinical Psychology*, *55*, 991–1004.

Hyer, L., & Intrieri, R. C. (Eds.). (2006). *Geropsychological interventions in long-term care*. New York: Springer Publishing Co.

Kazantzis, N., Pachana, N. A., & Secker, D. L. (2003). Cognitive-behavioral therapy for older adults: Practical guidelines for the use of homework assignments. *Cognitive and Behavioral Practice*, *10*, 325–333.

Knight, B. (2004). *Psychotherapy with older adults* (3rd ed.). Thousand Oaks, CA: Sage.

Knight, B. G., & Karel, M. J. (2006). National conference on training in professional geropsychology: The Pikes Peak model. *Adult Development and Aging News*, *34*, 10–12. Available at: http://apadiv20.phhp.ufl.edu/Fall06.pdf

Knight, B. G., & McCallum, T. J. (1998). Adapting psychotherapeutic practice for older clients: Implications of the contextual, cohort-based, maturity, specific challenge model. *Professional Psychology: Research and Practice*, *29*, 15–22.

Knight, B. G., Teri, L., Santos, J., & Wohlford, P. (Eds.). (1995). *Mental health services for older adults: Implications for training and practice*. Washington, DC: American Psychological Association.

Laidlaw, K., Thompson, L. W., Dick-Siskin, L., & Gallagher-Thompson, D. (2003). *Cognitive behavior therapy with older people*. London: Wiley.

Lichtenberg, P. A. (Ed.). (1999). *Handbook of assessment in clinical gerontology*. New York: Wiley.

Liese, B., & Beck, J. S. (1997). Cognitive therapy supervision. In C. E. Watkins, Jr. (Ed.), *Handbook of psychotherapy supervision*. New York: John Wiley & Sons, Inc.

Lopez, M. A., & Mermelstein, R. J. (1995). A cognitive-behavioral program to improve geriatric rehabilitation outcome. *The Gerontologist*, *35*, 696–700.

Mahoney, M. J. (2000). Training future psychotherapists. In C. R. Snyder & R. E. Ingram (Eds.), *Handbook of psychological change* (pp. 727–735). New York: Wiley.

Morgan, A. C., & Wieman, D. W. (1990). Mental health home visits to nonhomebound elderly. *Hospital and Community Psychiatry*, *41*, 1339–1341.

Newman, C. F., & Beck, J. S. (in press). Selecting, training, and supervising therapists in randomized controlled trials. In A. M. Nezu, & C. M. Nezu (Eds.), *Evidence-based outcome research: A practical guide to conducting randomized controlled trials for psychosocial interventions*. Oxford, UK: Oxford University Press.

Padesky, C. A. (1996). Developing cognitive therapist competency: Teaching and supervision models. In P. Salkofskis & D. M. Clark (Eds.), *Frontiers of cognitive therapy: The state of the art and beyond*. New York: Guilford Press.

Percevic, R., Lambert, M. J., & Kordy, H. (2006). What is the predictive value of responses to psychotherapy for its future course? Empirical explorations and consequences for outcome monitoring. *Psychotherapy Research*, *16*, 364–373.

Pinquart, M., & Sörenson, S. (2001). How effective are psychotherapeutic and other psychosocial interventions with older adults? A meta-analysis. *Journal of Mental Health and Aging*, *7*, 207–240.

Pretorious, W. M. (2006). Cognitive-behavioral therapy supervision: Recommended practice. *Behavioral and Cognitive Psychotherapy*, *34*, 413–420.

Rider, K., Thompson, L. W., & Gallagher-Thompson, D. (2004). *The California older person's pleasant events schedule: Computer version*. Stanford, CA: Older Adult and Family Center. Download available from website: http://oafc.stanford.edu.

Ronan, T., & Freeman, A. (2006). *Cognitive behavior therapy in clinical social work practice*. New York: Springer.

Satre, D. D., Knight, B. G., & David, S. (2006). Cognitive-behavioral interventions with older adults: Integrating clinical and gerontological research. *Professional Psychology: Research and Practice, 37*, 489–498.

Scogin, F., Welsh, D., Hanson, A., Stump, J., & Coates, A. (2005). Evidence-based psychotherapies for depression in older adults. *Clinical Psychology: Science and Practice, 12*, 222–237.

Shaw, B. E., & Dobson, K. S. (1988). Competency judgments in the training and evaluation of psychotherapists. *Journal of Counseling and Clinical Psychology, 56* 666–672.

Skovholt, T. M., & Ronnestad, M. H. (1992). *The evolving professional self: Stages and themes in therapist and counselor development*. Chichester: Wiley.

Skovholt, T. M., Ronnestad, M. H., & Jennings, L. (1997). The search for expertise in counseling, psychotherapy and professional psychology. *Educational Psychology Review, 9*, 361–369.

Sudak, D. M., Beck, J. S., & Wright, J. (2003). Cognitive behavioral therapy: A blueprint for attaining and assessing psychiatry resident competency. *Academic Psychiatry, 27*, 154–159.

Trepka, C., Rees, A., Shapiro, D., Hardy, G., & Barkham, M. (2004). Therapist competence and outcome of therapy cognitive therapy for depression. *Cognitive Therapy and Research, 28*, 143–157.

Weingardt, K. R. (2004). The role of instructional design and technology in the dissemination of empirically-supported, manual-based therapies. *Clinical Psychology: Science and Practice, 11*, 313–331.

Wetherell, J. L., Lenze, E. J., & Stanley, M. A. (2005). Evidence-based treatment of geriatric anxiety disorders. *Psychiatric Clinics of North America. Special Issue: Evidence-Based Geriatric Psychiatry, 28*, 871–896.

Young, J. E., & Beck, A. T. (1980). The Cognitive Therapy Scale. Unpublished manuscript, University of Pennsylvania, Phildelphia.

Zeiss, A. M., & Steffen, A. M. (1996). Interdisciplinary health care teams: The basic unit of geriatric care. In L. L. Cartensen, B. A. Eldstein, & L. Dornbrand (Eds.), *The practical handbook of clinical gerontology* (pp. 424–450). Thousand Oaks, CA: Sage.

Appendix

Resources: The following is a partial list of training courses in CBT, professional societies with a gerontological or CBT focus, and significant publications.

Training Courses

North America

- American Institute for Cognitive Therapy (New York, NY) www.cognitivetherapynyc.com

- Assumption College: The Aaron T. Beck Institute in Cognitive Studies (Worcester, lMA) http://www.assumption.edu/gradce/grad/coun_psych/beck.php)
- Atlanta Center for Cognitive Therapy (Atlanta, GA) http://www.cognitiveatlanta.com/)

- Beck Institute for Cognitive Therapy and Research (Philadelphia, PA) http://www.beckinstitute.org

  Center for Cognitive Therapy (Hungington Beach, CA)
  http://www.padesky.com/
  http://www.uphs.upenn.edu/psycct/edu/index.htm)

- Cleveland Center for Cognitive Therapy (Cleveland, OH) (http://www.behavioralhealthassoc.com/About_BHA/Educational_Programs educational_programs.html)
- Freeman Institute for Cognitive Therapy http://www.artcbt.com/Certification Program.html)

International

Flinders University Mental Health Sciences Postgraduate Coursework Program (Adelaide, South Australia)
http://www.flinders.edu.au/courses/postgrad/mentalhs.htm

- Institute of Psychiatry, Kings College (London, England) http://www.cbtcourse.org/)

  Massey University (Auckland, New Zealand)
  http://psychology.massey.ac.nz/grad/spec_pg_dipl.htm

- Mettaa Institute of Cognitive Behavior Therapy (Seoul, Korea) (http://www.mettaa.com/)
- Oxford Cognitive Therapy Centre (Oxford, England) (http://www.octc.co.uk/)
- Romanian Center for Cognitive and Behavioural Psychotherapies (Romanian Assoc. of CBT) (Cluj-Napoca, Romania) (http://www.psychotherapy.ro/)
- SIAD-Centro de Terapia Cognitiva in Guatemala (Guatemala ciudad, Guatemala) (http://www.geocities.com/skeqb/)

  UK Directory of Cognitive and Cognitive-Behavioural Training Courses
  babcp@babcp.com – http://www.babcp.org.uk/training/cbtcourses.htm

Professional Societies (Aging Focus)

American Association for Geriatric Psychiatry
kmcduffie@aagponline.org
http://www.aagpgpa.org/

American Geriatrics Society (AGS)
info@americangeriatrics.org
http://www.americangeriatrics.org

American Psychological Association (APA)
Division 12/II bradkarlin@neo.tamu.edu
http://geropsych.org/
Division 20 jmcdowd@kumc.edu

http://apadiv20.phhp.ufl.edu/
(Both 12/II and 20 Web sites contain useful training information such as examples of course syllabi and useful training videos)

Office on Aging Homepage publicinterest@apa.org
http://www.apa.org/pi/aging/
(resources include practice guidelines and resource guides)

Australian Psychological Society (APS)
Psychology and Ageing Interest Group (PAIG)
npachana@psy.uq.edu.au
http://www.groups.psychology.org.au/paig/

British Geriatrics Society
general.information@bgs.org.uk
http://www.bgs.org.uk/

British Society of Gerontology
info_at_britishgerontology.org
http://www.britishgerontology.org/

European Union Geriatric Medicine Society
secretariat@eugms.org
http://www.eugms.org/

Faculty for Old Age Psychology (British Psychological Society, Division of Clinical Psychology) (PSIGE)
http://www.psige.org/

Gerontological Society of America (GSA)
geron@geron.org
http://www.geron.org/

International Psychogeriatrics Association (IPA)
membership@ipa-online.org
http://www.ipa-online.org/

National Association of Social Workers (Aging Specialty Practice Section)
membership@naswdc.org
https://www.socialworkers.org/sections/default.asp

Royal College of Psychiatrists (UK) – Faculty of Old Age Psychiatry
rcpsych@rcpsych.ac.uk
http://www.rcpsych.ac.uk/college/faculties/oldage.aspx

Professional Societies (CBT Focus)

For a listing of CBT societies in 29 different countries, see:
European Association for Behavioural and Cognitive Therapies
http://www.eabct.com/

Association for Behavioral and Cognitive Therapies (ABCT)
(there is an interest group on Behavior Therapy and Aging within the society)
http://www.abct.org/sigs/sigs.html#BehaviorTherapy

Australian Association for Cognitive Behavioural Therapies
http://www.aacbt.org/

British Association for Behavioural and Cognitive Psychotherapies
http://www.babcp.org.uk/

International Association for Cognitive Psychotherapy
http://www.cognitivetherapyassociation.org/

International Directory of Professional Societies (CBT)
http://www.cbtarena.com/resources/societies.asp

Manuals and Publications of Note

American Association of Psychiatry Residency Training CBT Competencies
http://www.aadprt.org/training/tools/cbtc.pdf

APA Division 20 Guide to Graduate and Post-Doctoral Programs
http://apadiv20.phhp.ufl.edu/guide01.htm

APA Guidelines for Psychological Practice with Older Adults
http://www.apa.org/practice/Guidelines_for_Psychological_Practice_with_Older_Adults.pdf

Australian National University Mood Gym
http://moodgym.anu.edu.au/

Journal of Cognitive Psychotherapy: Special Issue on Supervision
http://www.cognitivetherapyassociation.org/journal/v12.aspx#number2

Laidlaw, K., Thompson, L. W., Dick-Siskin, L., & Gallagher-Thompson, D. (2003). *Cognitive behavior therapy with older people*. London: Wiley.

Older Adult and Family Research and Resource Center
(educational and research resources on CBT interventions with older individuals, families and caregivers)
http://med.stanford.edu/oac/

Reynolds, C. F., Arean, P. A., Lynch, T. R., Frank, E. (2004). Psychotherapy in old-age depression: Progress and Challenges. In S. P. Roose & H. A. Sackeim (Eds.). Late-life depression. London: Oxford University Press.

Yost, E. B., Beutler, L. E., Corbishley, M. A., & Allender, J. R. (1987). *Group cognitive therapy: A treatment approach for depressed older adults*. New York: Pergamon Press.

20
The Role of Positive Aging in Addressing the Mental Health Needs of Older Adults

R. D. Hill and E. Mansour

The utility of cognitive-behavioral therapy (CBT) in the treatment of psychiatric disorders among adults has been clearly established (Hollon, Stewart, & Strunk, 2006). The current volume covers the range of specific mental health issues that can be ameliorated through CBT in older adults. However, extending CBT to the broader issues of adaptation in old age requires new paradigms that encompass pathways to mental health through well-being and the skills needed to derive meaning in later life. This latter point is underscored in a recent review of the role of positive psychology in the future provision of mental health therapy. Duckworth, Steen, and Seligman (2005) note that:

> ...we believe that the job of the therapist of the future will not be simply to relieve the negative, but to help clients build the pleasant life, engaged life, and meaningful life. We call the techniques that build these three lives "positive interventions." (pp. 640–641)

Duckworth et al. (2005) suggest in their review that "positive interventions" in the future will have as a focus the whole person as opposed to narrowly defining the individual and her or his issues within a limited diagnostic category that includes a discrete set of symptoms or problematic behaviors. In fact, the preventive literature is clear that what most older persons who enter therapy want is not only to be free from the symptoms associated with psychiatric disease, but also to learn skills and to educate themselves in strategies that will help them cope better with circumstances of living that may arise in the future (Hill, Thorn, & Packard, 2000).

This notion of a positive intervention framework for counseling older adults fits well with a contemporary conceptualization of mental health, mental illness and its treatment that was operationalized by Keyes (2005; see also Keyes, 2002). In this scheme, a person's pursuit of better mental health would be considered flourishing (+) if she or he possessed high levels of positive emotion including active social engagement, the cultivation of meaningful personal relationships, and a positive psychological outlook. The absence of mental health, in contrast, Keyes labeled as languishing (–) or experiencing feelings of emptiness, loss, or stagnation associated with the lack of ability to engage one's environment to sustain a sense of purpose and meaning in life. In a sense, when a person experiences a lack of mental health he or she may report a perception of being cut off from internal resources that are needed to feel better (e.g., "I want to feel better, but I am so fatigued that I just don't have the energy to feel good.").

A person's effort to alleviate the symptoms of mental illness is qualitatively distinct from the pursuit of mental health in several important ways. First, this pathway involves the presence of active issues or reported symptoms of emotional distress and this distress is tied to a diagnosable psychiatric condition such as clinical depression or generalized anxiety disorder. Second, the presumption exists that one can "feel better" if those distressing symptoms are reduced which means that the effects of an intervention should focus on the symptoms of psychiatric disease that are diminished as a consequence of the intervention. Therefore, a person who is diagnosed with a psychiatric condition such as clinical depression would be in a state of floundering (–) if he or she felt at the mercy of the depressive symptoms. In contrast, the active effort to reduce depressive symptoms, as might emerge from an intervention such as CBT, provides the individual with a greater pool of resources to address such symptomatology. This process, Keyes labeled as struggling (+).

A critical element of this two-component heuristic is that those who are moving toward mental health (flourishing) and are actively working to diminish psychiatric symptoms (struggling) will more likely be engaged in coping or acting on their environment to enhance well-being or reduce distress. The opposite is the case for a person who may feel overpowered by her or his psychiatric symptoms and experience a sense of impotence that nothing can be done to improve the situation either through symptom reduction or finding ways to feel healthier.

Distinguishing mental health and its promotion from mental illness and its treatment opens the door to principles of positive aging, an extension of the positive psychology movement that focuses on the role of psychological resources in sickness and in health to optimize adaptation in old age (Hill, 2005). It could be assumed from a positive aging framework that therapeutic interventions designed to promote mental health would foster engagement and solution finding. Ameliorating negative symptoms (e.g., challenging automatic maladaptive thoughts) would also characterize an active effort to employ strategies that diminish symptom distress. This is not to say that symptom reduction is less important to well-being than finding ways to foster one's own mental health, but building the capacity to cope with one's challenges in either way is the key to positive aging. This is also consistent with a CBT approach to intervention which involves the learning of skills that can reduce psychopathology and/or engaging one's internal and contextual resources to experience overall well-being in old age (Laidlaw, Thompson, & Gallagher-Thompson, 2004).

The process of flourishing and struggling to preserve mental health in old age involves the recruitment potentialities that lay dormant (or unused) within the individual. The idea of accessing unused capacity to address age-related decline was articulated in the 1960s and 1970s as reserve capacity (Fries & Crapo, 1981). In this regard, reserve capacity, described as organ redundancy, is the explicit resource that allows the human physiology to maintain homeostasis even in the presence of substantial environmental stressors. Early proponents of this organ reserve hypothesis asserted that aging not only diminishes organ function but also erodes this reserve capacity, in effect making an older person more susceptible to disease and/or death than a younger person per se.

The notion of reserve capacity has been extended more recently to include cognitive reserve capacity or strategies that the brain employs such as recruiting neural pathways that are more resistant to age-related decline as a means to offset age- or disease-related cognitive loss (Stern, 2002). Of interest in this conceptualization is the notion of psychological plasticity or the ability of the "mind" to facilitate coping through the intentional generation of compensatory strategies or skills (Baltes, 1995). In fact, Baltes and Baltes (1990, p. 20) have used the term "adaptive capacity" to characterize coping by accessing psychological reserves. Psychological reserve capacity is a key element in positive aging because it raises the possibility that an older person can recruit latent psychological resources to preserve subjective well-being and mental health in old age.

Specific to positive aging is the proposition that older adults possess the ability to learn new skills and techniques which can work to offset the shift from gains to losses that is a predictable consequence of the aging process. For example, an individual dealing with memory loss may access reserve capacity to cope with such deficits by engaging specific strategies which could range from jotting down the location of her car in a parking lot prior to leaving the car or initiating a new routine to place her keys and cell phone in the same spot each time she returns home, or even learning to utilize mnemonic devices to enhance encoding and recall of material that she cannot write down such as personal identification numbers. The concept of reserve capacity also has implications for therapeutic interventions for treating mental illness and/or preserving mental health as well. The question here, however, is: What kinds of skills does a person need to acquire to recruit reserve capacity to deal with a mental health concern?

SOC and Reserve Capacity

Baltes and his colleagues (Baltes, Staudinger, & Lindenberger, 1999) have described a behavioral typology through which older persons access reserve capacity which they labeled as: (1) selectivity, (2) optimization, and (3) compensation or SOC. In brief, older adults who are better at adapting to the aging process may employ SOC in the following ways: first, they make numerous strategic discriminations to limit their options or life choices in order to optimize functioning as their faculties decline (selectivity). They also engage and/or rehearse behaviors through conscious repetition or practice to maintain the capacity so as to continue to perform tasks essential to everyday functioning and, in this way, stave off, through strategy use and rehearsal, the forces of loss through age-related performance efficiency (optimization). Finally, if limiting one's options or engaging in practice or rehearsal are insufficient to preserve functioning, an older person may seek out adjunctive support or instrumental assistance to help preserve functioning. Functioning through the utilization of such aids allows continued performance of the task (compensation) even though that performance is no longer possible without such assistive support. Examples of compensation may include such things as a walker or wheelchair to maintain mobility when one is not able to walk unaided, the use of notes or reminders to facilitate recall of items that one is not able to memorize, or the employment of a hearing aid or spectacles to preserve the ability to utilize sights and sounds when one is experiencing progressive hearing or vision loss (Bäckman & Dixon, 1992).

It is within this framework that notions of positive aging have emerged as an extension of SOC that is specific to the preservation of mental health. For example, positive aging highlights the possibility that in health or in illness a person continues to have the propensity to act on his or her environment to sustain quality of life (flourish) or alleviate symptom distress (struggle). In this regard, the evidence base for CBT strategies as sources of reserve capacity generation varies; however, problem solving (or active engagement of one's resources) point to specific techniques (or skills) that are inherent within CBT for enhancing one's ability to pursue mental health or to alleviate symptoms of psychiatric disease in order to feel healthy in later life (Karwoski, Garratt, & Ilardi, 2006). The special application of CBT as a positive aging approach to well-being merges with four capacities that Hill (2005) has identified as characterizing "positive agers" including the ability to: (1) deal with age-related decline through mobilizing personal resources to remain satisfied with life even in the course of decline, (2) make affirmative lifestyle choices that emphasize the preservation of psychological and physical resources, (3) invoke novel (or flexible) strategies of behaving and/or thinking that optimize adaptation, and (4) the propensity to focus on the "positives" versus the difficulties in life (e.g., seeing the glass as half full).

These characteristics represent skill pathways through which older adults access reserve capacity and there is a growing literature that has identified these in a variety of forms as they relate to coping in both health and illness among older persons (Hill, Thorn, & Packard, 2000). What is noteworthy about these characteristics is that they engage a mindset that is associated with reframing one's feelings and the context in ways that empower the individual to work through problems or difficulties versus to become stagnant or passive in the presence of challenges or as Keyes has described as languishing or stagnating. In this regard, not only is engagement key, but also finding novel ways to adaptively cope by utilizing one's environment is important.

With respect to psychotherapeutic interventions such as CBT and its strategic approach to ameliorating psychopathology, its techniques can be framed within a positive aging approach to mental health intervention. In CBT, the goal is to help a client learn alternative ways of thinking that will then influence her emotional state yielding positive behavioral consequences. Specific CBT strategies and how they can be employed to engender positive aging are described next.

CBT and Positive Aging Characteristics

Dealing with Age-Related Decline

The first characteristic of positive aging is manifested in a person's ability to directly deal with age-related decline and its effects. CBT can be applied to address general issues of age-related decline through cultivating of skills to facilitate how one interprets this universal old-age concern. The fact that age-related decline is a natural part of growing old and that it interacts with many features of the aging process, one's appraisal of the meaning and nature of decline is critical to preserving well-being as one ages. While it is true that decline is inevitable as one grows older, from a CBT perspective there are a number of automatic thoughts about age-related decline and loss that can block an older person from accessing reserve capacity for coping with decline. For example, the literature is clear that memory deficits are a consequence of age-related decline and that older people frequently complain about memory problems such as memory for faces and names. With this in mind, to deal with the predictable psychological consequences associated with memory (or general cognitive) decline and deficits due to aging, a CBT framework can be employed as a resource to help restructure stereotypical assumptions about cognitive loss for the purpose of making adaptation easier.

In his original writings, Beck (1976) asserted that, "...psychological problems can be mastered by sharpening discriminations, correcting misconceptions, and learning more adaptive attitudes." (p. 20) With respect to misconceptions about cognitive decline, the first faulty assumption is that it is uniform for everyone, and, once it begins, there is nothing that can be done about it. The facts related to decline and its mitigation, however, are much more complex than this generalized notion. For example, not all decline is unavoidable. While it is true that one's physical stamina and/or speed of responding may diminish in later life, there are ways to deal with this phenomenon through a variety of coping mechanisms.

Brandtstädter (1990) has proposed two types of coping based on the concept of assimilation or accommodation. Assimilative coping refers to one's ability to find new ways to maintain specific activities that become more difficult due to age-related decline. For example, an older man who has trouble remembering to take his daily insulin injections each morning, might decide to use a timer to cue him to administer the injection at the appropriate time each day (assimilation). Later, if his "timer cue" becomes more difficult for him to coordinate, he might simply let go of his responsibility for this behavior and employ an in-home service nurse to provide these daily injections at his home. In this latter case, the nurse assumes the responsibility so that he no longer needs to concern himself with the task. He is then free to let go of the worry or fear that he will not be able to manage administering this injection on time (accommodation). This shifting of meaning and emotion is a central component of CBT, and can be applied to address normative changes associated with age-related decline.

Making Affirmative Lifestyle Choices

The ability to make good decisions is a distinguishing characteristic of positive aging and is central in adaptive coping. Good decisions preserve resources by focusing on activities that are of the highest value while at the same time simplifying one's life by letting go feelings of guilt for those activities that are discarded (selectivity). The challenge in old age is how to make such decisions and then alter old routines and habits that are no longer meaningful or useful but are maintained by sheer force of will or duty. At its most basic level, CBT teaches a client how to construe such decisions so as to empower the client to engage her or his resources most efficiently (optimization). For example, holiday cards are a well-known Christmas tradition. Across the adult life span we collect names and addresses of our friends and associates and mail cards out during the holiday season. In old age, as friends become more distant, family members move away, and neighbors or co-workers change, the need to send this volume of cards to all of one's friends, family, and associates may no longer be needed or even desired, but may still continue because this is what one has done in the past. However, as a person moves into later life it may be that he or she decides to reduce the number of cards sent during the holidays and focus only on those individuals with whom one has an ongoing relationship. Such a decision might be difficult, especially in the face of a longstanding belief that everyone I ever knew "must" receive a card. However, this reasoning for sending cards may not match up with the meaning that holiday cards were intended to convey (sustaining an ongoing relationship). Engaging in a reevaluation process that modifies this task underscores the strategy of prioritizing one's motivations. Why send cards to sustain a ritualized behavior that has lost its meaning to the giver and receiver?

Invoking Novel Problem Solving Strategies

According to Lazarus and Folkman (1984) and Lazarus and Lazarus (2006), flexibility is a psychological coping mechanism that involves balancing one's own skills and resources against the demands of a given task. Baltes and Baltes (1990) use the term plasticity to represent the capacity to change in response to training, presumably through learning. In either case, a basis of adaptive coping is the ability to construe a problem or issue in such a way as to make novel solution strategies possible. This capacity for coping flexibility can be seen in examples of individuals who are able to "let go" of past regrets through reinterpreting them in a more constructive light. For example, reframing a personal error as a learning experience and an opportunity for improvement can free a person to employ reserve capacity with respect to the errant event.

A central goal of CBT is to build greater flexibility in approaching problems or issues through cognitive restructuring, that is, the conscious reconstrual of an event and its appraisal in a more optimistic light. A challenge in old age is to live in the present. It is all too often that older persons are caught in unpleasant past memories or in the negative construal loops of judging the future through the lens of past failures. A powerful component of CBT is learning the skill of living in the present. This strategy can be engaged through techniques such as systematic relaxation which focuses a person on her or his "present" situation and deals with emotions and feelings in the moment while setting aside fears or concerns about the future and resentments or regrets about the past. A novel approach to problem solving, therefore, would be to reduce worry about the future through a present-focused approach to living.

Focusing on the "Positives"

Positive aging involves cultivating an optimistic future outlook or, in a more direct sense, a bias in interpretation toward favorable outcomes. The old adage that "we get what we expect out of life" is a truism that fits positive aging in as much as those who are positive agers develop the ability to expect positive outcomes. This conscious construal of life through the lens of positive appraisal represents a skill in accessing coping reserves to generate such appraisals. The nature of optimism as an activating agent of positive emotion (and presumably positive aging) has been well documented (Brissette, Carver, & Scheier, 2002). In a similar way, one of the outcomes of CBT is to help a client learn how to anticipate positive outcomes, that is, the ability to optimistically evaluate one's self in the face of ambiguous future life circumstances (e.g., following the loss of a spouse; see Hill, 2005). Specific strategies embedded within CBT are designed to cultivate positive emotion and optimism that include techniques for increasing the frequency of positive events. Studies have documented that optimistically appraising one's self in relation to age-related change can enhance one's sense of identity as an older person leading to a greater sense of personal satisfaction for the "things that one can experience" in old age as opposed to focusing on experiences that are no longer possible (Freund, Alexandra, & Baltes, 1998).

Meaning-Centered Strategies and Coping Capacity

The role of CBT as an agent for cultivating positive aging characteristics has been examined under the implicit assumption that as one cultivates such characteristics, these in turn, lead to greater access to reserve capacity which can then be applied to address symptoms of mental illness (through symptom reduction *or struggling*) or to help an older person preserve positive mental health (*flourishing*). Techniques specific to CBT have been described as well as how these techniques address the general range of old age concerns highlighted above. These included increasing pleasant activities, focusing on success (as opposed to failure experiences), monitoring one's emotions (e.g., the frequency of positive versus negative self-construals), as well as learning and practicing present-focused living. Embedded in these techniques is the idea that there are strategies that are inextricably connected to meaning across the human life span, and that these strategies can be incorporated as part of a CBT framework to enhance coping. Meaning-centered strategies including gratitude, altruism, and forgiveness are described next along with how they can be incorporated as part of a positive aging CBT approach to therapy.

Gratitude

Gratitude is a commonly recognized positive emotion that is connected to feelings of appreciation that one experiences as the beneficiary of a gift or an altruistic act (Emmons & Paloutzian, 2003). The essence of gratitude is one's internal sense of appreciation as the "receiver" of a gift, and, at its most profound level, that the gift is undeserved. Appreciation and its psychological consequences, therefore, are at the heart of gratitude. Most people are familiar with not only the concept of gratitude, but also its impact on their emotional state. For the older adult, this emotion has likely been part of a lifespan of living well even in the presence of adversity.

The psychological benefits of gratitude are embedded not only in its religious traditions, but also in the role that gratitude plays in enhancing one's sense of well-being. It shares many of the same positive emotions as is found in hope, love, and forgiveness. Empirical evidence has documented that gratitude can be a powerful intervention strategy to mediate negative mood states as an emotional response that is incompatible with negativity. Gratitude is linked to positive aging as a meaning-centered life-span strategy that is predominantly cognitive (as opposed to altruism that is primarily based on the action of helping), which enables a person to reframe negativity. For example, to make one feel better after the experience of a loss, a counselor might encourage the sufferer to be grateful for what remains. In other words, "count your blessings." Whether gratitude works in the same way for all persons and under all circumstances is unknown; however, there is compelling evidence that for many people across cultural and ethnic groups, gratitude is uniformly viewed in traditions and beliefs and is considered to be a powerful source of positive emotion.

A central element of CBT is the notion of attribution or the propensity of a patient to construe certain stimuli in ways that precipitate a negative or catastrophic emotional response. If, for example, a patient becomes aware of an error or mistake that one makes during the course of one's daily routine, he or she might make the attribution that "I'm useless or I can't do anything right." In this instance, gratitude could be a powerful strategy to empower the patient to reframe the perception of a mistake by suggesting a healthier reattribution as follows: "Yes, I made one mistake, but I'm grateful for the 10 correct decisions that I made today." Learning how to be grateful is a natural way to prevent self-condemnation. Many older clients will readily understand the concept of gratitude even if it may be difficult to apply in some contexts (e.g., expressing gratitude for your challenges in life). Thus, gratitude can be an important positive aging skill in the counselor's CBT repertoire to foster reattribution of maladaptive automatic thoughts.

Altruism

The essence of altruism is that one acts in a positive manner toward others with no expectation of repayment or personal gain in return (Emmons & Paloutzian, 2003). Volunteerism is a well-known behavioral manifestation of altruism and studies have found that volunteerism engenders positive emotions and that those who volunteer are more likely to live longer than nonvolunteers (Oman, Thorensen, & McMahon, 1999).

Applying altruism as a meaning-centered strategy within a CBT intervention framework involves assisting the client in identifying opportunities for altruistic expression. This could be as straightforward as volunteering in a local service organization or in finding ways to help members of one's family who are in need. In fact, altruism includes not only other people, but animals and things as well. Because the benefits of altruism are almost entirely intrapersonally generated, the receiver need not even acknowledge an altruistic act for it to have a positive impact on the helper.

From a CBT framework, it may be that helping other people can be a powerful source of positive emotions that can enable access to psychological reserve capacity. It is important to be wanted and to feel like one's life is meaningful. This is often accomplished by an "other" focus or the developing of a frame of mind in which a person subordinates the needs of another person to one's own needs. Engaging in altruism through joining a volunteer group may be especially beneficial for those older adults who are isolated or are otherwise disengaged. Joining a volunteer group could allow for the benefits of engagement coupled with altruism to optimize well-being.

Forgiveness

Forgiveness involves absolving oneself of negative feelings toward an offender or an offense by means of a compassionate response. From the psychological literature, Robert Enright and Fitzgibbons (2000) have defined forgiveness as the "willful giving up of resentment in the face of another's considerable injustice and responding with beneficence to the offender even though the offender has no right to the forgiver's moral goodness." (p. 80). From a CBT perspective, forgiveness is a form of emotional reframing or reattribution of vengeful thoughts with positive emotions (Hill, 2005). In this regard, forgiveness could be conceptualized as a cognitive skill that enhances one's satisfaction by allowing a person to "let go" of resentment. Contemporary research on forgiveness has examined it as an adaptive coping strategy amenable to intervention. For example, in one study, a forgiveness intervention consisting of an 8-week course on "how to forgive" along with practice utilizing forgiveness skills was administered to 24 elderly women who reported that they were victims of some wrongdoing (Hebl & Enright, 1993). At the conclusion of the course, these women reported feeling less hatred and anger toward their wrong doers in addition to an improved sense of well-being including an increased willingness to help others become better forgivers.

In many respects, this form of intervention fits well within a CBT framework of helping clients reconstrue maladaptive thoughts (e.g., feelings of resentment or a desire to seek revenge) through a powerful, naturally occurring, cognitive restructuring technique, namely, forgiveness to generate new appraisals. As a strategy that is effective in providing an emotional correction for maladaptive thinking, whether it is real or perceived, forgiveness is a mechanism that can access psychological reserve capacity.

In summary, CBT is an action-oriented strategy for engendering positive appraisal by addressing maladaptive automatic thoughts, increasing pleasant events, and cultivating a present-focused orientation. In this regard, CBT is consistent with positive aging as a pathway for: (1) preserving mental health in old age and (2) dealing with psychiatric symptomatology as it occurs in later life. Meaning-centered life span strategies including gratitude, altruism, and forgiveness were highlighted as universal constructs that are part of a positive aging approach to emotional regulation. As facilitators of characteristics of positive aging, these meaning-centered strategies provide adjunctive support to CBT. The incorporation of these constructs within a CBT framework can increase the salience of a CBT with older clientele. These strategies may be a potent source of coping with age-related decline and the emotional consequences of advanced aging including, bereavement, cognitive decline, physical disability, and a loss of purpose for living.

The case below highlights how these strategies might be conceptualized in CBT with an older couple. A noteworthy aspect of the case is that the couple is from an ethnic minority group. Because gratitude, altruism, and forgiveness cut across cultural, economic, and religious boundaries, they increase the likelihood of extending CBT to older adults who vary in ethnic and cultural backgrounds. Such an approach may be more acceptable to older clients who might distrust a Western-oriented approach for assisting them in addressing a mental health concern. Couching CBT through these universally accepted meaning-centered strategies may produce greater impetus for personal change regardless of the older person's living situation, personal and/or cultural characteristics as noted in the case below.

Case Presentation

Mary (62 years old) and Omar (64 years old) recently immigrated to the United States leaving behind the place that they had called home from childhood and for most of their lives. The economic hardship in their home country had left them with little money and not enough savings for retirement. Omar felt hatred toward the occupying force in his country and blamed their hardships on its policies. Having secured a job in the USA, Omar and Mary relocated with the hopes of being able to secure a better retirement. Although all of their children had immigrated to the USA many years earlier, these children did not live in the same state as Omar and Mary, and so they were not able to have regular contact with their parents. This was of some concern to the children because they were concerned that Mary and Omar would have difficulty adjusting to living in the USA given their limited English speaking ability.

Mary and Omar did successfully immigrate and had settled into a semi-comfortable living routine, but as the couple grew older, Omar began to experience persistent memory problems that became more progressive over time. For example, he began to occasionally forget family members', including his own grandchildren's, names and telephone numbers. He also experienced great difficulty with word finding and would frequently substitute inappropriate words, including expressions from his native language, which made much of his communication with others, including Mary and his adult children, incomprehensible.

Omar was also experiencing some changes in his personality – becoming very irritable and reporting frequent bouts of anxiety especially when he was confronted with family activities that had been common for him to engage in (e.g., attending birthday parties with his grandchildren). Recently, while shopping in a grocery store with Mary, he became lost and confused and was unable to find the exit to the store.

Mary was exceedingly worried about Omar's condition and mentioned it to her children. Mary was convinced that Omar should return home to his country of origin where things, she believed, would be more familiar to him. The family, however, did not have the financial means or the extended relationship connections for this to happen. One of Mary's children suggested that Omar visit her family doctor. When this visit happened, the doctor diagnosed Omar with Alzheimer's disease (AD). The diagnosis put Mary in a panic. Omar was the center of her life and she felt that if he carried the diagnosis of AD, then their life was essentially over. This feeling of being overwhelmed was further compounded by the fact that her children lived in different states and she had no regular friendships or neighborhood contacts. Mary's intense despair troubled her children. Although they were concerned about their father's safety, if Mary was in such poor mental health, she could be at risk for self-harm. Eventually, the children persuaded Mary to enter counseling even though Mary stubbornly objected to this because she felt that her situation was a family matter and outsiders would just make things worse.

The counselor who worked in the medical complex where Omar had received his diagnosis was sensitive to Mary's cultural background and treated the situation and treatment as a family issue. In doing so, he invited two of the daughters to attend Mary's sessions monthly. He taught the daughters how to engage in simple tasks to assess Omar's changing cognitive and physical condition. At first Mary was relieved to have the support and help that she felt she was lacking. However, as time went on, Mary began to feel guilty that her children were too involved in her problems and began asking them not to help because Mary believed that she was "robbing" them of their own lives. This was exaggerated to the point where Mary began to think that her children secretly wished that she and Omar should have never immigrated to the USA in the first place and that their move had precipitated Omar's medical condition. This issue was directly addressed in counseling. The counselor helped Mary realize that her children's motive was to express gratitude to their parents for helping them begin life in the USA. One mechanism for showing their appreciation was to engage in altruistic service on behalf of Mary and Omar. As part of therapy, the counselor encouraged the children to each draft a letter that expressed their gratitude for Mary and Omar. Sessions followed during which these letters were shared with Mary. Over time, Mary began to reframe her beliefs in the direction of allowing her children to help in this way as an opportunity for them to develop into better persons through altruistic service to their parents.

Although Mary began to feel more comfortable accepting help from her children, she was troubled by her persistent fear that her life was "over" because of Omar's condition. He had been her long-term companion and she questioned how she could continue living as Omar's spouse. The counselor helped Mary reconstrue this global attribution. He focused on Mary's new role as a caregiver for Omar and through several counseling sessions they explored areas where Omar and Mary could continue to maintain their marital relationship even though Omar could not adequately express his love for Mary. Counseling also focused on helping Mary disengage from Omar and learn how to look after herself in order to be a better caregiver for her husband. Sessions followed where the counselor explored Mary's hesitancy to place Omar in respite care.

Working from the adage that was consistent with Mary's cultural tradition, that is, Mary's view that "Trouble shared is trouble halved," Mary joined a local Alzheimer's caregiver support group. At this support group she was able to meet people who lived near her and even one member of the group who was from Mary's home country who was also dealing with the same problems and issues. This decreased Mary's sense of loneliness and gave Mary a chance to make new friends, thus increasing her social support network. Mary's counselor also encouraged her to engage in volunteer work through her church even though Mary felt she had no time for this. He suggested that Mary consider placing Omar in respite care for longer periods of time so that Mary would have more hours to volunteer. Through this process, Mary became even more active in her church and was able to strengthen her social support network from which she drew spiritual strength.

One result of the increased social support was Mary's growing confidence in her ability to successfully negotiate caregiving tasks such as bathing and dressing Omar. Counseling then began to focus on ways in which Mary could keep her mood positive by encouraging Mary's children to become even more involved in Omar's care. Because of the previous counseling sessions where the issue of her children's altruism had been a focus, Mary willingly accepted increased help from her children and in this way her relationship with her children became even more meaningful and Mary began to understand that this involvement helped her children maintain close ties with their cultural heritage and traditions.

One of the main purposes of counseling was to allow Mary to approach Omar's care from a place of love instead of regret and/or fear. Eventually, Mary was able to approach caregiving from a more optimistic perspective. Mary realized that Omar's disease had changed her life and their relationship, but that Omar remained important to her. Mary was able to establish new goals including prioritizing her health so that she could remain strong and capable of helping Omar. Mary learned over time to appreciate the time that she and Omar had remaining (gratitude) and accepted life as it unfolded. She was no longer paralyzed by the future.

In summary, this case study demonstrates how a CBT approach, blended with aspects of positive aging, can be employed with a couple whose values and world view represented substantial cultural diversity. The role of the family, church, and one's neighborhood were important components in this intervention. Mary learned to alter her attributions related to her changing situation and she eventually was able to derive personal benefits from the caregiver role. Gratitude and altruism were essential components in this intervention as they were active strategies employed by the therapist to alter Mary's cognitive schema for caregiving.

References

Bäckman, L., & Dixon, R. (1992). Psychological compensation: A theoretical framework. *Psychological Bulletin*, *112*, 259–283.

Baltes, M. M. (1995). Dependency in old age: Gains and losses. *Current Directions in Psychological Science*, *4*, 14–19.

Baltes, P. B., & Baltes, M. M. (1990). Psychological perspectives on successful aging: The model of selective optimization with compensation. In P. B. Baltes & M. M. Baltes (Eds.), *Successful aging: Perspectives from the behavioral sciences* (pp. 1–34). Cambridge, MA: Cambridge University Press.

Baltes, P. B., Staudinger, U. M., & Lindenberger, U. (1999). Lifespan psychology: Theory and application to intellectual functioning. *Annual Review of Psychology*, *51*, 471–507.

Beck, A. T. (1976). *Cognitive therapy and the emotional disorders*. New York: International Universities Press.

Brandtstädter, J. (1990). Tenacious goal pursuit and flexible goal adjustment: Explication and age related analysis of assimilative and accommodative strategies of coping. *Psychology and Aging*, *5*, 58–67.

Brissette, I., Scheier, M. F., & Carver, C. S. (2002). The role of optimism in social network development, coping, and psychological adjustment during a life transition. *Journal of personality and Social Psychology, 82*, 102–111.

Duckworth, A. L., Steen, T. A., & Seligman, M. E. P. (2005). Positive psychology in clinical practice. *Annual Review of Clinical Psychology*, 1, 629–651.

Emmons, R. A., & Paloutzian, R. F. (2003). The psychology of religion. *Annual Review of Psychology*, *54*, 377–402.

Enright, R., & Fitzgibbons, R. P. (2000). *Helping clients forgive*. Washington DC: American Psychological Association.

Fries, J. F., & Crapo, L. M. (1981). *Vitality and aging*. San Francisco: W. H. Freeman.

Freund, A., Alexandra, M., & Baltes, P. B. (1998). Selection, optimization and compensation as strategies of life-management: Correlations with subjective indicators of successful aging. *Psychology and Aging*, *13*, 531–543.

Hebl, J. H., & Enright, R. D. (1993). Forgiveness as a psychotherapeutic goal with elderly females. *Psychotherapy*, *39*, 658–667.

Hill, R. D. (2005). *Positive aging: A guide for mental health professionals and consumers*. New York/London: W. W. Norton & Company.

Hill, R. D., Thorn, B. L., & Packard, T. (2000). Counseling older adults: Theoretical and empirical issues in prevention and intervention. In S. D. Brown & R. W. Lent (Eds.), *Handbook of counseling psychology*. New York, NY: Brown & Little.

Hollon, S. D., Stewart, M. O., & Strunk, D. (2006). Enduring effects for cognitive behavior therapy in the treatment of depression and anxiety. *Annual Review of Psychology*, *57*, 285–315.

Karowski, L., Garratt, G. M., & Ilardi, S. S. (2006). On the integration of cognitive-behavioral therapy for depression and positive psychology. *Journal of Cognitive Psychotherapy: An International Quarterly*, *20*, 159–170.

Keyes, C. L. M. (2005). Mental illness and/or mental health? Investigating axioms of the complete state model of health. *Journal of Consulting and Clinical Psychology*, *73*, 539–548.

Laidlaw, K., Thompson, L. W., & Gallagher-Thompson, D. (2004). Comprehensive conceptualization of cognitive behaviour therapy for late life depression. *Behavioural and Cognitive Psychotherapy*, 32, 389–399.

Lazarus, R. S., & Folkman, S. (1984). *Stress, appraisal, and coping*. New York: Springer Publishing Company.

Lazarus, R. S., & Lazarus, B. N. (2006). *Coping with aging*. New York: Oxford University Press.

Oman, D., Thorensen, C. E., & McMahon, K. (1999). Volunteerism and morality among the community dwelling elderly. *Journal of Health Psychology*, *4*, 301–316.

Stern, Y. (2002). What is cognitive reserve? Theory and research application of the reserve concept. *Journal of the International Neuropsychological Society*, *8*, 448–460.

21
How Medicare Shapes Behavioral Health Practice with Older Adults in the US: Issues and Recommendations for Practitioners

Paula E. Hartman-Stein and James M. Georgoulakis

The Medicare system is the largest healthcare insurance plan in the United States, covering older adults of age 65 and older, individuals with disabilities, and patients with end stage renal disease. Although Medicare went into effect in 1965, the mental healthcare services provided by psychologists and social workers became covered benefits as a result of the Omnibus Reconciliation Act (OBRA) of 1987, but were initially restricted to settings such as community mental health centers. As of July 1990, under OBRA 1989, the Health Care Financing Administration (currently called the Centers for Medicare and Medicaid Services (CMS)) allowed psychologists and licensed independent social workers to submit claims directly for clinical services under Part B of Medicare. Lack of interest by the leaders of the American Psychological Association in the mid-1960s was the primary factor that delayed inclusion of psychologists into Medicare for 25 years (Cummings & O'Donohue, in press).

At the present time, psychologists can easily become Medicare providers. Any licensed psychologists and licensed independent social workers can apply for Medicare provider status by downloading and printing the application form from the main government Web site, http://www.cms.gov. The applicant must sign an agreement to follow the regulations of the Medicare system. By doing so, the professional agrees to provide Medicare with accurate information on the care provided to the beneficiary. Failure to provide this information subjects the provider to both civil and criminal sanctions at the federal level.

Following a straightforward credentialing process, the regional insurance company that has a contract with CMS, called the Fiscal Intermediary (FI), assigns the applicant a provider number. Unlike membership in managed mental healthcare panels, there is no obligatory number of years of practice after licensure required before becoming an independent provider.

The enrollment process is where the easy part stops. Medicare regulations are complex and frequently change with variability by geographic region. Prerequisite elements to be a successful Medicare provider include periodic allotment of time to learn the regulations, attention to details, commitment to strong professional ethics, patience, and perseverance.

Despite the ongoing effort and time required to master the rules, inclusion in the Medicare system has catapulted psychology to major league status in American healthcare. If they so choose, psychologists can function in a broader role as healthcare providers, no longer restricted to the role of mental health specialist. The development of the health and behavior codes is one powerful example of how Medicare has shaped clinical practice.

In 2003, CMS adopted a series of procedures, both assessment and intervention, called Health and Behavior codes that reimburse psychologists for treating patients who do not meet criteria for psychiatric diagnoses but who need help following their medical regimen and coping with acute or chronic medical illnesses. These procedures are appropriate when designing and implementing disease management strategies, with individual patients or in small group settings, illustrating an effective way to better integrate medical care and behavioral healthcare (O'Donohue, Naylor, & Cummings, 2005).

Commercial payers are slowly following suit in adopting these codes, allowing psychologists to provide behavioral health services to children and adults of all ages throughout most regions of the country.

Along with the status of Medicare provider comes sobering responsibilities. We have observed that substantial numbers of psychologists and social workers view Medicare regulations as intimidating and difficult to comprehend, and therefore avoid learning them, or at the other extreme, minimize their importance. Some clinicians have the cognitive distortion that knowledge of empirical research findings, clinical expertise, and professional reputation are sufficient to ensure payment from Medicare without the need to carefully document treatment and learn the nuances of billing codes.

Negligible information is provided about Medicare coding, billing, and documentation in graduate school and pre- and postdoctoral training curricula or through continuing education course offerings, serving to reinforce naïve attitudes and misinformed views.

This chapter will provide basic background information of how Medicare works for behavioral/mental healthcare, make pragmatic recommendations to increase compliance of the regulations, and highlight the dynamic, evolving nature of the system, illustrating how payment regulations impact the delivery of clinical services while being shaped by empirical research, current economic realities, and political policies.

Much of the information originates from interviews conducted by the first author while researching articles for *The National Psychologist*, as well as 24 years of experience from her own consulting and clinical practice with older adults. The second author has been a consultant to HCFA and CMS on reimbursement for psychological services since 1989. Over the last 10 years, he has served as the Chief Consultant to the American Psychological Association's (APA) Practice Directorate on Medicare Reimbursement Issues as well as APA's representative to the American Medical Association's (AMA) Relative Value Update Committee (RUC). The RUC is responsible for recommending work values and policies regarding medical services under the Medicare program. In 1993, both authors were members of the Harvard Technical Advisory Group (TAG), which helped to develop the first "work values" for psychological services.

Why Bother to Become a Medicare Provider?

After learning the complexities of the regulations and stringent sanctions for erroneous billing, clinicians may question the wisdom and usefulness of becoming a provider. We maintain that if a clinician wants to provide clinical services to a substantial number of older Americans and get paid for doing so, it is advantageous to become a Medicare provider. The federal Medicare system remains the primary source of payment for the majority of older adults' health and behavioral/mental healthcare costs. There are senior managed care insurance products, but the numbers of beneficiaries enrolled in these systems are relatively small.

Based upon the first author's experience, only a small percentage of older adult patients choose to pay out of pocket for behavioral/mental health services. If an older adult disallows his/her claim to be sent to the Medicare carrier, then the clinician must abide by this request, and the clinician can collect privately for the service. When an older adult or his/her family request services that are not covered under Medicare, such as lifestyle coaching after retirement or determining legal decision-making capacity to sign a prenuptial agreement, the clinician is obliged to inform the older adult and/or family member requesting the service that submitting a claim for such services constitutes fraud.

Legal hassles can potentially occur for the physician or psychologist with a busy practice who refuses to accept Medicare patients. Blanket refusals can result in the filing of a class action lawsuit that alleges discrimination of Medicare beneficiaries. Clinicians may close their practice temporarily to *new* Medicare patients, but they are opening themselves up to legal issues prompted by a disgruntled patient who is refused treatment.

Some contend that the stringency of Medicare rules may discourage mental health clinicians from working with older adults. Because Medicare sets the precedent for other healthcare insurance companies across this country, practitioners who learn and follow its documentation and coding rules will easily pass chart audits from any third party payer. Increasing numbers of commercial insurance companies are using fraud detection software that picks up patterns of submission of incorrect claims (Appleby, 2006). Healthcare law defines fraud as the submission of an incorrect claim regardless of intent; a pattern is two or more incorrect claims. The private insurers follow the lead of Medicare regarding individual prosecution following testing of CMS policies in the courts.

A little known fact is that even patients who choose not to use their health insurance benefits can file complaints against practitioners for fraud and abuse. A federal prosecutor can investigate such accusations and if there is sufficient evidence, he/she can bring charges against the clinician, based on the tenets of Constitutional law that it is the federal government's responsibility to promote the general welfare of its citizens. Therefore, in the current healthcare environment, no behavioral healthcare practitioners are free from outside scrutiny. Therefore, we conclude that there is no discernible advantage for a psychologist who works with older adults to forego Medicare provider status.

Essential Sources of Information

By virtue of their contractual arrangement, we advise all Medicare practitioners to become familiar with the CMS Web site, http://www.cms.gov. Other resources include the APA practice Web site, http://www.apa.org, and organizations such as Psychologists in Long Term Care (PLTC), the National Alliance of Professional Psychology Providers (NAPPP), and the National Academy of Neuropsychology (NAN). Additionally, practical articles about Medicare changes impacting behavioral health are published regularly in *The National Psychologist* and are available on the Web site, http://www.centerforhealthyaging.com.

The most important source for news about Medicare regulations is found on the Web sites of the FI that functions as the Medicare payer in the region where the behavioral/mental health clinician practices. This is especially critical because as of this writing, June 2007, there is no uniform carrier mental health policy in effect. The Medicare carriers have the authority to interpret federal directives that originate from CMS. For example, the FI can legally set a limit on the number of units that will be paid for neuropsychological testing. However, the FI does not have the authority to set a local policy governing the eligibility for Medicare services.

There are approximately 20 Medicare Part B Contractors in the country and 10 CMS Regional Offices to contact for answers to questions from providers.

A Cautionary Tale for Medicare Providers

The ramifications of ignorance about the regulations are sobering and serious. For example, audits that uncover incorrect diagnostic and procedural coding with accompanying poor clinical documentation can result in demands for payment back for completed professional work, monetary penalties of $12, 500 per claim plus triple damages, potential loss of Medicare provider status, the closing of nonprofit clinics and private practices, and criminal prosecution. In 2005, about 170 psychologists received jail time for fraudulent billing under Federal Sentencing Guidelines.

One commonly held Medicare myth is the conspiracy theory about mental health services, i.e., the notion that "the feds are out to get us." What is true is that the federal government has a limited budget, and for every dollar Congress puts into monitoring claims, they get four dollars in return.

About 90% of the cases that are tried by the Department of Justice (DOJ) result in felony convictions with providers repaying money, paying fines, and receiving jail time. Under the Federal Sentencing Guidelines, the psychologist can literally be tried under civil law (e.g., fined) and under criminal law (incarcerated) for the same offense.

The Resource-Based Relative Value Scale, Basis of Reimbursement

At the end of each calendar year, the regional Medicare carriers publish the new fee schedule that will go into effect the first of the year. Medicare pays for a percentage of the allowed fees, and providers cannot bill the patient for more than the Medicare allowable charge for each clinical service.

National economic and political factors impact the fee schedule. Congress allocates a pot of money in the federal budget to pay for estimated Medicare services. A formula developed by Harvard economists called the Resource-Based Relative Value Scale (RBRVS) determines the fees for each and every covered medical service. Fees are based on the "work value" of the code, the practice expense, and the malpractice costs. Work value of a service is based upon the clinician's technical skill and physical effort, the mental effort and judgment put forth, and the stress associated with risk to the patient or others. The original scale against which the relative work value is judged is a therapy hour with a patient suffering from panic attacks that is given a value of 100 (Hartman-Stein, 1993).

Fees vary by region to cover differences in overhead and malpractice costs. For behavioral/mental health services, Medicare reimburses the clinician 80% of the allowable fee for the initial diagnostic interview, psychological and neuropsychological testing, and health and behavior interventions. The provider is required to attempt to collect the remaining 20% from the patient, except in the case of indigent patients who are covered under state funded Medicaid. Private coinsurance policies frequently cover the remaining costs. For psychotherapy services, Medicare covers 50% of the allowed fee. The 50–50 split is the hot button issue triggering parity debates across the country.

Congress mandates budget neutrality for Medicare. When the value of one service goes up, other services decrease in monetary value to maintain the budget. CMS alters regulations as well as reimbursement per clinical service based upon the US economy, political pressure, and feedback from individual providers and specialty professional groups.

The adoption of the RBRVS system in 1992 resulted in dramatically lowering fees for psychological testing, because the work values had been determined only for psychotherapy services at that point. Reimbursement for psychological testing was based upon overhead and malpractice costs only and remained so until January 2006 when work values had finally been assigned. This scenario is a striking example of the dynamic of Medicare shaping practice. Many psychologists either stopped offering psychological testing services and emphasized psychotherapy because of the drastically lower market value of assessment services after the RBRVS system was adopted.

The Medicare fee system directly impacts what other insurance companies will pay. According to the American Medical Association (Center for Medicare and Medicaid Services, 2006), more than 95% of the private payers utilize the Medicare fee schedule in some fashion, e.g., as a benchmark for payment and or policy.

Advocacy Efforts Make a Difference

As of January 2001, CMS required Medicare contractors to post drafts of their new policies on their Web site and have a time open for comment by individual clinicians or groups. This is a major avenue available that can influence the regional carriers by demonstrating how policies should reflect limits of clinical service backed up by both anecdotal and empirical evidence.

An example of a successful national advocacy effort that changed Medicare regulations was that of the Alzheimer Association's push to stop automatic denial of services such as psychotherapy for depression to patients with dementia. Leslie Fried, attorney and advocate, highlighted the research evidence that led to CMS changing its previous policy that allowed Medicare contractors to deny psychological or physical therapy services to patients with dementia (Hartman-Stein, 2001). This landmark change exemplifies that Medicare is a fluid, changing system.

Another example is the success of the efforts of Joseph Casciani, PhD, to change the policy of National Heritage Insurance Company that initially restricted health and behavior interventions to 2 h of treatment *per year* for any medical condition. Using numerous case examples and available clinical research findings, he advocated in writing and during meetings with the Medical Director that successful behavioral interventions required more than 2 h of treatment. The carrier agreed to reimburse the Health and Behavior interventions for its California beneficiaries for 1 h of assessment and up to 12 h or 48 units of treatment sessions per year, resulting in a significant policy change (Hartman-Stein, 2005).

Medical Necessity

One of the core factors that determine whether a claim for a clinical service is legitimate and acceptable is its medical necessity. The American Medical Association developed the gold standard for this concept that is utilized by both private payers and CMS. According to the AMA, a medically necessary healthcare service or product is one "that a prudent provider of practice would provide to a patient for the purpose of preventing, diagnosing, or treating an illness, injury, disease, or its symptoms in a manner that is in accordance with medical practice, clinically appropriate in terms of type, frequency, site, and duration; and not primarily for the convenience of the patient or provider" (Smith, 2006). The overall purpose of treatment must be the alleviation or mitigation of a medical condition. Chart documentation must reflect that each and every treatment or service meets the standard of being medically necessary.

In addition to meeting the criteria for medical necessity, the service provided must be a covered service. For example, if the services are primarily for personal growth or prevention, such as cognitive fitness programs for socialization purposes (such as support groups for new residents in an assisted living setting), or to answer a legal question, they are not deemed medically necessary and therefore are not Medicare covered services.

On occasion what we were taught in graduate school as to what is medically necessary is at odds with payment policy. For example, 10 or more hours of neuropsychological testing to determine a diagnosis of dementia or twice weekly psychotherapy sessions per year for treatment of depression may not be deemed medically necessary by current standards. If a patient requests counseling for personal growth purposes after retirement, or memory enhancement classes, for example, the prudent clinician should inform the client that the requested services do not meet the criteria for payment under their program but can be obtained through private payment arrangements.

Documentation Should Reflect Observable Symptoms and/or Problem Behaviors

Observable symptoms or problem behaviors must be noted in the clinical record when explaining the reason for the clinical intervention as well as the progress the patient is making in treatment, another way that Medicare impacts clinical practice. For example, although Medicare does not require that specific modalities of treatment be utilized (such as cognitive behavioral treatment), the documentation of symptoms must be in clear-cut, behavioral terms.

Audits Impact Clinical Practice

Receipt of a letter from the Medicare intermediary can strike fear in the heart of any well-meaning clinician. The fact is that most if not all Medicare providers will be audited at some point. Moreover, these audits are not limited to Medicare and Medicaid but include private payers as well. At the time of this writing, more than 55 private insurers who paid nearly 230 billion dollars in benefits in 2005 utilized or were in the process of utilizing antifraud software. Malcolm Sparrow, a professor at Harvard's John F. Kennedy School of Government and author of *License to Steal: How Fraud Bleeds American's Health Care System*, has noted that "pattern recognition is a growing field in health fraud detection" (Sparrow, 2000). Pattern recognition in the most basic sense is comparing one provider's profile (pattern of services) to the peer group.

Audit triggers may include the following: use of a code that provides a higher reimbursement rate when documentation supports a lower level of code, showing a high percentage of the same code (e.g., only billing for initial interviews), excessive visits to nursing home patients, and excessive testing hours for a single day.

An individual provider's profile of claims can also trigger an audit. For example, if a clinician has a procedure profile of billing a code well over the average of his/her peers, that clinician's work will be audited.

The Use of Modifiers in Correct Billing

The National Correct Coding Initiative (NCCI) determines numerous specific rules about coding known as edits. Information on NCCI edits can be found at http://www.cms.hhs.gov/physicians/cciedits/. A recent edit change went into effect in October 2006 that requires a modifier, a two-digit number (59), to be added to the claim form when a clinician conducts, interprets, and writes a psychological evaluation on the same day that a technician performs face-to-face testing or when the patient completes a psychological test using a computer. Regional Medicare carriers received the directive to require the edit in billing testing codes, but the best recommendation is always for the clinician to seek guidance from the local carrier. Additionally, the clinician should make a memorandum for the record that includes the date, the person with whom the issue was discussed, and the guidance provided. This is especially critical because the carrier can change policy during the course of the year. Some of these changes can be retroactive.

Another edit of importance to clinicians is modifier 52, indicating a reduced service. For example, when a patient cannot stay sufficiently alert to engage in psychotherapy or a health and behavior intervention due to a reaction from sedating medication, and the service is terminated prematurely, modifier 52 should be added to the claim form. The therapist will be paid for his/her time at a reduced rate, and appropriate documentation needs to be recorded in the medical record (Hartman-Stein, 2006). Modifiers should not be viewed as a penalty but as a way for psychologists and social workers to ensure that they are meeting their contractual agreement with Medicare in providing accurate information on the care provided to the patient.

Expansion and Change of Clinical Procedural Codes

The procedural codes for psychological testing were completely changed and revamped for 2006. However, at the time of this writing (December 2006), the code descriptors were being reviewed by the AMA Editorial Staff for clarification. As a result of this review and the potential for changes in the code descriptors, clinicians should consult the most current edition of AMA Current Procedural Terminology (CPT).

Documentation of Clinical Service

In order to pass an audit, thorough documentation is required for every claim, with each chart note standing on its own merit. Following a strict research protocol or a manualized cookbook approach may conflict with the Medicare requirement of medical necessity for each treatment session. In psychotherapy research studies, for example, the numbers of sessions are generally predetermined such as a 16 session group format for treating posttraumatic stress in senior Veterans (Meuser, Clower, & Padin-Rivera, 1998). If a provider bills for the clinical services under the Medicare system, each and every treatment session must be necessary, with corresponding documentation that shows whether there is progress toward the lessening of the symptom severity. If the patient demonstrates a significant reduction of symptoms by the sixth session, for example, then a continuation of ten additional treatment sessions billed to Medicare or any third party payer may be questioned during an audit.

The following are recommendations for the charting of treatment sessions:

a. Type your notes whenever feasible because legibility of chart notes is one of the most critical elements in the event of a Medicare audit. The easier the record is to audit, the less likely it will be audited in great detail.
b. Match the date of service in the clinical record to the date on the claim form. Mistakes in this detail can easily occur, so check dates carefully.
c. Track and record the number of the treatment sessions; e.g., session number 4 for this episode of treatment.
d. Document start and stop times of the treatment session. This is especially critical for timed procedures, e.g., the therapy codes and testing codes. Regardless of whether the Medicare carrier insists upon this, documentation of the times are helpful in an audit.
e. Include the diagnosis code number as well as descriptor. Check that the diagnosis on the claim form matches the diagnosis on the chart note. Some computer billing software has the original diagnosis recorded as the default diagnosis. Billing personnel must check that the diagnosis the clinician charted in case progress notes match the diagnosis on the claim form.
f. Use the diagnostic code numbers from the *International Classification of Diseases, 9th Revision (ICD-9-CM) 2006* (World Health Organization, 2005). The cross walk from the *ICD-9-CM* to the *DSM-IV* (American Psychiatric Association, 1994) is fairly consistent but is not a straight one-to-one. All psychologists would be better off in submitting claims if they utilized the *ICD-9-CM* for all diagnoses; e.g., using the mental disorder section codes for psychotherapy and the physical disorders with the Health and Behavior codes. A general rule is that when coding the diagnosis for neuropsychological testing procedures, use diagnostic codes from the nervous system section of the *ICD-9-CM*.
g. Record the global assessment of functioning (GAF) score from the *Diagnostic and Statistical Manual of Mental Disorders, 4th Revision (DSM-IV)* for both psychiatric and Health and Behavior Interventions.
h. On chart notes and assessment reports, use only abbreviations that are commonly recognized.
i. Include a description of treatment, operationalizing treatment in behavioral terms.
j. Document medications reported by the patient.

k. Record outcome measures, prognosis, and homework assignments.
l. In order to authenticate the record, include a signature, title, and date on every chart note.
m. Chart notes need to be individualized, clinically rich descriptions of the patient and the treatment provided. Auditors generally frown upon canned, "cookie cutter" notes.

Develop a Medicare Compliance Plan for Your Practice

Although having a compliance manual for your practice is technically voluntary, a viable, active compliance plan is considered to be the best single insurance in the event of an adverse audit (Hartman-Stein, 2000). Additionally, the DOJ has stated that this is the *only* resource that can be utilized as a mitigating factor in a fraud and abuse case. To begin developing a compliance plan, obtain a copy of the guidelines published in the Federal Register (http://www.oig.hhs.gov/authorities/docs.physician.pdf), and place it in a three ring binder. Read through the elements required in the manual, as outlined below, and designate a section of the binder for each of the component parts.

Compliance manuals are to reflect the uniqueness of an organization, describing the services the practice offers as well as what risks these services may entail. The content in the manual should contain evidence that chart entries and claims are checked periodically, including documentation of phone conversations or written correspondence with insurance company personnel. The manual should contain documentation of any questions, problems, or actions that result from the monitoring process. We recommend conducting an internal audit of charts monthly or at least every quarter and to hire an external reviewer to examine approximately 30 charts per year for each clinician in the practice for proper documentation and coding. Remember to review the OIG Web site annually.

The Office of Inspector General's Compliance Program Guidance for Individual and Small Group Physician Practices published in the Federal Register (2000) includes the following components of an effective compliance program:

1. Internal monitoring and auditing, including what is being monitored, how it is done, and who is doing the monitoring.
2. Implementing compliance and written practice standards.
3. Designating a compliance officer (e.g., the owner of the practice or the office manager). Include a typed page indicating who the compliance officer is and under whose authority the person was appointed.
4. Conducting training and education of all staff in the practice, both professional and clerical. The training can be an in-service or a continuing education program.
5. Responding appropriately to detected offenses and developing corrective action. For example, include a statement in the manual that a monthly staff meeting will occur to discuss any changes in coding, problems with billing, or any newly published regulations.

6. Ensuring that open communication is the standard in the practice, i.e., the director or owner of the practice supports discussion and disclosure of any compliance problems without any fear of retribution.
7. Enforcing disciplinary standards through well-publicized guidelines for employees to follow so they take seriously their compliance responsibilities. For example, include in the compliance manual a letter to each employee that states the practice is committed to exceptional patient care and adherence to policies that govern that care.

Pay-for-Performance: A Future Trend in Medicare Reimbursement?

In recent years, policy makers and leaders in healthcare have explored controversial payment arrangements in which a portion of payment for medical services is based upon defined measures of quality (Hahn, 2006). Pay-for-performance programs are an attempt to bring the relationship between prices and quality care services into a closer balance. The passage of the Tax Relief and Health Care Act passed by Congress in December 2006 introduces performance-based reimbursement for the first time into the Medicare system on a large scale. It includes a provision for a 1.5% bonus incentive payment for Medicare providers who voluntarily report quality measures borrowed from an existing CMS program that can be modified to better fit the professional specialty. It is apparent that there is exciting work ahead for behavioral health specialists to work with CMS, proposing additional measures in the future that have been used in empirically based psychological outcome research. According to Cummings (Hartman-Stein, 2006), the future of healthcare reimbursement will become increasingly performance-based, something that effective clinicians do not have to fear. This initiative is yet another example of how Medicare influences clinical practice and is driven by economic and political forces.

In closing, although Medicare regulations are complex and can resemble a moving target, striving to understand and then comply with them is essential and eminently possible. All providers must realize and accept that there are limitations and constraints in the Congressional budget; therefore, working with patients in a cost-effective manner by adhering to the guidelines of medical necessity and striving to demonstrate effectiveness through routine use of quality measures are paramount for working within the Medicare system.

References

American Psychiatric Association. (1994). *Diagnostic and statistical manual of mental disorders.* Washington, DC: American Psychiatric Association.

Appleby, J. (2006). Computer program help flag insurance fraud before payment. *USA Today,* November 7.

Center for Medicare and Medicaid Services, U.S. Department of Health and Human Services. (2006). Improper Medicare FFS payments long report (Web version) for November 11,

2006. Retrieved November 25, 2006, from http://www.cms.hhs.gov/apps/er_report/preview_er_report.asp?from = public&which = long&reportID = 5&tab = 5.

Cummings, N. A., & O'Donohue, W. T. (in press). *Eleven blunders that cripple psychotherapy in America: A remedial unblundering.*

Hahn, J. (2006). Retrieved December 12, 2006, from http://www.vascularweb.org/_CONTRIBUTION_PAGES/Government_Relations/PDF_Doc/CRS%20report%20on%20P4P.pdf Pay-for-Performance in Health Care. *Congressional Research Report for Congress.*

Hartman-Stein, P. E. (1993). Psychologists' fees to be affected by Medicare RBRVS study. *The National Psychologist, 2,* 1–2.

Hartman-Stein, P. E. (2000). Inspector General urges psychologists to set up voluntary compliance procedures. *The National Psychologist, 9,* 1–2.

Hartman-Stein, P. E. (2001). Federal agency to begin reimbursing long-denied diagnosis for dementia. *The National Psychologist, 10,* 16.

Hartman-Stein, P. E. (2005). Persistent psychologist wins change in Medicare regulation. *The National Psychologist, 14,* 17.

Hartman-Stein, P. E. (2006). Billing code modifiers ensure payment for testing services. *The National Psychologist, 15,* 9.

Meuser, T., Clower, M. W., & Padin-Rivera, E. (1998). Group psychotherapy: Ideas for the managed-care environment, In P. Hartman-Stein (Ed.), *Innovative behavioral healthcare for older adults: A guidebook for changing times.* San Francisco: Jossey-Bass Publishers.

O'Donohue, W. O., Naylor, E. V., & Cummings, N. A. (2005). Disease management: Current issues. In N. A. Cummings, W. T. O'Donohue, & E. V. Naylor (Eds.), *Psychological approaches to chronic disease management.* Reno, Nevada: Context Press.

Office of Inspector General's Compliance Program Guidance for Individual and Small Group Physician Practices, 65 Fed. Reg. 59, 435–59, 452 (October 5, 2000).

Smith, S. L. (Ed.). (2006). *Medicare RBRVS: The physicians' guide.* Chicago, Illinois: American Medical Association.

Sparrow, M. K. (2000). *License to steal: How fraud bleeds America's healthcare system.* Boulder, Colorado: Westview Press.

World Health Organization. (2005). *International classification of diseases, 9th revision.* Los Angeles: Practice Management Information Corporation.

Index

Printed in the United States of America